CLIMATE CHANGE

AND THE ROAD TO

NET-ZERO

By Mathew Hampshire-Waugh

Crowstone
Publishing

First Published 2021

By Crowstone Publishing

www.crowstonepublishing.com

Originally published by Crowstone Publishing in 2021. Crowstone Publishing 160 City Road, London, UK, EC1V 2NX. Crowstone Publishing is a trading name of Net-Zero Consulting Services Ltd.

Graphics by Mr Paul Bean and Mathew Hampshire-Waugh.

British Library Cataloguing-in-Publication Data

A catalogue record for this book is available from the British Library.

ISBN: 978-1-5272-8796-9

Acknowledgments

I would like to thank the many experts who gave up their time to read, review, and fact-check relevant parts or all of the draft book, including: Mark Campanale (founder of Carbon Tracker Initiative who conceived the 'unburnable carbon' capital markets thesis), Michael Liebreich (CEO of Liebreich Associates and founder of New Energy Finance, now BloombergNEF), Mark Z. Jacobson (Professor of Civil and Environmental Engineering, Director of the Atmosphere/Energy Program at Stanford University, and author of *100% Clean Renewable Energy and Storage for Everything)*, Syukuro Manabe (senior meteorologist in the Program in Atmospheric and Oceanic Sciences at Princeton University, a pioneer in the use of computers to simulate global climate change, and author of *Beyond Global Warming*), Spencer R. Weart (formerly Director of the Centre the History of Physics of the AIP and author of *The Discovery of Global Warming*), Andrew Dessler (Professor of Atmospheric Sciences at Texas A&M University, former senior policy analyst in the White House Office of Science and Technology Policy, author of *Introduction to Modern Climate Change* and *The Science and Politics of Global Climate Change*), Andrew Hagan (Head of Chemistry and Advanced Materials at the World Economic Forum), Vincent Gilles (Chief Investment Officer at Clim8 Invest), Glen Peters (Research Director at the Center for International Climate Research), Philip E. Tetlock (Annenberg University Professor at the University of Pennsylvania and author of numerous social science/policy books including his most recent *Superforecasting: The Art and Science of Prediction*), Vincent McCarthy (founder of ESG Ireland ®), Hannah Ritchie (Head of Research at Our World in Data), Richard Tol (Professor at the Dept. of Economics at the University of Sussex, Professor of the Economics of Climate Change at the Institute for Environmental Studies, Vrije Universiteit, Netherlands, and author of *Climate Economics*), Nick Steinsberger (the pioneer of modern fracking), Robert S. Pindyck (Professor of Economics and Finance at the Sloan School of Management MIT, author of *Microeconomics and Investment Under Uncertainty*), Geetu Sharma (founder of AlphasFuture, sustainable asset manager), David G. Victor (Professor of Innovation and Public

Policy at the School of Global Policy and Strategy at UC San Diego, co-director of the Deep Decarbonisation Initiative, and author of numerous climate change books including his latest *Making Climate Policy Work*), Thomas Frölicher (SNF Assistant Professor, Climate and Environmental Physics, University of Bern), and the multiple corporations which prefer to remain anonymous.

Finally, I would like to thank my friends and family for all their advice, support, love, and patience. I couldn't have started and certainly wouldn't have finished this book without you.

Contents

Preface: A Story Unwritten

Planet Earth & Humans

Net-zero: a balance between the emission and removal of greenhouse gases, to and from the atmosphere, by human activity.

What follows is a story.

A story of how humanity has broken free from the shackles of poverty, suffering, and war and for the first time in human history grown both population *and* prosperity.

It's also a story of how a single species has reconfigured the natural world, repurposed the Earth's resources, and begun to re-engineer the climate.

Our story begins with a bitter argument over the future of humanity between two heavyweight eighteenth century scholars, and as it unfolds, we will encounter a cast of colourful characters. From a pioneering female scientist, largely forgotten in history, who first warned about global warming, to a famous chemist who formed the scientific basis for the greenhouse effect as a distraction from his ongoing divorce. We will meet the wealthiest man who ever lived, a computer worm designed to limit the rise of new nuclear nations, and a fictional walrus swimming in the Gulf of Mexico. From fossil fuel tycoons detonating nuclear bombs in the Colorado Mountains to oil executives drinking fracking fluid to appease shareholders. And from the stolen head of an eighteenth century reformist, a drove of pigs living a life so luxurious they brought down a government, to a gaggle of Cold War physicists spreading disinformation and doubt.

At the heart of this story is a wager over the future of our planet. A bet between a celebrity biologist who predicted the impending doom of civilisation unless increases in population and consumption were limited, and a subversive economist who resolutely believed in the power of free markets, technological innovation, and infinite boom.

Where these disparate narratives converge is with the protagonist of our story, carbon dioxide (CO_2).[i] This once innocuous by-product of industrial revolution is now the very definition of anthropic change, threatening to undo two hundred years of progress.

But this is a story still being written. We will learn how the fundamental science, economics, and technology that underpin climate change forecasts can be adjusted to create drastically different outcomes. Throughout the book I will refer to these three sensitivities as the 'dials of doom and boom': the *climate sensitivity* is a science-based value which governs how quickly the temperature will change as we load the atmosphere with CO_2; the *discount rate* is an economic factor which defines how much value we ascribe to future generations; and the *experience curve* is the technological rate of learning which predicts how quickly zero carbon solutions become cost competitive. By tweaking these three dials of doom and boom, turning them up or down, we can completely re-write the narrative and create diametrically opposed visions of the future using the same science, economics, and technology.

And yet there is an alternative ending. An ending where there is no trade-off between economy and environment, present and future generations, or rich and poor. A net-zero carbon economy. I argue we are at a juncture where the possible range of values for these dials are sufficiently narrow to assert that a rapid transition to net-zero will create a win-win outcome for all sides of the argument. The information is not perfect, the details are not final, but the road is illuminated.

Net-zero offers not only a solution to climate change and air pollution, but an opportunity to create a cheaper, more resilient energy system, a more productive economy, and a better quality of life for everyone on the planet. Net-zero accelerates the ideals of human progress and creates a sustainable society living within the planet's boundaries and fulfilling our moral obligations to nature.[1]

I wrote this book as somewhat of an outsider to the argument, an impartial observer with a background in science and technology and an expertise in financial modelling and forecasting the future. I spent the best part of the last 20 years researching many of the sub-plots in our story; from publishing scientific papers and patents on energy materials as part of my engineering doctorate to forecasting market trends in renewables, electric cars, batteries, and biofuels, during a decade issuing share recommendations at a major investment bank.

---

i Carbon dioxide or CO_2 is an invisible, odourless gas composed of one carbon bonded to two oxygen atoms.

Like many others, I too am guilty of climate apathy. I always thought the world would solve global warming, just as "we" banned leaded fuel or fixed the hole in the ozone layer. But we are nearly 30 years on from ratifying the United Nations' framework on climate change and we continue to pump more CO_2 into the atmosphere than ever before. By most measures little progress has been made, and the warnings from scientists grow ever louder and ever more troubling. The world is aligning with long-term net-zero pledges, but the pace of change today is slow and there is little consensus on exactly *how* to reach zero emissions.

I began writing this book to illuminate all sides of the climate debate, to better understand exactly how big an issue climate change is. How does it compare with other global problems? Is there a workable solution? How much will the solution cost? Are there better ways to use our resources? Is it worth doing and if so, how quickly we should act? And finally, why has climate change proved such an intractable problem? I finished writing this book convinced that a rapid transition to net-zero is undoubtedly the best direction of travel.

Predicting the future is inherently uncertain, so I ask you not to read this as a rigorous scientific proof, economic theory, or political white paper, but as a back-of the-envelope take on the science, technology, economics, and politics of climate change, air pollution and the road to net-zero. Enrico Fermi was the father of nuclear power; he was also famous for his back-of-the envelope estimates. Using incomplete information he could calculate the blast strength of an atomic bomb by dropping pieces of paper as it exploded, or he could estimate the possible number of alien civilisations in our galaxy with no more than a handful of astronomical observations. Throughout this book I hope to provide enough information to frame the big picture and enough insight to project multiple visions of the future without getting lost in the technicalities.

I'm not a writer, journalist, or political commentator. I'm a scientist and an analyst. Which is why I want to begin by giving away the end. What follows is a synopsis of the book.

Synopsis: Net-Zero

Empowering Action Towards a Sustainable Future

Human Prosperity and Planet Earth

If the history of the Earth were condensed into one day, the story of human civilisation would start just 0.2 seconds before midnight, or 12,000 years ago, as the planet emerged from the last ice age. At this moment in time, concentrations of CO_2 in the atmosphere stabilised at just below 300 parts per million (ppm)[i] and average temperatures on Earth settled at a comfortable 14ºC. Humans advanced from nomadic hunter gatherers to build permanent settlements supported by a newly dependable agricultural yield. We domesticated animals, repurposed land for growing crops, and transformed the Earth's natural resources to support an ever-expanding human empire.

Populations grew, but prosperity did not. Agricultural production slowly increased but any gains were met with more mouths to feed. So, for most of human history, most humans have experienced the same levels of disease, hunger, and violence as all generations before them. Humanity was stuck in what would become known as the Malthusian Trap.

Thomas Robert Malthus, an eighteenth century scholar, asserted how any improvements to human existence would prove short lived because linear resource increases would always be met with exponential[ii] population growth. He believed humanity would remain forever imprisoned by the shackles of poverty unless action was taken to limit reproduction. And he was right for most of human history, all the way up until he *actually* published his theory when, two centuries ago, at 0.004 seconds before midnight, along came the industrial revolution.[1]

---

i 300 parts per million means there are 300 particles of CO_2 for every million particles of all the gases making up the atmosphere or 0.03% concentration.

ii A linear relationship means that if you double one variable the other variable will also double. Straight line change. Exponential growth is where the rate of increase in one variable increases with the other. Creating ever larger change. A line curving ever steeper.

This was the moment modern humans mastered combustion, just as Homo erectus had mastered fire. We unlocked millions of years of the sun's energy stored in fossil fuels[iii] and used it to power the machinery of the modern world: transport, industry, agriculture, and convenience living. Both population *and* prosperity not only expanded for the first time in human history; they exploded. And so did carbon dioxide.

The industrial revolution has created unprecedented population growth *and* prosperity. Average incomes have increased from less than $1,000 to $18,000 per person per year (in the equivalent of today's money).[2] Life expectancy has doubled, child mortality and extreme poverty are ten times lower, and deaths on the battlefield have declined twenty-fold.[3]

We are living through a period of unprecedented wealth, quality of life, and relative peace. Yet not everybody is on board. Half the population still has no basic sanitation, clean cooking, or electricity. Nearly one billion people have no access to safe water or sufficient food.[4] Population and consumption will continue to grow as developing countries rightfully strive for modern standards of living. So, unless we fundamentally change how our system is powered, CO_2 emissions will also continue to climb, and ever higher concentrations will accumulate in the atmosphere.

Over the last 150 years, modern science has unravelled the complex relationship between CO_2 and temperature. A relationship that dates back hundreds of millions of years into the deep geological record, through the last million years of ice ages, and into the modern archives of the twentieth century. This is a relationship that has held constant, unwavering through time. The data is clear, the scientific basis sound, and the predictive models have proven accurate so far. Change either CO_2 or temperature and the other will mutually follow; and we now understand how humans are changing CO_2 concentrations in the atmosphere faster than any other planetary event in at least 66 million years.[5]

Where eighteenth century scholars debated the perils of an expanding population and the future of the human race, now carbon dioxide, rising temperatures, and climate change have reignited the argument for the twenty-first century. On one side of the dispute sit the neo-Malthusians who foresee a dystopian vision for the future should we not limit population, consumption, and emissions. On the other side of the quarrel are those we might call the cornucopians, drinking from the horn of plenty, convinced that human innovation, technology, and free markets can weather the coming storm. So, who is right? Where should we set our dials of doom and boom as we forecast the future? Are we stumbling into disaster or striding towards ever greater prosperity?

---

iii Fossil fuels are formed from dead plant and animal matter compressed and heated in the Earth's crust over millions of years.

Understanding Climate Futures to Empower Action

Temperatures are already over 1°C higher than in pre-industrial times. Make no change to our existing systems and most likely we will warm the planet by at least 3°C by the end of this century and by 4.5°C by the middle of the next. Sea levels will rise, and the ocean will become increasingly acidic. Amplifying feedback mechanisms not yet captured in climate models – such as melting tundra, burning forests, or weaker than expected carbon dioxide uptake by land and ocean – may accelerate change even faster.

We are already experiencing three times the number of climate related not-so-natural catastrophes compared to 40 years ago. The last 20 years have recorded 19 of the hottest in the last two millennia, and with half of the strongest hurricanes on record.[6] If we don't change the way we fuel the economy, over the next 80 years 40 million people may die from hunger and natural catastrophes, and 500 million more could be forced to migrate due to extreme heat and rising seas. That would cost the economy $5 trillion per year.

But the biggest impact won't be from headline grabbing destruction, but from hidden death, suffering, and economic toll. With rising heat, melting ice, and changing weather, more than one third of the global population will lack sufficient clean water and nutrition. The air pollution from burning fossil fuels already kills at least eight million people every year.[7] By 2100, one billion lives will have been cut short from respiratory disease and heart failure. The hunger, water shortages, poor health, and heat, will force labour productivity declines across the economy, bringing total damages to more than $40 trillion every year or 5% of GDP by 2100.

Compare the death, suffering, and economic toll of climate change and air pollution to other global issues and they already rank in the top three of major world problems, alongside poverty in the developing world and the prevalence of heart disease in rich nations. Continue burning fossil fuels and climate change quickly takes the top spot, before spilling over into nearly every other human problem.

But for our cornucopians the forecast needn't look so bleak. The future isn't simply a bigger, warmer version of the past, and humanity won't just sit back and let rising temperatures and rising seas destroy their food supply, homes, work, and health. If GDP[iv] grows at just half the rate of the last 40 years, average annual income per person will still rise to over $70,000 by the end of this century. Suddenly, losing 5% of future income becomes a much smaller concern. Why can't future generations simply spend their way out of the problem? Humanity can adapt to a changing climate using an increasingly large pot of wealth to not only minimise associated death and suffering, but also to lessen its economic burden. With the benefit of advanced technologies, more information, and greater wealth, surely future generations are better placed than we are to solve the problem. If done well, climate change could cost the world less than 0.5% of GDP by 2100.

This is where the plot takes another twist. Thus far our story has assumed that climate change follows a manageable or predictable pathway. However, we know that not all change is linear. We know that dangerous tipping points exist where risks grow exponentially and sub-systems may collapse, quickly leading to complete systemic breakdown, whether of the environment, society, or the economy: or all three. The acceleration of melting ice, biosphere feedback, global pandemics, a financial Carbon Crunch, or conflict could trigger a tipping point moment where change is unpredictable and sudden. Damages grow exponentially. The problem may become unmanageable no matter how much money is thrown at it. Tipping points represent a known-unknown event that risks throwing humanity back into the Malthusian Trap unless we take action.

The Earth's systems are currently playing catch up with the rapid increase to the concentration of CO_2 and that means nearly 1.5ºC warming is already locked in. Further change is inevitable but should prove manageable. Adaptation will be necessary, but other tactics must be deployed if we are to avoid higher temperatures and unmanageable tipping points.

At first glance, a technical fix like climate engineering seems the easy, cheap option. But as with most quick fixes, it only solves part of the problem. Limiting solar radiation using space mirrors, aerosols, or cloud whitening, might keep temperatures at bay, but such measures don't solve air pollution or ocean acidification and they run a seriously high risk of going wrong. Negative emissions technologies to remove CO_2 from the atmosphere are lower risk, but more expensive, or can only cover a fraction of emissions. Mitigation of carbon dioxide and the transition to a net-zero carbon economy offers the only complete and low risk path.

---

iv Gross Domestic Product (GDP) is a monetary measure of the market value of all the final goods and services produced in a specific time period (usually one year). Divide global GDP by the global population and you get average GDP per capita which is very close to average income per person (ignoring a few adjustments).

Mitigation assumes prevention is better than cure, and yet we are still presented with a range of neo-Malthusian and cornucopian ideals. Greenhouse gases[v] are the product of population, consumption, energy efficiency, and emissions intensity; so which factor should we target to produce the best outcome? The answer, of course, is all four, but not in equal measure. Attempting to control human reproduction is slow, morally difficult, and practically challenging and the global population should self-limit with growing wealth. Consumption limits are effective but demand a complete re-wiring of socio-economic behaviours and are perhaps best left to activities with little other option. Energy efficiency buys time but never offers a complete solution. The bulk of the change must come from slashing emissions intensity and reaching net-zero.

Systems Thinking to Build Net-Zero

A net-zero transition requires fundamental changes to both our energy supply and our energy demand. Wind and solar electricity coupled with pumped hydro and battery storage can form the backbone of a net-zero economy providing cost competitive, safe, reliable, distributed energy, with no carbon emissions or air pollution. We must electrify everything where possible and where not, switch to hydrogen, biofuels, or carbon capture. Electrification will replace oil-powered transport, gas-powered heating, and coal-powered industry. Global agriculture will need to be modernised to cut waste and lower carbon intensity. Global eating habits will require cutting back on meat consumption to free up enough agricultural land to regrow forests and offset stubborn residual emissions.

Change will be all-encompassing but requires no compromise on quality of life: simply the breaking of old habits. Done well, the transition will barely register for the top 10% of the global population already accustomed to the highest quality of life in developed countries. The remainder of the world will quickly be enabled to raise living standards towards those in the top bracket. The poorest 10% will for the first time connect to the modern world with electricity access, clean cooking, and basic amenities. A just and equitable transition for all.

Electricity prices will decline, but as electricity must replace cheap coal and gas heating, so the overall blended price of energy may rise by one third. However, electrification brings large efficiency gains – heat pumps use four times less energy than gas boilers, electric cars three times less energy than combustion engines, and electrified industrial manufacturing can halve energy consumption. A net-zero economy will

v Greenhouse gases or GHGs cover all gases that act to warm the atmosphere with the same mechanism as CO_2.

demand less than half the energy of the equivalent fossil fuel system with no compromise on travel, goods, food, or amenities. Spending on the energy supply declines by one third, despite higher blended prices. The whole system, including the higher upfront cost of equipment, will end up 25% cheaper than the fossil fuel alternative.

Wind, solar, and battery costs are already shrinking by 10-35% every time the number of installations doubles. Hydrogen and biofuels will soon join this commercial experience curve. It's not the passing of time that lowers costs, but the scale of production. Sitting on our hands and waiting for a breakthrough is not the way forward. We have already commercialised 80% of the technology we need so the faster we push, the cheaper it gets. The other 20% merely requires incremental development. The faster we reach net-zero, the sooner we save over $5.5 trillion per year on our energy system and we avoid worst case climate impacts: we get paid to breathe cleaner air, save lives, avoid climate change damages, and create a sustainable world for future generations.

If that isn't enough, net-zero also cuts air pollution which could save nearly eight million premature deaths per year and create more jobs by directing our energy spending into labour rather than exorbitant profits for petro-states. Net-zero side-steps conflict over dwindling resources and avoids the worst impacts of climate change. A net-zero world will be cheaper to run, cleaner, safer, more reliable, more sustainable, and create more employment than a world bound to fossil fuels.

But let's pause for a moment. Let's sit back, take stock and remember the words of British statistician George Box, that "essentially, all models are wrong, but some are useful".[8] The model we will build through the pages of this book is no exception. We can't predict the exact values for our dials of doom and boom: the response of the Earth, the desires of future generations, and the rate of technological progress are inherently uncertain sensitivities. But, I believe that we now have enough information to sufficiently narrow their possible values so that regardless of whether we turn our dials up high or down low, a rapid transition to net-zero will create the best possible outcome for the environment and for the economy. We can avert disaster *and* drive greater human prosperity.

Net-zero is inevitable. But that doesn't mean the speed of the transition can't be optimised. Continuing to run a fossil fuel economy is inefficient and, just like any other market inefficiency, it can be corrected to drive greater prosperity. Think about the value investor Warren Buffet who buys companies trading below their intrinsic value and profits as they rise, or Silicon Valley venture capital firms who recognise the potential of tech enthusiasts running makeshift operations from their parents' garages. These market players profit from inefficiencies and in doing so they also correct them. Warren Buffet's undervalued investment portfolio appreciates, and Apple, Google and Amazon are born. In the same way, markets will undoubtedly move towards net-zero as the best solution; however, by providing an early helping hand to overcome market inefficiencies, the speed of the transition can be optimised, generating greater wealth, improving health, and accelerating prosperity.

Getting to net-zero will require $70 trillion of upfront investment, $46 trillion more than sticking with fossil fuels, but it will create an energy system more than $5.5 trillion cheaper to run each year. A twenty-year transition provides the best balance with enough time for governments, businesses, and individuals to adjust whilst bringing the system cost savings forward and more than offsetting any losses from fossil fuel assets left stranded or from crowding out other investments. The headline numbers sound daunting but break the transition down over the global population and it seems far more manageable. Over the next two decades, we install just four solar panels, plant 27 trees and develop 16 square metres of biofuel cultivation for each person on the planet. We add the equivalent of one (radiator sized) battery in every house and one wind turbine between every 7,000 people. One quarter of car owners shift to ride sharing and the remaining three quarters replace their old vehicles with electric. Half of broken boilers are replaced with heat pumps. It starts to feel far more achievable. Part way through the transition, we decide whether renewables can cost-competitively carry us all the way to net-zero, or whether support from next generation nuclear or fossil fuels with carbon capture is needed. Either way, the best course of action for the next decade is abundantly clear.

Finding $3.5 trillion of annual investment also sounds like a daunting task. But the world already invests $1.75 trillion in mostly fossil fuel energy each year and wastes another $0.5 trillion on fossil fuel subsidies. We need to redirect this money and find another $1.25 trillion per year (1.5% of GDP) to deploy in net-zero solutions at a competitive market rate of return. The money is ready and waiting to go. It just needs clear direction from governments and private savers. Net-zero financing will create new and productive investment opportunities, boost jobs, and stimulate the economy whilst creating a cheaper energy system with a ten-year payback.

Bridging the Climate Divide to Accelerate Change

Despite the abundant benefits, progress towards net-zero has so far been slow. Since the International Panel on Climate Change formed in 1990, we have emitted as much CO_2 into the atmosphere as all human history before and in 30 years net-zero energy supply has grown from 12% to just 15% of total energy.[9] The hurdles to action range from apathy, misaligned economic interests, inequality, freeriding, and political wrangling to deliberate sabotage: each and all has slowed progress. But ultimately the balance of fear has stopped humanity transitioning away from fossil fuels. The perceived threat of climate change hasn't been great enough to overcome the political, social, and economic risk of transitioning the energy system. But the longer we wait, the bigger and more immediate the threat becomes, the more people die from air pollution and, although the cost benefits of a zero-carbon system will become increasingly clear, our window of opportunity for the best possible outcome will inevitably narrow.

The time has come for governments around the globe, corporates, and individuals to stop stalling and to push towards a better system. We are now lining up, ready to clear the bar, and if we can only recognise the benefits of a net-zero system and act on that recognition, we will leapfrog most of the remaining hurdles. Are you a sceptic who doesn't believe in climate change? Or perhaps you don't care about other countries? Well, net-zero will be cheaper, and it will reduce air pollution in your own country. Do you think net-zero technology is too expensive? Well, total buying and running costs will become cheaper once we are just one quarter of the way through the transition and will end up 25% cheaper than fossil fuels once fully scaled. Don't want to risk change? Once the transition is underway, to remain bound to fossil fuels will become uneconomic, unethical, and unimaginable.

Awareness and willingness to act are certainly on the rise. The unified scientific voice grows louder. 26 countries have declared a climate crisis and protests across the globe are mobilising young and old alike.[10] Investors are beginning to push companies for change and media outlets are spreading an ever-greater message of urgency. Over one hundred countries around the world, covering well over half of all emissions, have made net-zero pledges for the middle of this century. However, today's average world leader will be nearly 100 years old when those targets come due.[11] Clearly translating the long-term pledges, promises, and good-will into short and mid-term action is crucial. We must create change, not just talk about it. Science will continue to be constrained by burden of proof, business limited by burden of profits, and politics limited by popular votes, but recognise the benefits of net-zero and all become aligned. We already have the tools to expedite change; from left leaning command and control regulation to right leaning market economics and consumer choice.

Individuals can electrify end uses and source zero carbon electricity, forcing demand away from fossil fuels. Vote with your ballot paper and your wallet, choose credible policy and low carbon products. Politicians will soon respond to the changing polls and companies will soon react to falling revenues. Cut back on meat, the one area of consumption where there is no easy substitute solution (the average diet requires one football pitch of agricultural land, whereas the vegan equivalent requires just the penalty box). Reduce consumption where you can, but this should remain a personal choice. Do we need cars, planes, or washing machines to survive and reproduce? No. But they are culturally and economically embedded in our social systems – giving them up is a perceived hardship and we need to avoid the idea that solving climate change requires hardship or a trade-off between our current and future needs.[12] The clear message is that net-zero is not a zero-sum-game, but a win-win, and we have everything needed to transition without compromise. As ecological economist C.S. Holling states, sustainable development is "not an oxymoron but a term that describes a logical partnership".[13]

Individuals must recognise their role in changing energy demand; not just blaming government inaction or big business profiteering. Corporates should strive for net-zero as the best sustainable economic outcome; greenwashing[vi] should not be used as a marketing tool. Politics should be used to transition the energy system; climate change should not be used as an excuse to change politics. Yes, there may be significant shortcomings with traditional economics, GDP as a measure of progress[14,15] and globalised capital markets, but once we understand that a net-zero system will create a more favourable outcome for both right and left politics, big and small business, the individual, and society, then we can accelerate change for the better.

The last decade of stimulus following the Credit Crunch transformed the economics of wind and solar technology which are now competing with fossil fuel power. No analyst, energy agency, or government predicted just how quickly renewables prices would decline compared to their fossil fuel predecessors. Everyone underestimated how much faster technology moves compared to geology. With the unprecedented stimulus being

---

vi Greenwashing is considered an unsubstantiated claim to deceive consumers into believing that a company's products or services are environmentally friendly.

deployed following COVID-19, we now have the opportunity to transform the economics of supporting technologies such as hydrogen and batteries and to begin the full-scale transition to net-zero. Supply must be deployed five times faster and demand switched nearly ten times quicker to reach an optimal outcome. The solutions are ready: we just need to push.

Our story began with CO_2 as the protagonist and air pollution quickly emerged as an equally dangerous accomplice. But the real twist in the tale is that net-zero can solve both problems whilst accelerating underlying wealth, health, and human progress. Net-zero will be cheaper, cleaner, safer, more reliable, more sustainable, and will create more employment than if we remain bound to fossil fuels. By understanding and acting on this statement we can redefine the argument and create a better future for all on planet Earth.

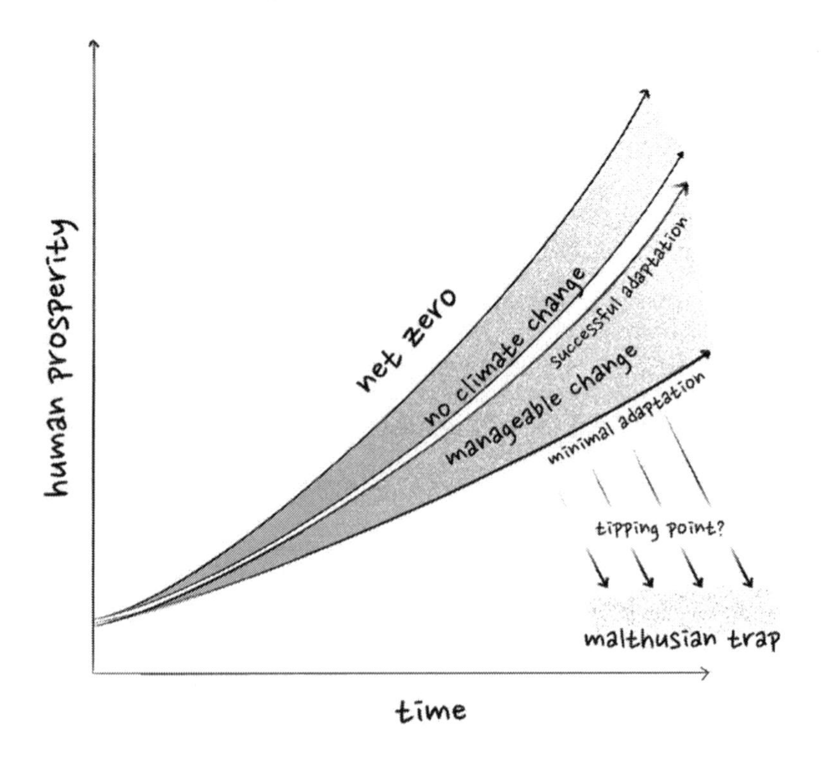

"We have come this far because we are the smartest creatures that have ever lived. But to continue we require more than intelligence. We require wisdom." David Attenborough, A Life on Our Planet, 2020.

Chapter 1: Third Rock from the Sun

Human Prosperity and Planetary Pressures

In this chapter we will explore how, in such a short amount of time, humans have come to dominate planet Earth. We will chart the rise of civilisation to understand how fossil fuels have supercharged progress, with planet changing consequences and how the industrial revolution has freed humanity from poverty, and yet provides the highest quality of life for just ten per cent of the global population. We will end the chapter by delving into a 200-year-old argument over the 'limits to growth' and why climate change has dragged this debate into the twenty-first century.

A Brief History of Planet Earth and the Rise of Civilisation

The universe started with the Big Bang, some 14 billion years ago. Another 9.5 billion years passed before our solar system settled and planet Earth formed from the gravitational pull of hot swirling gas and dust. Earth cooled, lava turned to rock, water condensed into oceans, and the first single celled organisms appeared four billion years ago.[1] These simple lifeforms eventually chanced upon a route to harness and store energy from the sun; photosynthesis released huge quantities of waste oxygen. Earth's systems were overwhelmed, and its atmosphere irreversibly changed.

This Great Oxidation event, two billion years ago, led to the evolution of increasingly complex life.[2] Early plants formed nearly one billion years before present day.[3] The colonisation of land, with the expansion of dense forests, dates back 500 million years. Dinosaurs were roaming the planet 230 million years ago, swiftly followed by the first mammals on Earth, shrew-like creatures which inhabited a planet 10-15ºC warmer, with sea levels over 200 metres higher, great forests stretching to the poles, and crocodiles swimming in the Arctic.

Early humans didn't evolve until the temperature of planet Earth cooled by 10-15ºC. We first mastered walking on two legs, allowing us to cover vast distances, then Homo habilus – "The Handy Man" –

developed stone tools around three million years ago. Our earliest close ancestors, Homo erectus, are the longest lived of all human species and first appeared nearly two million years ago, surviving multiple ice ages until extinction 1.7 million years later.[4] Homo erectus learnt to hunt, to fashion stone tools, and to cook food.

Human existence is a mere tick on the geological clock. Homo sapiens first evolved just 300,000 years ago: if the existence of planet Earth were condensed into one day, our species would appear just 6 seconds before midnight. Nonetheless, we have made good use of our time with early Homo sapiens developing language, art, and social cooperation. These attributes no doubt helped our nomadic hunter gatherer ancestors to survive another series of ice ages where average temperatures dropped 4-8ºC below today and vast ice sheets, thousands of metres thick, extended across Europe, North America, and one third of all the land on Earth.[5]

The Egyptians formed the first human settlements 13,000 years ago but volatile temperatures and unpredictable weather forced their return to nomadic life. Human civilisation began in earnest 12,000 years ago (0.2 seconds before midnight) when the temperature of the planet settled into a stable range similar to today. We became the dominant species on planet Earth and asserted our authority over the natural world by clearing forests for crops, rearing animals for food, and re-purposing the Earth's natural resources. Humans developed agriculture, writing systems, philosophy, religious order, cities, empires, and global trade.

Despite all the achievements of humanity, for most of human history the majority of humans have experienced very little improvement in quality of life. Records of GDP or income per person dating back 2,000 years show the average human earnt the equivalent of less than $1,000 per year (in today's money) through most of this period.[6] The total sum of global GDP may have increased or decreased with population, and the distribution of wealth certainly changed depending on the whims of the ruling elite, but productivity was stuck at $1,000. So, although the experience of life would have been quite different for your average human living in sixteenth century Shakespearean England compared to Cleopatra's Egyptian empire in 50 BC, the prevalence of disease, war, and hunger were pretty similar.[7] Humans have a potential life span of 70 years but for most of human history life expectancy was pegged at 20 to 40 years due to the prevalence of child mortality, disease, malnutrition, and violence.[8,9] Humanity was stuck in the Malthusian Trap.

In 1798, Thomas Robert Malthus, an English cleric and economist, published his book *An Essay on the Principle of Population*[10] asserting how any increase in resources, such as an abundance of food, will inevitably lead to population increase. Improved living standards don't last long because the rising population soon begins to exhaust the extra nourishment, so quality of life deteriorates and famine and disease once again rise so that human prosperity is *once again* reset. Malthus considered increases in available resources to be linear and unchecked population growth to be exponential – as the maths seemed incompatible, so famine, disease, and war would inevitably hold back the human species.

Let's pause for a moment and dig into the difference between linear and exponential growth as these concepts will prove a recurring theme throughout our story.

Linear increase is the more intuitive notion, where one variable will change by a given amount when compared with the change in a second variable. For example, if you invest £100 and earn 10% interest on this original deposit each year, after ten years you have £200 in your account. You make a linear increase of £10 interest for every one year of time.

Exponential growth, on the other hand, is where the rate of change in one variable depends on the magnitude of the other variable. An example in finance is called compound growth, where the interest accumulated in your account earns more interest. Imagine you deposit your £100 at 10% annual compound interest, then after the first year you have £110, but in the second year the extra £10 also earns interest: you end up with £121 by the end of year two. If you let your investment compound at 10% per year for ten years you end up with £259 in savings, which is significantly more than in our linear interest example. This is the power of exponential growth.

There are many aspects of modern life where we experience exponential growth. Economists expect GDP to grow at 1-3% per year compound, we have learnt the hard way that viruses spread exponentially, and, as Thomas Malthus pointed out, the human population itself has followed an exponential path through history.

At the start of the agricultural revolution there were less than five million humans on the planet. It took two thousand years to reach 10 million people, a further 1,500 years to reach 20 million, and the population kept on doubling every 1-2 thousand years: a compound growth rate of 0.05% per year.[i] By the year 1,500 AD, half a billion humans inhabited planet Earth and the exponential change kicked up a gear: in just 350 years the population doubled again to just over one billion people.

---

i You can estimate the exponential growth rate by dividing 70 by the time to double. And vice versa.

Another 75 years went by and we hit two billion people with the compound growth rate running at over 1% per year. By 1974, the population hit four billion and is now nearly eight billion people today.[11] Currently, every year we are adding around eight million people or the equivalent of one extra Greater London to the global population.

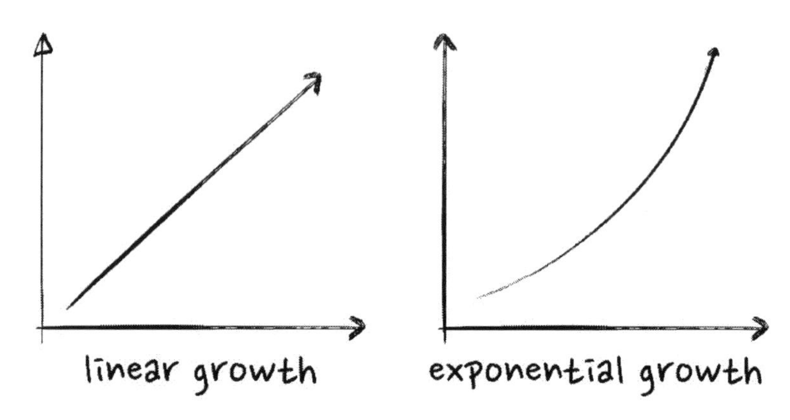

Charts of linear and exponential growth. "Population when unchecked increases in a geometrical ratio and subsistence for man in an arithmetical ratio." Thomas Robert Malthus, An Essay on the Principle of Population, 1798.

As Malthus put it, "The power of population is so superior to the power of the Earth to produce subsistence for man that premature death must in some shape or other visit the human race. The vices of mankind are active and able ministers of depopulation. They are the precursors in the great army of destruction, and often finish the dreadful work themselves. But should they fail in this war of extermination, sickly seasons, epidemics, pestilence, and plague advance in terrific array, and sweep off their thousands and tens of thousands. Should success be still incomplete, gigantic inevitable famine stalks in the rear, and with one mighty blow levels the population with the food of the World."[10]

The Industrial Revolution: Unlocking a Vast Energy Surplus from Fossil Fuels

Malthus' theory of doom was not accepted by all and many prominent scholars of the eighteenth and nineteenth centuries disagreed with his theories. The English political philosopher and novelist William Godwin thought Malthus' idea of unbounded population growth was "founded upon nothing" and asserted that the Earth could sustain nine billion people with little improvement to technology. Godwin, who hated aristocratic privilege and believed the monarchy to be corrupt, is considered the first modern proponent of anarchism. He believed in the innate good of people and advocated minimal state intervention and equality for all. He acknowledged population increases may follow increased standards of living but asserted that technological advance would reduce the amount of time individuals spend on production and labour and allow more time to develop "intellectual and moral faculties". He believed technology would fundamentally reshape human nature, leading to an "eclipsing of the desire for sex by the development of intellectual pleasures", limiting population, and driving human progress towards a utopian ideal.

Certainly, throughout human history, the relationship between accumulated knowledge, technology, and our ability to harness ever greater energy supplies has proven a fundamental driver of human progress. A virtuous circle of more surplus energy, creating more time to improve technology, and leading to more efficient ways of recording and transferring knowledge through the generations. From the spoken word of early Homo sapiens to writing and accounting developed by the ancient Sumerians, or Johannes Gutenberg's printing press, we have gradually improved our recordkeeping and accumulated more and more information. We are no longer limited to just recording and transferring information in our genes: the slow process of biological evolution has been augmented by language, writing, and mathematics.

However, for most of human history, the recording of knowledge was stuck within small networks of humans, prone to error, and its accumulation was linear and slow. Most human time was spent hunting, gathering, or growing food energy, leaving little spare time for knowledge building.

Our earliest ancestors, 3-4 million years ago, relied on their hands as the primary means of production and on the power of the sun, harnessed by plants, to provide food energy. But the system was inefficient, for plants capture less than 1% of the sun's energy and humans convert just 25% of food calories into useful work.[12] Early humans had to expend nearly all available energy hunting and gathering for more food, leaving little spare time to figure out how to do things better.

It took 1-2 million years to develop stone tools and to harness fire, but once we had mastered cooking, humans could grind and heat otherwise inedible or hard to digest food. Homo erectus had unlocked our first energy surplus. Less time and less energy were spent gathering and digesting food and eventually Homo sapiens developed primitive art and language to further disseminate our newfound knowledge.

Progress was slow but eventually humanity accumulated enough shared knowledge to begin to tend the land, grow crops, and rear animals. The agricultural revolution created a larger, more dependable source of energy which supported the first settlements and freed up more time for other endeavours. The development of permanent agricultural settlements also meant, for the first time, humans could accumulate more possessions than they could carry. So, to keep count of our accumulated stuff the Sumerians of Southern Mesopotamia (now Iraq) developed basic accounting techniques and writing systems.

The agricultural revolution continued to increase the energy surplus, but growth remained linear and slow. It took another ten thousand years before the next energy surplus from mechanised wind and water wheels was developed by the Romans 2,000 years ago.

The Roman Empire came remarkably close to breaking free from the Malthusian Trap, establishing a constitutional republic with a two-tier magistrate and senate political system, a large, well trained army, and a respectful, pragmatic approach to integrating conquered people of different races and religions. The stability of the Roman system coupled with organisational efficiency, record keeping, and expanding infrastructure meant that by the year 14AD the average GDP per person had increased to just over $1,000. Unfortunately, the increasing wealth peaked around the time of Emperor Caligula who was famed for his extravagant orgies, tyrannical grip on power, and lavish expenditure. By the end of the second century, plague, overspending, and civil war had weakened the Roman armies and barbarian invaders would occupy Italy in the fifth century AD.[13]

Nonetheless, Roman windmills and water wheels provided a new way to harness the power of the sun and created an even bigger energy surplus. Yet, the amount of energy available to support and supplement human labour was still constrained by the suitability of soils, availability of water ways, or the strength and direction of the prevailing breeze. Even the Romans couldn't harness energy fast enough and use it efficiently enough to grow production faster than population.

In 1543, the Polish astronomer Nicolaus Copernicus published his work *On the Revolutions of Heavenly Spheres.* This was a pioneering theory based on meticulous observations of the stars and suggested that the Earth rotated around the sun and was not, in fact, at the centre of the universe. Copernicus' work, though controversial at the time, laid the foundations for the scientific revolution which would transform our understanding of nature and our ability to manipulate the world.[14] In turn, science supported endeavours such as shipbuilding and navigation which, when combined with new systems for land rights and financing, led to the beginnings of large scale global trade allowing the transfer of new crop varieties to new agricultural lands where they were best suited for growing. This once again boosted agricultural yields and our energy surplus. Population growth began to meaningfully accelerate but still humanity was unable to break free from the Malthusian Trap.

Just three hundred years ago, most energy was still supplied by people, animals, biomass (wood), and basic wind or water wheels. Every one-hectare of farmland provided around half a tonne or 1,000 kWh[ii] of food energy per year. This was enough to support one to two humans but most of their efforts still had to be channelled back into tending the land.[15]

However, humanity was on the cusp of change. The development of the steam engine and the ability to harness the power of fossil fuels would break through linear productivity increases and set us on a path of exponential growth in energy, technology, and knowledge.

Thomas Newcomen invented the first steam powered machine in 1712, using the combustion of coal to heat water into steam to drive a mechanical pump. The machine was used on Lord Dudley's estate in the West Midlands, UK, to pump water from the bottom of a mine to help extract more coal. His early designs harnessed just 0.5% of the energy from combustion which limited wider practical use and it wasn't until 1770 when James Watt's redesign boosted the efficiency (to 5%) that steam power started to garner serious interest. Watt's more efficient design could generate 20-100 kW of power which was comparable to the biggest and best water wheels of the time.

Watt's patents expired in 1805 and a new generation of high-pressure steam engines came to market capable of even higher power output and more compact designs. These small, powerful coal-fired steam engines were used to power agricultural and industrial machinery, boats, and railways. By 1890, coal had overtaken biomass as the energy source of choice.[16] The production of food, wood, textiles, and metal products was mechanised. Agricultural productivity doubled, and every hectare of land could now generate one tonne of crop or 2,000 kWh of food energy per year, enough to support up to four people.

---

ii 1 kWh (kilowatt hour) is a measure of energy equivalent to 860 kcal (calories) or 3.6 million joules. The kW is a measure of power. Run a 1 kW steam engine for 1 hour and you generate 1 kWh of energy. Think of filling a sink with water as an analogy: power is the rate at which the water flows from the tap, and energy is the amount of water which ends up in the sink.

In 1858 Edwin Drake spent the best part of a year using a steam engine to drive a metal stake 21 metres into the ground. Though some thought he had gone mad, he was actually on the hunt for oil. A year in and just as his crew were starting to lose faith and his funding about to run out, he struck black gold. Drake successfully created the world's first commercial oil well.[17] Oil was first sold as a replacement for whale blubber in oil lanterns, but the market really took off with the invention of the two-stroke engine by Nicholas Otto in 1866. Otto's engine used the combustion of oil to drive the wheels of the first mechanised transport carriage. By 1900, further design refinements by Benz, Daimler, Maybach, and Lavassor, and supported by the invention of the sparkplug by Bosch, led to the commercialisation of the internal combustion engine, the beginnings of the car industry, and the birth of many of the world's largest companies still trading today.

Following steam power and the combustion engine, electrical power was the third major technology breakthrough of the industrial revolution. Minor electrical applications had been developed through the 1800s, but Thomas Edison was the first to build a full-scale electric power station in London in 1882. Holborn Viaduct station boasted 93 kW of power – enough to support up to 1,000 filament lights in the surrounding area – the design based around a steam turbine developed by Charles Parsons in Newcastle upon Tyne through the mid-1800s.[18] Parson's turbine used the combustion of coal to heat water into steam, driving the rotation of large blades which spun an electrical generator. By 1910, the design was 25% efficient with up to 25,000 kW power output (equivalent to 1,000 water wheels). Electricity quickly became the premium energy source for many applications, providing the flexibility to create heat, motion, light, or sound, clean at the point of use and requiring no mechanical linkage other than flexible wires to connect the equipment.

Whilst engineers were busy developing steam, combustion, and electrical power through the 1800s, chemists were busy developing additives to boost the yields of agriculture. In 1813, Humphry Davy published his bestselling book *Elements of Agricultural Chemistry*, which encouraged the use of guano for stimulating plant growth.[19] Guano is the accumulated excrement of seabirds or bats and is rich in the essential plant nutrients nitrogen, phosphate, and potassium.[20] "Amongst excrementitious solid substances used as manures, one of the most powerful is the dung of birds that feed on animal food, particularly the dung of sea-birds." The use of guano soon caught on and through the 1800s ever larger quantities were shipped from Peru, Namibia, and Bolivia to Great Britain, Germany, and the US. But the insatiable appetite for more fertiliser soon exhausted the limited supply of guano, and interest turned to deposits of nitrogen containing saltpetre in the Atacama Desert in Chile. But by the early 1900s, the industrialised world realised there was not nearly enough naturally occurring fertiliser to supply future needs. The focus turned to developing an artificial alternative and by 1910 Fritz Haber had figured out the chemistry and Carl Bosch, at German industrial giant BASF, had designed the industrial process.[21] The Haber-Bosch reaction could generate seemingly unlimited quantities of ammonia fertiliser using natural gas which was combined with nitrogen from the air.

Whilst coal-powered steam engines kick started the industrial revolution in the 1800s, the addition of oil-powered internal combustion engines and electricity generating steam turbines led to the second wave of industrial progress through the early 1900s. Electrical energy allowed for the development of more diverse industrial processes and combined with better transport networks and mass production techniques (pioneered by Henry Ford), led to rapid increases in productivity. The discovery of synthetic nitrogen fertiliser unlocked huge agricultural productivity gains: nearly 50% of the nitrogen in your body now originates from a factory running the Haber-Bosch process. By the early twentieth century, agricultural productivity had once again doubled, so that one hectare of agricultural land could now yield two tonnes or over 4,000 kWh of energy per year and support 5-8 humans. We had unlocked such a vast surplus of energy it was no longer just a handful of philosophers who had time to ponder big abstract questions, but growing numbers of workers had the opportunity to innovate and solve everyday problems, driving ever greater productivity and exponential progress.

The third phase of industry began with the invention of the integrated circuit, by Jack Kilby of Texas Instruments in 1958, which unlocked the potential of computing power.[22] The initial design was quickly improved and the number of transistors packed into an integrated circuit has doubled every two years since (an observation known as Moore's Law). Computers allowed for centralised control of industrial operations and better monitoring to boost productivity. Early applications developed into complex enterprise planning tools and automation, with machines replacing humans in repetitive or dangerous jobs. The internet opened the world of e-commerce and revolutionised communication. Global networks enhanced international trade which led to the opening of borders and many of the trans-national companies we know today.

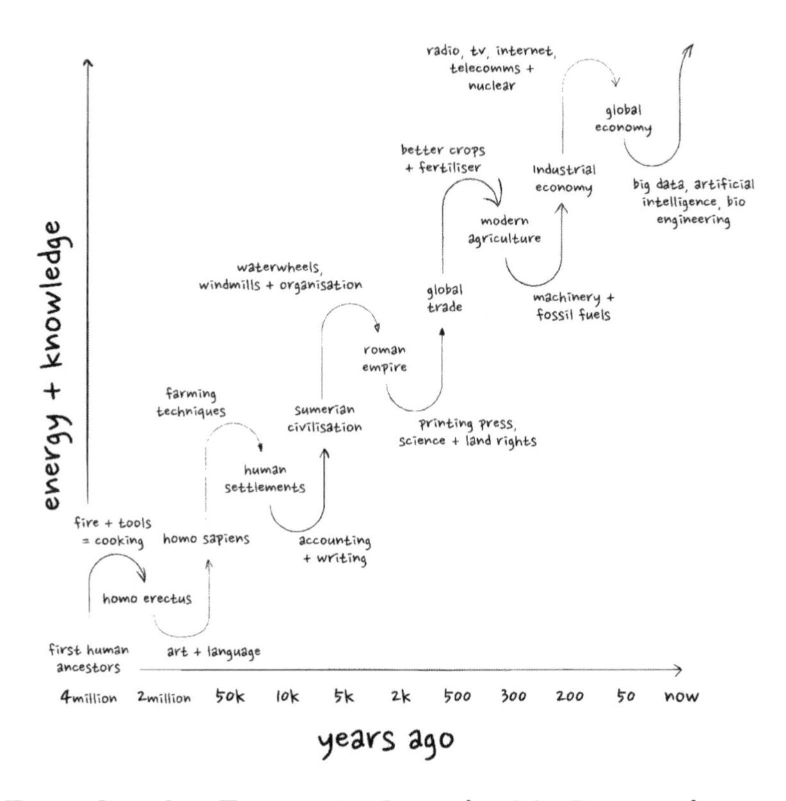

From Surplus Energy to Sustainable Prosperity

When Newcomen first discovered how to harness the combustion of fossil fuels, he started humanity on a path which would transform relatively linear progress into exponential growth. Today, we no longer rely on the direct power of the sun but have been able to unlock the vast stores of the sun's energy captured by plants over millions of years, energy that is buried, concentrated, and stored in fossil fuels. One barrel of oil contains 100 tonnes of concentrated ancient plant matter and can deliver as much energy as nearly a decade of manual labour[iii] and with this the power to deploy what seemed a limitless energy supply wherever and whenever it was needed.[23] Fossil fuels unlocked the potential of human capital. The number of industrial machines per person rose more than 200 times through the twentieth century.[24] The ability to supplement human labour with mechanised work meant more time could be spent developing other endeavours such as art, science, and politics, connecting ever greater networks of human beings. In turn these exploits have sped

iii 1 barrel of oil contains 1,700 kWh of energy. Humans can sustain 75W power for 8 hour shifts per day, the equivalent of 219 kWh work per year.

up the accumulation of knowledge and this has increased innovation across all aspects of life. For the first time in history, sustainably improved prosperity was made possible for an increasing number of people. We had escaped the Malthusian Trap.

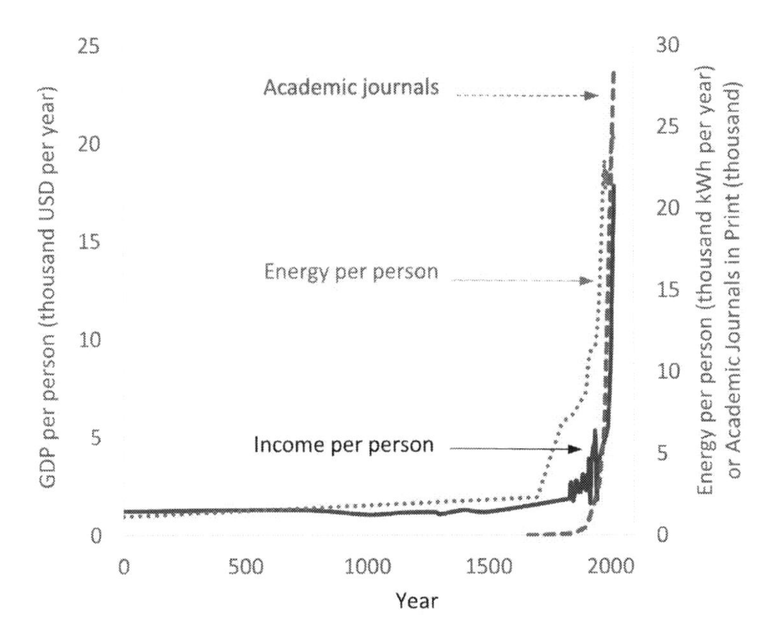

Solid line shows global average GDP (Income) per person for the last 2,000 years with scale on the left axis, inflation adjusted to 2011 USD purchase price parity (comparable spending power).[7,25] *Dotted line shows energy consumption per person per year in kWh (right axis).*[26,27,28] *Dashed line shows the number of academic journals actively publishing articles (right axis).*[29]

Energy consumption per person has increased ten-fold, from less than 2,000 kWh to over 20,000 kWh per capita per year. Whereas the first academic journal was not established until 1665, there are now nearly 30,000 of them publishing close to two million papers every year. The average income or GDP per person has broken through $1,000 and grown to nearly $18,000 per person per year. Agricultural employment now accounts for just a quarter of global working life.

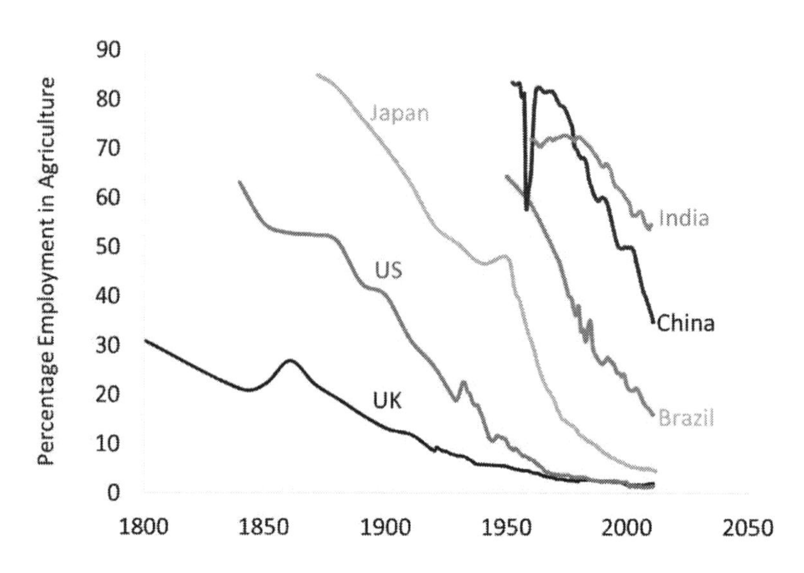

Agricultural employment as a percentage of total employment by country. Data from Our World in Data. [30]

Life expectancy has more than doubled from 30 to 70 years, battlefield deaths have declined from 10-20 per 100,000 people per year to below one. Child mortality has fallen from 43% to 4%. Extreme poverty, measured as monetary or equivalent income less than $1.90 per day (in today's money), has fallen from 90% in 1800 to just 10% today.

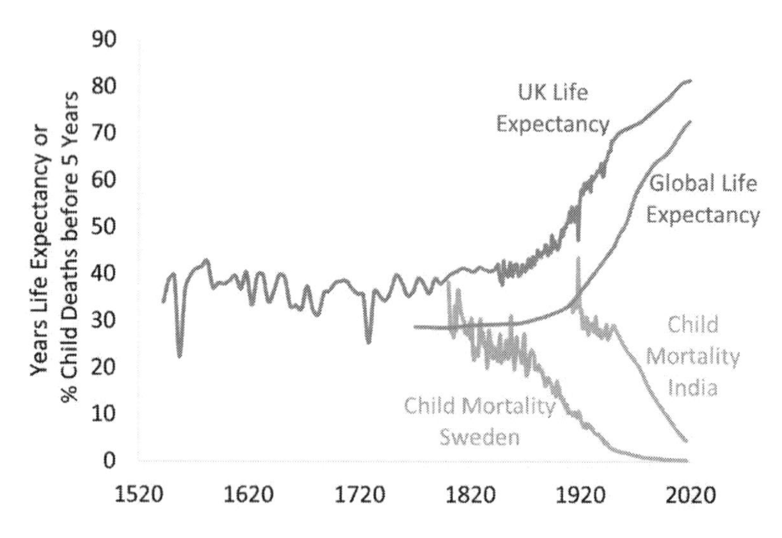

Life Expectancy for a new-born in years. [31] *Child Mortality measured as % of deaths before the age of five. Data from Our World in Data.*[32]

Though it may not always feel like it, we are living through a period of unprecedented wealth, quality of life, and peace. Life for your average human being is better than it has ever been and yet we remain a long way from completing our journey.

Prosperity for some, but not for all: Global Inequality and Rising Consumption

The statistician who sleeps with his head in the oven and feet in the freezer may be the perfect average temperature, but clearly averages don't tell the full story. Improved prosperity has come for some, but not for all, and there remains huge inequality across the world. Average incomes may have risen 18 fold, but three quarters of the world earns less than the average (the remaining quarter earns significantly more) and 10% of people still earn less than $1,000 per year.[33] The distribution is heavily skewed.

A significant proportion of the world's population still spends most of their time labouring in the field, earning little income, and isolated from the energy surpluses of the modern world. For many, life expectancies are short and child mortality rates high due to poor diet and lack of access to medicines and professional healthcare. Income inequality, corruption, and crime are prevalent. Education, basic human rights, and self-reported life satisfaction are low.

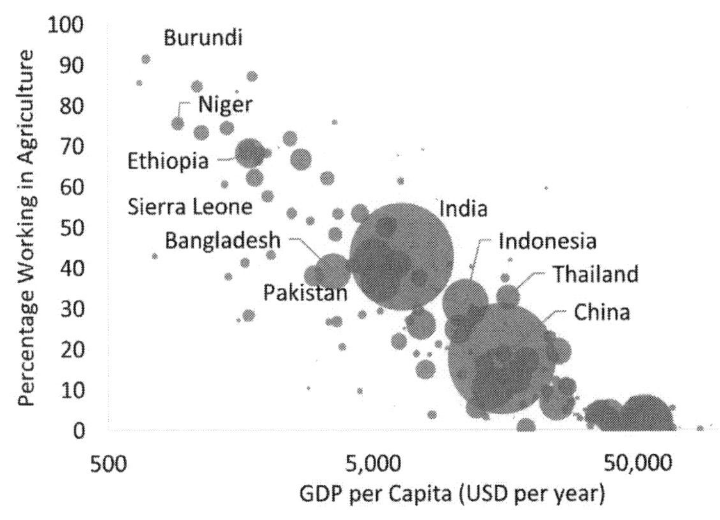

GDP per capita (using a logarithmic scale) versus the percentage of adults working in agriculture. Data from Our World in Data.[34] Each data point is the average for the country. The size of each bubble represents the population of the country.

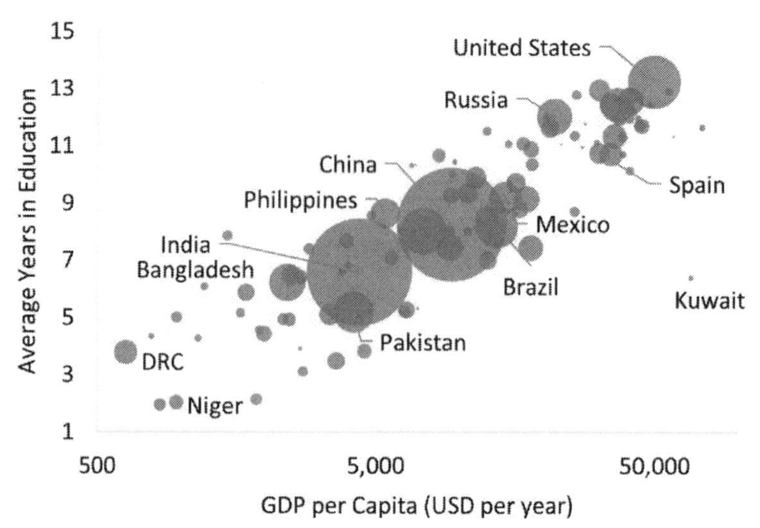

GDP per capita (using a logarithmic scale) versus the average number of years spent in education. Data from Our World in Data.[35] *Each data point is the average for the country. The size of each bubble represents the population of the country.*

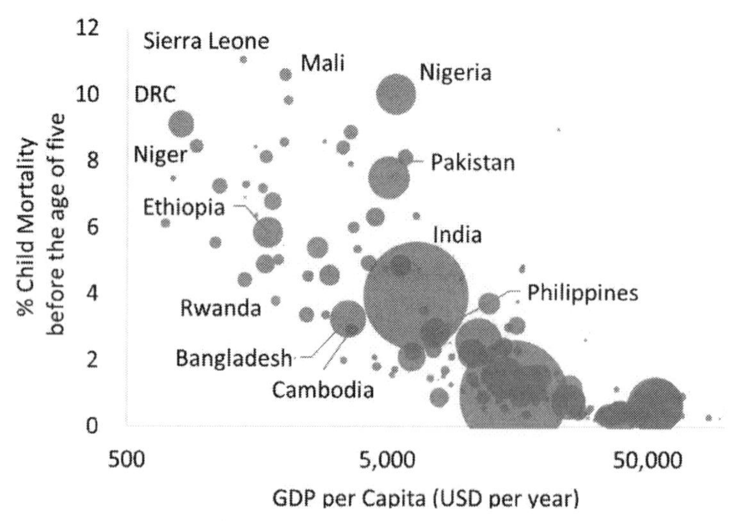

GDP per capita (using a logarithmic scale) versus the percentage deaths before the age of five. Data from Our World in Data.[36] *Each data point is the average for the country. The size of each bubble represents the population of the country.*

In their book, *Factfulness*,[37] Hans, Ola, and Anna Rosling break down the global population into four income levels: if you have enough leisure time to be reading this book, you are probably in the top 10%. You most likely earn over $16,000 per year, you went to school, drive a car, have access to electricity, hot running water, and you eat pretty much whatever you want.

But for every person in the top 10% there is someone in the bottom 10% who consumes the equivalent of less than $1,000 per year and remains stuck in the Malthusian Trap. Work is done by hand and you travel by foot. Your small plot of land must provide enough food for you and your family. Most of your time is spent in the field or walking to the nearest well as you have no access to safe water or sanitation. You cook and heat your house on an open stove which makes it difficult to breath. You hope that you or your family don't get ill as you have no income to afford medication. Everything you produce is required to sustain your family with nothing left over to invest in farm equipment or basic amenities which could make life easier, more productive, and help break free of the poverty trap. You are largely isolated from the modern world.

The remaining 80% of the global population fall somewhere in between but importantly are on the path towards the top 10%. Faster transportation, better sanitation, more nutritious diets, and clean cooking and heating lead to better health, growing wealth, and a virtuous circle of increasing productivity.

As the majority of the population rightfully strive for a better quality of life, so global consumption will increase. More miles travelled, food eaten, and products made. But what strain will this exert on the planet? Are Malthusian concerns about limits to growth warranted?

Images courtesy of $Dollar Street, Gapminder.org. [38] *Annual income bands estimated based on the future of Worldwide Income Distribution.*[39] *Note: income per capita is slightly less than GDP per capita quoted elsewhere because of depreciation, cash retained by companies, certain government spending and under reporting on tax returns. Amount quoted in 2011 USD purchase price parity including the value of self-production and consumption (e.g. food growing).*

Planetary Pressures and the Argument over Limits to Growth

Thomas Robert Malthus published his book *An Essay on the Principle of Population* in 1798, claiming that unless we intervened, humanity was destined to remain bound to population-led poverty forever.[10] But just seven years later, James Watt's steam engine patents expired, ushering in a wave of mechanisation and productivity improvements across agriculture and industry. The global human population has swelled from 1.2 billion people in 1850 to nearly 8 billion people today, but for the first time in human history it appears we have broken free of the Malthusian Trap. The industrial revolution has achieved the impossible: squaring the demands of population growth *and* increasing per capita wealth.

By most measures, the industrial revolution has been of great benefit for humanity, however it has also changed the planet as never before. For the first time in the four billion years of life on Earth, a single species has begun to permanently reconfigure the natural landscape, repurpose the Earth's natural resources, re-write the rules of evolution, and re-engineer the climate.

Agricultural land has grown from five million to nearly 50 million square kilometres. Oil-powered, mechanised, agricultural machinery has turned one third of all dryland from forests into fields of regimented crops and animal pastures.[40] Humans account for just 0.5% of animal flesh on the planet and yet we use 23% of the primary production of land-based plants.[41] Domesticated animal biomass for feeding humans now outnumbers large wild animals by seven to one with half a billion pigs, one billion cows, and over 20 billion chickens on the planet at any one time.[42,43,44]

Extraction of fossil fuels has been used to power mining and for purification and chemical transformation of natural minerals to build sprawling cities covering 2% of Earth's land. Oil-powered travel across land, sea, and air has brought once disparate geographies into a single ecosystem and redistributed once distinct species and diseases.

We humans began our training as subsistence farmers[iv] and two hundred years later we have graduated top of the class: the first species on planet Earth to yield the power to transform planet Earth itself. The exponential rise in human population, consumption, and changes to

---

iv Subsistence farming provides enough food to meet local requirements with no surplus.

the natural world have led many after Malthus to repeatedly question the limits to increasing prosperity and population growth. His fear of exponential growth would inspire the works of many famous nineteenth century scholars; from British economist David Ricardo who warned of land scarcity and rising rent creating unsustainable inequality, to the German philosopher Karl Marx who feared the capitalist system would concentrate industrial production in the hands of the few, leading to limits on growth and even revolution: "This law of capitalist society would sound absurd to savages, or even civilised colonists. It calls to mind the boundless reproduction of animals individually weak and constantly hunted down".[45] These ideas were carried into the twentieth century by Marxist revolutionaries: "Once we have overthrown the capitalists, crushed the resistance of these exploiters with the iron hand of the armed workers, and smashed the bureaucratic machinery of the modern state, we shall have a splendidly-equipped mechanism, freed from the parasite", Vladimir Lenin, State and Revolution, 1917.

Fast forward half a century or so and, following the post-war boom, concerns over the rate of population increase resurfaced in the 1950s and 1960s, the *Zeitgeist*[v] captured by Stanford University Professor Paul R. Ehrlich's *The Population Bomb*, published in 1968.[46] The book predicted global famine in the coming decades. Ehrlich and his wife argued that population had already doubled from two to four billion people in just one generation and, unless something was done, was on track to do so again. They asserted that the impact on Earth's natural systems was a combination of population, affluence, and technology (I=PAT). The Ehrlichs were convinced that limiting population was the only way to live sustainably on our allotted sources of natural income and without squandering Earth's natural capital of non-renewable energy, minerals, and soils.[47] Ehrlich asserted that food production was already near limits and the choice was stark: either "population control or the race to oblivion".

As the Ehrlichs were worrying about population growth, many others at the time were preoccupied with concerns over peak oil and the limits to fossil fuel production. M. King Hubbert, a geophysicist working at Shell in the 1940s and 1950s, devised a mathematical model which linked fossil fuel reserves, discovery rates, and output, to calculate peak production rates of finite resources. His models correctly predicted the peak of traditional US oil production in the 1970s and he went on to estimate that global oil would peak by the year 2000. In 1973 the Organisation of Petroleum Exporting Countries (OPEC) declared an

---

v Zeitgeist: the defining spirit or mood of a particular period of history as shown by the ideas and beliefs of the time.

oil embargo on nations supporting Israel in the Arab-Israeli war. This coincided with global oil wells running at full capacity and plunged the western world into crisis. The price of oil quadrupled, gasoline was rationed, and the US imposed a 55mph driving speed limit to curb consumption. Worries moved from food shortages to energy shortages.

Concerns were further heightened after the publication of *Limits to Growth* in 1972.[48] The report was researched and written by four scientists at MIT for The Club of Rome, an international collection of businessmen, scientists, and statesmen trying to better understand global problems as part of larger interconnected systems. The team of four scientists built an early computer model called World3 which linked changes in population, resource use, and pollution, using the ideas of economic investment and consumption.[49] The scientists input the rate at which technologies improved, to try and understand not only the limits to ***sources*** of energy, food, and minerals but also the ability of soil, water, and the atmosphere to accommodate pollution: limits to the Earth's ***sinks***.[50] The idea of sources and sinks was a more nuanced take on Ehrlich's ideas, which the scientists used to estimate the carrying capacity of the planet. The team estimated that the quality of life for humanity would rapidly deteriorate at some point over the next 100 years. They calculated that the only way to avoid disaster was through a combination of limiting further population expansion, ending consumption growth, and pushing sustainable technologies as hard as possible.

By the end of the 1970s, limits to growth concerns had reached the highest levels and US President Jimmy Carter commissioned Global 2000 to better understand the problems.[51] The report surmised, barring a revolutionary advance in technology, life for most people on Earth would be more precarious in the year 2000 compared to 1980, unless the nations of the world acted decisively to alter ongoing trends. The growing body of work by scientists and environmentalists was impressing an ever more dire vision of the future in the public consciousness. The possible response to these limits started to attract the attention of liberal economists who saw a rising threat to free markets and unhindered consumption.

As Paul Sabin describes in his book *The Bet*, Julian Simon, a professor at the University of Illinois, was one of those economists. He had gained an MBA in 1957 at the University of Chicago business school around the time that Milton Friedman was challenging Keynesian ideas of big government in favour of deregulation, free markets, and printing money. These ideas no doubt influenced Simon and his early work focussed on the declining cost of food, and on demographic economics which led him to challenge the Malthusian account, asserting instead that a larger population is of benefit to everyone because more people equals more innovation. He was convinced that although material supplies are physically limited on a finite Earth, they are economically infinite because technology creates more efficient use, enables resources to be recycled, and markets will ultimately find substitutes, where necessary, with the action dictated by price signals. Should resources run into short supply, prices increase and attract more efficient use, new supply, or alternatives: "we find new lodes, invent better production methods, and discover new substitutes".

In collaboration with Herman Khan, who was a prominent US Cold War strategist, a futurist, and the inspiration for Stanley Kubrick's title character in the film, *Dr Strangelove*, Simon published a book in 1984 called *The Resourceful Earth*,[52] a direct attack on the ideas of the Global 2000 report. Simon and Khan asserted that by the year 2000 the world would be more populated but less crowded, less polluted, more economically stable, and less vulnerable to resource supply disruption. Yes, problems would arise, but capital markets and society would find solutions and leave the world a better place. The book was pro-nuclear in the long run, anti-big-government and, at its core, made the case for the capacities of market substitution to create plentiful resources.

Animosity between Simon and Ehrlich had been simmering since the publication of *The Population Bomb*. The book had become an international bestseller and Ehrlich's celebrity status was consolidated with the publication of opinion pieces everywhere from the *Washington Post* to *Playboy* magazine. In his regular appearances on 'The Tonight Show', he forcefully delivered his message of population control or impending doom to large audiences. He dismissed Simon's theories of infinite growth and referred to Simon as the "leader of a space age cargo cult".

Infuriated by the widespread attention being showered upon Ehrlich, and following Ehrlich's claim that he would take even-money odds on England not existing by the year 2000, Simon decided to call him out.[53]

Simon challenged Ehrlich to choose any natural resource he wanted and wagered that the price of those commodities would decline over any period of time greater than one year because technology would improve, and substitution would occur. Ehrlich took him up on the offer and in 1980 the two academics agreed to bet $1,000 on whether the price of five metals would rise or fall over the next decade.[54]

The rivalry between Ehrlich and Simon represented the growing division between scientists or environmentalists and liberal economists. The limits-to-growth-thinkers reasoned that Earth's resources were finite and therefore consumption or population must be regulated. As we have seen, these are the constituency that became known as the neo-Malthusians. Those liberal economists who believed in the power of capital markets, substitution, and indefinite growth became known by the derogatory term 'Cornucopians', after the mythical horn of plenty from Greek mythology.

American economist Kenneth Boulding, giving evidence at a 1973 congressional hearing on the Club of Rome report, famously quipped: "Anyone who believes that exponential growth can go on forever in a finite world is either a madman or an economist".[55] US journalist and author Max Lerner meanwhile evidently disagreed, writing in mockery: "Ashes to ashes. And dust to dust. If the bomb doesn't get you. Exponential curves must".[52]

So, who is right? Are we heading for Dystopia or Utopia? Doom or Boom?

The Ongoing Argument and Concerns over Climate

Herman Khan wrote, in an essay in 1976, "our disagreement with advocates of the limits-to-growth positions sometimes is that they raise false, non-existent or misformulated issues; equally often, perhaps they pose as being basically insoluble real problems for which we believe rather straightforward and practical solutions can be found in most cases".[56] So, is the cornucopian mantra correct? Can we find practical solutions to all our problems?

The Ehrlichs viewed the human population as no different from populations in the natural world, both ultimately controlled by the actions of predators, the prevalence of disease, and the availability of food. Early humans learnt to fashion tools and hunt, which quickly took them to the top of the food chain, nullifying the threat from predators. Modern humans mastered microbiology, so reducing the impact of disease. The industrial revolution and modernisation of farming opened up greater food supplies, thus reducing the prevalence of famine and starvation. The Ehrlichs may have correctly predicted a population explosion, but the race to oblivion has, so far, not come to pass. Expanding agricultural land, improving yields, and other efficiencies have meant that, despite the world population doubling from four to nearly eight billion people over the last 50 years, the prevalence of hunger has rapidly declined in percentage terms and has made small progress in absolute numbers.[57] Availability of food has yet to constrain the population explosion.

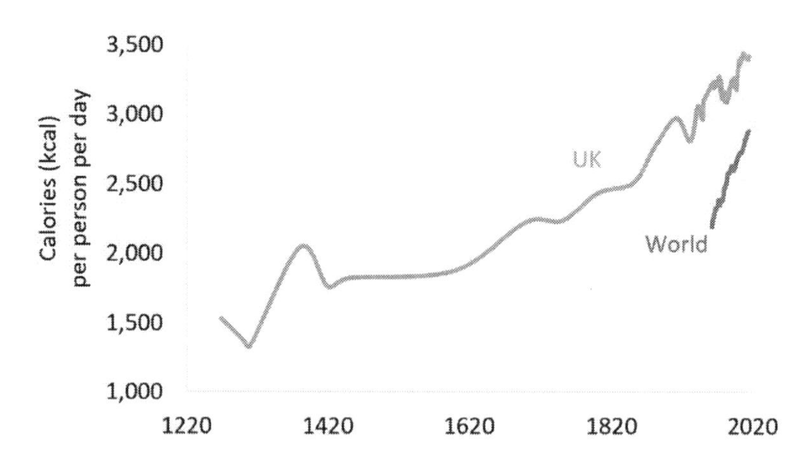

Daily average calorie supply per person in the UK and in the World. [58]

M. King Hubbert correctly predicted the peak in supplies of traditional oil in the late twentieth century.[59] However, his calculations failed to reckon on improving technologies, global discoveries of new sources and the substitution of traditional oil for non-conventional deposits like deep sea oil, shale gas, or tar sands. Despite the energy shocks of the 1970s, fossil fuel consumption has only continued to expand into the twenty-first century. Based on current consumption there is enough coal, gas, and oil to last another 200 years at a similar cost as that of today.[60] Add the deposits not yet located, or more difficult to reach, and resources more than double again.[61] Enough to power the planet for another 500 years.[vi] Availability of fossil fuel energy is yet to hold back growth.

Thus far we *have* found practical solutions to many of our biggest problems. The cornucopian view of technological advancement driving better yields and substitution of resources seems to have prevailed, allowing for an increasing population and greater levels of consumption. Yet that progress has required significant governmental intervention, including the nuclear non-proliferation, acid rain, and ozone treaties.

vi There are an estimated 1,000 billion tonnes of coal, 200 trillion cubic metres of gas and 1,700 billion barrels of oil reserves which have been located and can be extracted at reasonable cost with today's technology.

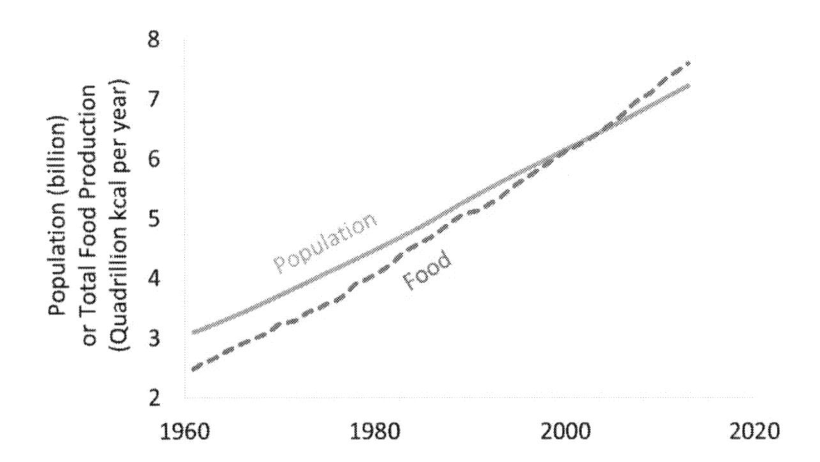

Global population in billions and global food calories (measured in 10^15 kcal).[56]

So, perhaps it's the Club of Rome's sinks, not sources, which pose the more imminent threat to humanity. Could the ability of the Earth to accommodate growing pollution prove the ultimate limit to humanity?

The concern is no longer availability of fossil fuels, but the socio-economic damage that combustion of oil, gas, and coal may inflict on civilisation and the natural world. CO_2 was once considered an innocuous by-product of combustion, but we have come to understand that CO_2 plays a fundamental role in the temperature and climate of Earth.

CO_2 concentrations in the atmosphere have held steady at 250-280 parts per million (ppm) throughout the last 12,000 years of civilisation.[5] In the last 150 years, however, humans have pushed concentrations to over 410ppm.[62] The last time atmospheric CO_2 was at this level was 3 million years ago when temperatures were 3-4ºC hotter, the Arctic was ice free, sea levels were 20 metres higher, and humans were yet to evolve.

The world's emissions of greenhouse gas have grown to over 50 billion tonnes per year, a by-product of the global economy and modern life.[63] Fossil fuels power the world but provide the highest quality of life for just 10% of people. Half of new annual emissions are removed by the land and oceans, but the remainder accumulate in the atmosphere, building ever higher concentrations which will persist for centuries or millennia, even if we could cut annual emissions to zero tomorrow.

If we remain bound to our existing fossil fuel system as the population rises to over 10 billion people, who will rightfully want to increase consumption towards the highest standards of living, then 50 billion very quickly becomes 100 billion tonnes of emissions per year. CO_2

concentrations could reach 1,000ppm by the start of the next century. The last time CO_2 reached this level was deep in geological time, 50 million years ago, when temperatures were over 10°C warmer. There was no ice on Earth, sea levels were 200 metres higher, palm trees grew in Alaska, and 200-pound penguins as tall as modern humans roamed New Zealand.[64,65]

Through a virtuous circle of accumulated knowledge and innovation, humans have unlocked ever greater stores of surplus energy. We have figured out how to control and direct the combustion of fossil fuels to drive productivity increases across transport, industry, and agriculture. By most measures prosperity for your average human is better than ever, and yet 90% of the global population are still striving for a better quality of life. As living standards rightfully increase around the world, so too will the rate at which we release carbon dioxide. This has shifted the 'limits to growth' arguments from sources to sinks. We are no longer asking when will fossil fuels run out, but rather, how much can we burn before climate change destroys the improvements to wealth, health, quality of life, and global peace created over the last 200 years? Are the cornucopians right? Can we just leave it to free markets and technology to solve the problem? Or do we listen to the neo-Malthusian view and limit growth or run the risk of humanity falling back into the Malthusian Trap? Where are the dials of doom and boom set?

In the next chapter we will explore the fundamental science of global warming, examine the evidence of human influence on the climate, and start to understand the limits of the Earth's atmosphere.

Chapter 2: Carbon Dioxide and Climate

Earth's Geological Past to the Epoch of Humans

In this chapter we will trace the history of climate science back over 150 years to better understand the scientific theory of the greenhouse effect, the relationship between CO_2 and temperature, and the natural cycles which control Earth's climate. We will journey through 500 million years of proxy data to understand how scientists reconstruct the past atmosphere, and why when combined with the direct records of the last 150 years it provides irrefutable evidence than human activity is warming the planet.

Carbon Dioxide: The First Understanding of the Atmospheric Thermostat

Eunice Newton Foote was an American scientist and women's rights campaigner in the nineteenth century. She was a key signatory of the first women's rights convention in 1848 and six years later become the first person to link atmospheric temperature with CO_2.[1] By measuring the temperature rise of jars containing different gases, she predicted how carbon dioxide and water vapour concentrations could modulate solar heating: "An atmosphere of that gas would give to our Earth a high temperature; and if as some suppose, at one period of its history the air had mixed with it a larger proportion than at present, an increased temperature... must have necessarily resulted".[2] Her work was presented at the Annual Meeting for the American Association for the Advancement of Science in 1856 by a male colleague, Joseph Henry, but her results never made it into the conference proceedings. Her pioneering insights went unrecognised for 150 years until, in 2010, retired petroleum geologist Ray Sorenson came across her work in an 1857 volume of *Annual Scientific Discovery.*[3]

Further laboratory work on CO_2 and a range of other gases, published by Irish physicist John Tyndall in 1860, further supported Foote's work.[4]

By 1890, Swedish chemist Svante Arrhenius had used the experimental data to make the first climate predictions. Arrhenius estimated that if the concentration of CO_2 doubled from 280 to 560 ppm[i], temperatures would increase by 5°C.[5] At the time, though, his work was regarded as no more than scientific curiosity, nobody imagining that humans might *actually* change the atmosphere so drastically. Yet half a century later, a select few scientists started to take serious interest in Arrhenius' work. In 1931 American physicist E. O. Hulburt refined the CO_2 absorption values and estimated a 4°C sensitivity to the doubling of CO_2.[6] Then in 1938, Guy Stewart Callendar presented the first evidence that Earth might already be warming. He used his own climate model to predict that if the concentration of atmospheric CO_2 doubled, the planet would be 2°C warmer.[7]

These early predictions have proven remarkably accurate, with CO_2 concentrations now over 410 ppm and temperatures already over 1°C higher. The best estimates from modern climate models, which use super-computers and masses of data, suggest that doubling CO_2 concentrations will lead to around 3°C total warming with the range of estimates between 1.5 and 4.5°C.

The Greenhouse Effect: How CO_2 Warms the Planet

Understanding the link between CO_2 and temperature requires we start with the all-important relationship between our planet and the sun. Planet Earth is kept warm by the sun's constant release of energy in the form of electromagnetic radiation. The most familiar and largest part of the electromagnetic spectrum is visible light with wavelengths between 0.4 and 0.7 micrometres (less than one thousandth of a millimetre) which our eyes use to see the world. Other parts of the spectrum include high energy, short wavelength UV light, and X-rays, alongside lower energy, longer wavelength infrared, microwaves, and radio waves. It's this spectrum of the sun's energy which warms your face on a summer's day, and which keeps the Earth temperate enough for life.

The sun's radiation energy hits the Earth's atmosphere with an average

i The temperature increase due to concentrations of CO_2 doubling in the atmosphere from pre-industrial times is now known as the **climate sensitivity.**

intensity of 340 watts[ii] of power per square metre.[iii] Atmospheric gases, clouds, snow, ice, oceans, and the Earth's surface reflect 30% of this incoming energy back into space and the remaining 240 watts per metre is absorbed by the atmosphere, land, and oceans.

The portion of the sun's energy which is absorbed is re-radiated and sent back towards space as lower energy, longer wavelength, infrared radiation. Water vapour and greenhouse gases like CO_2 in the atmosphere are more effective at absorbing this outgoing infrared and in turn re-radiate the energy in all directions. This serves to trap more heat before it escapes back into space, heating the lower atmosphere (troposphere), and cooling the upper atmosphere (stratosphere). This is the familiar phenomenon known as the "Greenhouse effect" coined by mathematician Joseph Fourier in 1824 before the process was fully understood.[8]

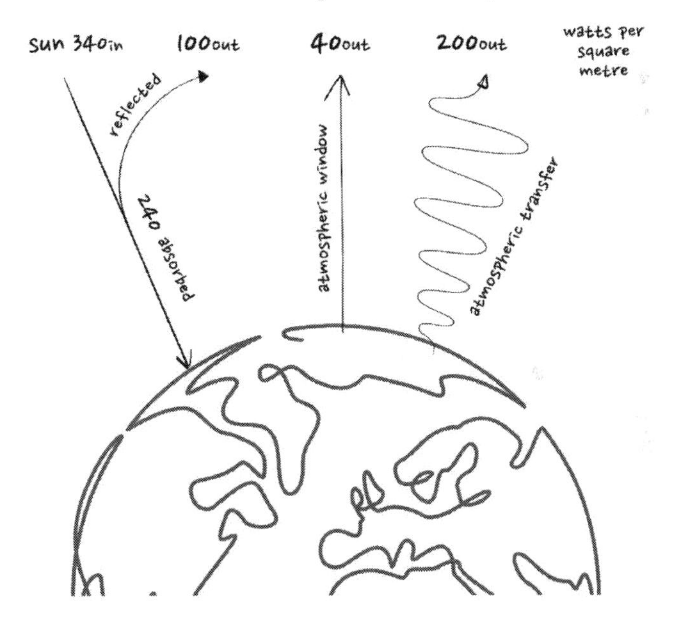

Earth's Energy Budget showing watt per square metre flow of energy between space and Earth. Adapted from NASA.[9] Total outgoing energy matches incoming energy at 340 watt per square metre when the temperature is stable.

ii Watts (W) are units of power - the rate of energy change equivalent to 1 joule per second.

iii The sun radiates 380 trillion-trillion watts of power which reaches Earth's orbit with 1,360 watts per square metre power. Multiplied by the area of Earth's shadow (130 trillion square metres) and 173,000 billion kW of power shines down on the planet. Divide the total power by the Earth's surface area (510 trillion square metres) and 340 watt per square metre average power reaches the planet.

But the physics of the greenhouse effect are a little more nuanced than was initially thought. To better understand the phenomenon, first start by imagining that the Earth has no atmosphere. It simply heats up and re-radiates energy with nothing to slow the outgoing radiation, and all the energy is lost into space from the Earth's surface. The temperature of that surface can be calculated using the equation for a black-body[iv] – the Stefan-Boltzmann law – which says the rate of energy loss equals 5.67×10^{-8} multiplied by the temperature to the power of four.[10]

Power Radiated = Stefan-Boltzmann Constant x Temperature^4

We know 240 watts per square metre is absorbed by Earth so, if the temperature is stable, 240 watts per square metre must re-radiate back to space to balance the energy flows. Plugging these numbers into the equation tells us the temperature of the Earth would be 255 kelvin[v] or -18ºC with no atmosphere.

But luckily our planet isn't floating through space completely naked. Earth is covered in a blanket of atmospheric gases like water vapour and CO_2. About 40 watts per square metre of the energy re-radiated from Earth is emitted at a frequency which avoids being re-absorbed by atmospheric gases and escapes unhindered into space: this is known as the atmospheric window.[vi] But the remaining 200 watts per square metre of outgoing energy is absorbed and re-radiated over and over again, bouncing around the atmosphere until it reaches a point high enough where gases like CO_2 and water vapour are few and far between and the radiation energy can finally escape: the average height at which energy escapes the atmosphere to space is called the 'emission level' and it is about 5 km up.[11] The temperature at the emission level is cooler than

---

iv A black-body is an idealised body that absorbs all incoming radiation.

v Scientists use kelvin as temperature units. 1 kelvin is equal to 1 ºC. But zero degree C or the freezing point of water is 273 kelvin. Zero kelvin is absolute zero where there is no energy whatsoever.

vi Wein's Law states that the wavelength (the inverse of frequency) where the greatest amount of energy is emitted from a black-body equals 2,898,000 ÷ temperature (in kelvin). The sun is 6,000 kelvin and emits most energy at 500 nanometres or green light. The Earth is about 300 kelvin and emits most energy around 10,000 nanometres (infrared). However, there is a large range of radiation emitted around these values. There is enough water vapour in the Earth's atmosphere to block most energy from escaping directly to space apart from a narrow window at 8,000 to 13,000 nanometres where water cannot absorb, called the atmospheric window. CO_2 can absorb in part of this atmospheric window, but concentrations in the atmosphere are so low the absorption is only partial, and is not saturated, therefore adding more CO_2 partially closes the atmospheric window and forces the radiation energy to bounce around the atmosphere until it escapes high up in the colder atmosphere.

the surface of the Earth because the gases of the atmosphere expand and become less dense at higher altitudes. Think about how cold it is at the top of a mountain or the temperature of an expanding aerosol spray on your skin.

Now think back to our naked, cold Earth and imagine that we instantaneously added an atmosphere to the planet. Most of the energy is no longer lost straight from the surface but from the colder emission level 5 km up in the new atmosphere and where a lower temperature emits less radiation energy (following the Stefan-Boltzmann law). The lower radiation creates an energy imbalance, more energy comes in than goes out, and the planet and atmosphere heat up until the temperature of the emission level reaches 255 kelvin (-18ºC) again and the balance of energy in and out is restored.

Without the warming impact of greenhouse gases, the surface temperature of Earth would be 18ºC and there would be no proliferation of complex life. Atmospheric gases create a warmer surface. Water vapour is the strongest infrared absorber and adds 20ºC to the Earth's average temperature. Carbon dioxide, with the help of other greenhouse gases such as nitrous oxides, ozone, methane, and fluorocarbons, adds another 13ºC. This brings the average surface temperature on Earth to 14-15ºC which has proven optimal for modern humans and civilisation.

Now you can start to understand why adding even more CO_2 to the existing atmosphere creates a warming effect. Firstly, because CO_2 is at a relatively low starting concentration, the frequencies where the gas absorbs infrared are not saturated and can close off parts of the atmospheric window. This prevents an extra portion of energy from escaping directly into space from the Earth's surface. Instead, the energy is emitted from the colder atmosphere creating a small energy imbalance and initial warming. This initial temperature increase allows the atmosphere to hold more water vapour which raises the height of the emission level. A higher emission level is a colder emission level and even less radiation makes it to space until the whole system heats up enough to rebalance the energy flows.

In short, adding greenhouse gases to the air partially closes the atmospheric window, creates a thicker blanket, and warms the planet.

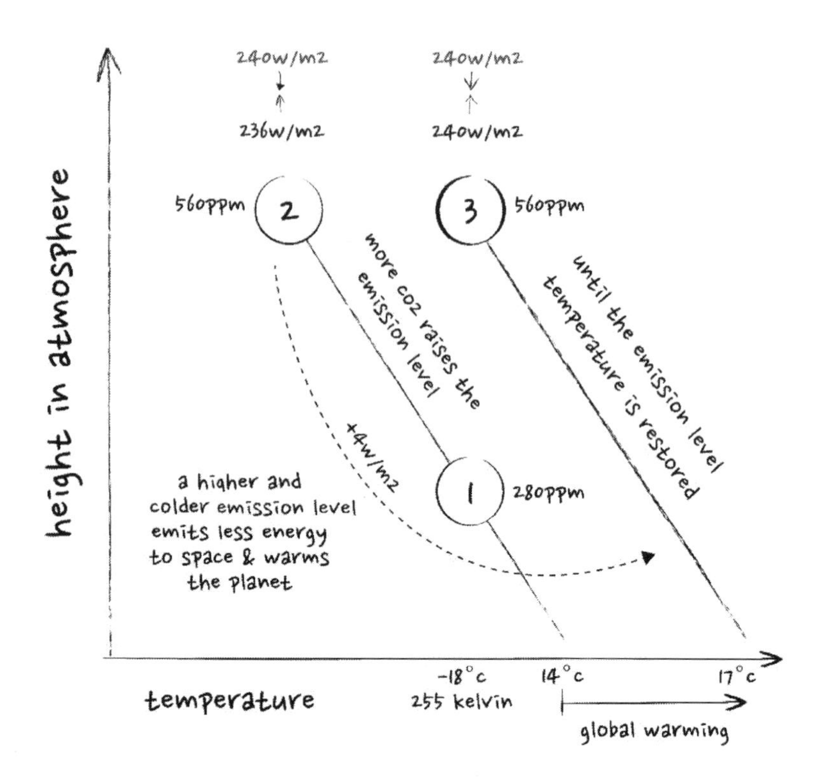

Diagram of the Greenhouse Effect: (1) Pre-industrial CO_2 concentrations (280 ppm) double to 560 ppm. This increases absorption in the atmospheric window and increases the average altitude that energy is radiated to space (emission level). (2) The new higher/colder emission level emits less energy (due to Stefan-Boltzmann) and a net energy flow of 4 watts per square metre enters Earth. (3) The extra incoming energy warms the surface and atmosphere until the emission level returns to 255 kelvin (18°C) and the energy balance is restored. If CO_2 doubles the surface temperature must warm by 3°C to rebalance the energy flow and reach a new, warmer steady state. Adapted from Andrew Dessler.[12]

Despite water vapour contributing the greater impact on warming overall, CO_2 plays the most important role in controlling atmospheric temperature. Water vapour is self-regulating because if you add more water vapour to the atmosphere with no immediate change to temperature, the water vapour will condense back to liquid. Water is in a balance between vapour and liquid for a given temperature. Greenhouse gases are different because they remain in the atmosphere as gas long enough to warm the planet. Only then can the lower atmosphere hold more water vapour, which further amplifies the initial CO_2 driven heating until no more water vapour can be accommodated.[13]

Carbon dioxide acts as the planetary thermostat, whilst water vapour is the boost button.

Setting the Thermostat: Energy Flows and the Carbon Cycle Controlling the Climate

If carbon dioxide is the atmospheric thermostat which regulates the temperature of Earth, what controls the CO_2? This is where climate science gets even more complex. The Earth's temperature and the concentration of CO_2 in the atmosphere are determined by several biological, chemical, and physical factors; many of which interact and may have both positive and negative or amplifying and dampening feedback cycles. Amplifying feedback reacts to an initial change in temperature or CO_2 by further increasing the change. Dampening feedback is the opposite and reduces the magnitude of the original change.

Here are some of the natural drivers of the CO_2 concentrations and temperature:

Rock weathering traps CO_2 from the atmosphere in the Slow Carbon Cycle: Carbon dioxide gas in the atmosphere dissolves in rainwater forming carbonic acid (H_2CO_3) which slowly dissolves rock. The net impact of rock weathering is to remove CO_2 from the atmosphere, trap it in chemical solution, and transport the carbon down river to the ocean where, either directly or through the formation of biological shells, the CO_2 becomes buried as sediment in the deep ocean. Rock weathering is the first part of the slow carbon cycle taking half a million years to move carbon from atmosphere, land, and ocean to rock.[14] The slow carbon cycle will act to rebalance temperatures over geological time – whereby higher temperatures and more water vapour drives faster weathering of rock, reducing levels of CO_2 and gradually lowering temperatures – a dampening mechanism but over hundreds of millions of years.[15]

Tectonic and volcanic activity re-release CO_2 from rock in the Slow Carbon Cycle: The Earth's crust consists of seven large and many smaller tectonic plates which have been moving for the last 3.5 billion years. As the plates shift, at a few inches per year, this creates subduction zones (one plate moves under the other) and divergent margins (where new sea floor forms). This activity reshapes continents, creates mountain ranges, and leads to earthquakes, sea vents, and volcanic activity. The amount of tectonic activity affects the rate at which subducted rock chemically transforms under heat and pressure to release silicate and CO_2.[16] Volcanic eruptions and continental rifts release the silicate and CO_2 back into land and atmosphere. This is the second part of the slow carbon cycle which returns the carbon trapped in rock back into the atmosphere over millions of years.[17]

The ocean physical carbon pump – the not so slow carbon cycle: The ocean can absorb CO_2 directly through diffusion from the air. The CO_2 reacts with water to form hydrogen and bicarbonate ions making the ocean more acidic. The physical carbon pump is the process where cold dense water sinks and transports the dissolved bicarbonate into the deep ocean before up-swelling currents return the carbon to warmer water and then return it to the atmosphere over thousands of years. Wind, currents, salinity, temperature, and CO_2 concentrations in the atmosphere all affect the net direction and rate of flow through the physical carbon pump. Typically, higher atmospheric CO_2 concentrations will increase ocean uptake, but warmer water holds less CO_2, so the ocean physical carbon pump acts as a dampener of initial CO_2 led change, but amplifies an initial temperature led change. Every 1°C increase in ocean temperature reduces CO_2 uptake by 4%, every 100 ppm concentration increase in CO_2 reduces the potential for further ocean uptake by 15%.[18]

The land based biological carbon pump – the Fast Carbon Cycle: Plants utilise the sun's energy to grow through photosynthesis which takes CO_2 from the atmosphere, combines the CO_2 with water, and produces energy-rich carbohydrates and by-product oxygen. Plants will use this energy through respiration which takes oxygen and carbohydrates and releases CO_2. The combined effect of photosynthesis and respiration is to uptake CO_2 in growing plants which store the carbon in their trunks, stems, and roots. Once the vegetation dies, the carbon enters the soil where it either remains bound with a mix of rock minerals or is decomposed by microbes which release the carbon back into the atmosphere. Atmospheric gases, climate, and temperature all impact the flux of carbon in and out of vegetation and soil. Typically, plants will grow faster at increasing CO_2 concentrations and up to 20-25°C average temperatures, beyond which lack of water will become detrimental to growth leading to decomposition and net carbon release.[19] At higher temperatures, soil respiration also increases releasing more CO_2. Vegetation will therefore act as a dampener of initial CO_2 led change but may amplify larger temperature changes.

The ocean's biological carbon pump – the Fast Carbon Cycle: Algae or phytoplankton in the ocean serves as the start of the marine food chain. Residing in the upper 100 metres of water these single-celled organisms or seaweeds grow using sunlight and CO_2 for photosynthesis, trapping atmospheric carbon in their biomass. This biomass ends up eaten by other marine life, decays, and returns to the atmosphere, or it can sink to the bottom of the ocean and eventually form sedimentary rock. Although marine biomass does not store much carbon itself, it is critical for moving carbon from the upper ocean to the deep ocean. Water temperatures play a role in the carbon flux. If the ocean becomes warmer, the upper 100 metres becomes increasingly unable to mix with the deeper ocean as the temperature gradient, or thermocline layer, increasingly separates the two (typically at above 10°C). This limits the mixing of nutrients and oxygen for algae to grow, creating dead zones. Higher temperatures favour carbon release and colder temperatures favour carbon uptake. More atmospheric CO_2 creates faster algae growth. The ocean biological pump acts as a dampener of initial CO_2 led change, but acts as an amplifier of initial temperature led change.

Earth's proximity to the sun: Day and night come and go as the Earth spins on its axis. Seasons change as the tilt of the planet directs each hemisphere closer or further from the sun throughout its annual orbit. But neither the spin, the tilt, nor the orbital path around the sun are constant. The planet's spin undergoes a 26,000-year wobble around its axis which changes the intensity of seasons for the two hemispheres. The planet's tilt switches between 21.8° and 24.4°, every 41,000 years, changing the intensity of summer and winter. The Earth's orbit around the sun is slightly elliptical with the sun off centre, creating a 7% difference in solar irradiance over 100,000-to-400,000-year cycles. Together the impact of these changes are known as Milankovitch Cycles and can trigger heating or cooling events over thousands of years which in turn impact the carbon flux between rock, vegetation, and ocean, further amplifying the warming or cooling.

Solar forcing variation: The sun's intensity changes through geological time. When the Earth first formed, the sun's intensity was 30% less than today (1.36 kW per square metre). In one billion years the sun will be 50% hotter than today (2 kW per square metre). On a shorter timescale the sun's intensity changes through the 11-year solar cycle as the magnetic poles swap and the number of sunspots, flares, and coronal loops change the energy output by about 1 watt per square metre or 0.1% intensity peak to trough. This impacts temperatures by ±0.1°C.[20]

El Nino southern oscillation: This is the periodic cycle of warming (El Nino) and cooling (La Nina) sea surface temperatures that occur in parts of the Pacific Ocean every two to seven years. The changing sea temperature impacts rainfall patterns and can drive extreme weather events and reverse low-lying trade winds, impacting global temperatures by up to ±0.2°C.[21]

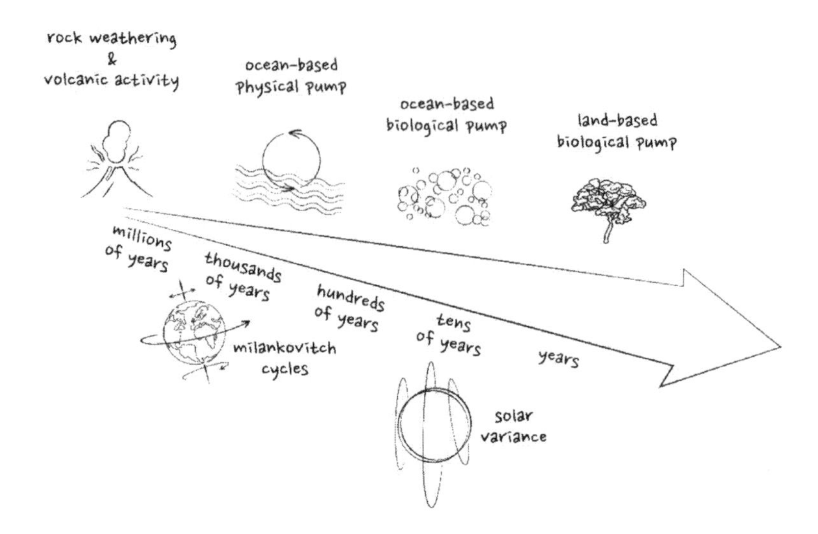

Diagram showing natural processes which control CO_2 concentrations and the temperature of the planet over different timescales.

Combining the impact of the slow and fast carbon cycles with the changing proximity of the Earth to the sun alongside the changes to atmospheric chemistry, ocean chemistry, cloud cover, ocean circulation, and solar variation – including all feedback loops between each process – and you start to understand why the Earth is a complex system. Broadly, CO_2 will regulate over geological time but changes in volcanic activity, solar variation, and ocean/land systems create fluctuations on more human timescales.

Proving the Theory: Recording and Reconstructing a History of the Earth's Atmosphere

To test the theory of global warming, scientists have been able to record and reconstruct the Earth's past temperature, CO_2 concentration, and sea levels, using a range of direct and proxy data.[20] The data shows the historical relationship between CO_2 and temperature. It also gives us an insight into our possible climate futures.

The data is collected using many different techniques:

Historical: Direct global records date back to 1850 for temperature, 1870 for sea levels, and to 1959 for atmospheric CO_2. The data must be carefully vetted for changing measurement techniques, localised disturbances, and natural cyclical variations.

Biological: Proxies include tree rings[vii], pollen, coral, and the fossils of plants, animals, insects, and plankton. These samples can be dated using radiometric analysis of unstable isotopes which decay at a known rate. Once the date of the sample is known the temperature can be estimated by measuring the ratio of stable isotopes in the sample, which evaporate or condense at different rates to their standard atom equivalents depending on the past temperature.[viii]

Cryological: Ice cores are drilled from ice sheets and glaciers providing trapped bubbles of the past atmosphere as far back as 800,000 years. Ice cores can be dated using the distinct layers of annual snowfall and using unstable isotopes. The CO_2 from the atmosphere trapped in tiny bubbles can be measured directly. Temperatures are again inferred from the ratio of stable isotopes of hydrogen or oxygen.

Geological: Various types of rock and sediment can be dated by radioactive decay all the way to the start of life on Earth. The ratio of stable isotopes of carbon, magnesium, lithium, and boron found in the shells of ancient plankton, called foraminifera, provide temperature and CO_2 proxy information that paleoclimatologists can use to reconstruct the climate deep into the geological past.

---

vii Tree rings form once per year because trees grow faster in summer than winter. Many trees in the tropics have no discernible rings because they grow at a similar rate all year round.

viii Isotopes are atoms that are heavier or lighter than usual, due to the number of uncharged neutrons. Unstable isotopes decompose at a known rate. Measure the ratio of remaining isotopes compared to their decay products and the age of the sample can be estimated. Carbon-14 can date back 50,000 years and uranium-235 as far back as 700 million years. Stable isotopes like oxygen-18 or deuterium weigh slightly more than normal oxygen or hydrogen and so will evaporate slower depending on the temperature.

The Deep Geological Record: Earth's Climate Over the Past 500 Million Years

Paleoclimatology is the study of the climate over all of Earth's history. Reconstructing climate deep in the geological past is still a relatively new branch of climate science and one where the range of estimates continues to show large variation. But the reconstructions of temperature, sea level, and CO_2, are continually being refined with ever increasing data and more sophisticated analysis, allowing us to see deeper into the Earth's climate past.

A synthesis of hundreds of studies by Ross J Salawitch and his team from the University of Maryland suggest that nearly 400 million years ago carbon dioxide concentrations ran higher than 2,000ppm and temperatures were more than 10ºC above today's average.[5] Conditions were perfect for the mass expansion of forests across the single super continent, Pangea.

As the forest spread, the deep tree roots likely broke apart the silicate rock, exposing it to the elements and speeding up carbon sequestration through chemical weathering. Levels of carbon dioxide began to fall, and temperatures dropped 10ºC. By 360 million years ago, 70% of all species had been wiped out, unable to adapt to the rapid temperature changes.

Roughly 300 million years ago, global temperatures and CO_2 concentrations stabilised, at levels not too dissimilar from today, before once again beginning to climb. The next 250 million years experienced a volatile warming trend as currents in the Earth's liquid mantle broke apart the supercontinent Pangea, drove increasing volcanic activity, and released CO_2 and methane into the atmosphere. Through the early temperature increases, up to 90% of all sea creatures and 65% of land creatures were driven to extinction. The dinosaurs emerged as the dominant species on Earth.

Carbon dioxide concentrations peaked at over 1,000ppm around 50 million years ago, with temperatures over 10ºC warmer than today. The next 50 million years were characterised by a gradually cooling planet as volcanic activity slowed. The Antarctic ice sheets formed 30 million years ago, and Greenland glaciation took place 3 million years ago. The evolution of our early ancestor, Homo habilus, "the handy man", followed soon after, before Homo erectus, "the upright man", first appeared nearly 2 million years before the present day.

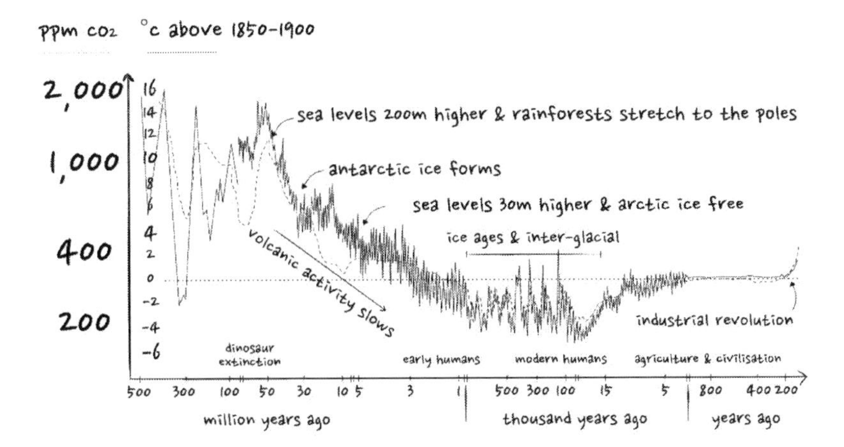

CO_2 ppm (dashed) and temperature (relative to pre-industrial period from 1850-1900) for the last 500 million years. Data based on the average of hundreds of studies: Adapted from Paris Climate Agreement: Beacon of Hope by Ross J. Salawitch et al.[5]

The record of rock and ice is both a complex tapestry of the flourishing life on planet Earth and, at the same time, a reminder of the extreme vulnerability of species over geological time.[22] The records show five mass extinction events over the last 500 million years, each of which wiped out over three quarters of all species on the planet and each caused by rapid changes to the Earth's climate.[23]

The Ordovician-silurian extinction 440 million years ago wiped out most small marine organisms as the planet rapidly cooled and iced over before once again thawing. The Devonian extinction, 365 million years ago, destroyed many tropical marine species as Earth cooled, sea levels dropped, and the oceans ran low on oxygen.

The Great Dying, 252 million years ago, was Earth's most severe extinction: rapid warming was caused by volcanic activity spewing CO_2 into the atmosphere and the sudden release of methane from the sea floor. Ocean waters at the equator rose to hot tub temperatures, seas became starved of oxygen, and 96% of all marine life plus over 70% of terrestrial life was obliterated.

The last major extinction was 65 million years ago when an asteroid struck the Gulf of Mexico, vaporising rock into atmospheric aerosol, blocking the sun, and rapidly cooling the planet. The asteroid-induced climate change decimated food chains, wiped out the dinosaurs, and ushered in a new age of mammals.

Humanity has taken full advantage of the recent stable climate but to think of our survival as guaranteed is, at best, short-term thinking, or misplaced hubris. We have existed as a species less than 3 million years: that is 0.06% of the planet's life, 0.6% of the time of trees, and represents just 2% of the length of time the dinosaurs roamed the Earth.

We have a long way to go.

The Rise and Fall of Ice: A Reconstruction of Ice Ages Over the Last Million Years

Ice cores provide a more detailed record of climate over the last 800,000 years; a period characterised by alternating ice ages and warmer interglacial periods. Serbian mathematician Milutin Milankovitch theorised in 1941 that tilts, wobbles, and the slight elliptical shape of Earth's orbit around the sun, could create periodic changes in solar intensity and push Earth in and out of ice ages.[24] It wasn't until 1976 when Jim Hays, Nick Shackleton, and John Imbrie published their seminal work on marine sediment, that so-called Milankovitch cycles were proven to match the Earth's climate record.

Ice ages may be initiated or ended by the amount of sunlight reaching the planet, but the correlation between temperature and CO_2 has still held tight. Small solar induced warming or cooling cycles are amplified by the response from rising or falling carbon dioxide concentrations.

The exact process is still uncertain but, broadly, solar intensity declines, temperatures start to drop, ice sheets grow and reflect more of the sun's energy, thus cooling the planet further (ice-albedo). The upper and deep oceans start to mix and increasingly more CO_2 is taken from the atmosphere by a stronger biological (growing plankton) and physical carbon pump. Two thirds of the cooling is led by less sunlight and ice-albedo and one third by lower CO_2 and greenhouse gases. Ice ages have seen polar temperatures as low as 10ºC below current averages. Through these glacial periods ice sheets extend from the poles, covering up to one-third of land on Earth and lowering sea levels by over 100 metres. Much of North America and Europe would have been under hundreds of metres, if not kilometres, of dense ice.

Once Earth passes its orbital low, the solar intensity starts to slowly increase and temperatures rise once again. Ice sheets recede and reflect less sunlight leaving more heat energy in the Earth's lower atmosphere. Melting ice releases decaying vegetation and warmer oceans hold progressively less life and less carbon. Atmospheric CO_2 increases and further amplifies the initial warming. The Earth enters a warmer interglacial period, as we are in today. Ice sheets retreat to the poles, the tropics expand, and oceans turn from green to blue as up to 80% of surface waters reduce their mixing with the deep ocean and support less life.

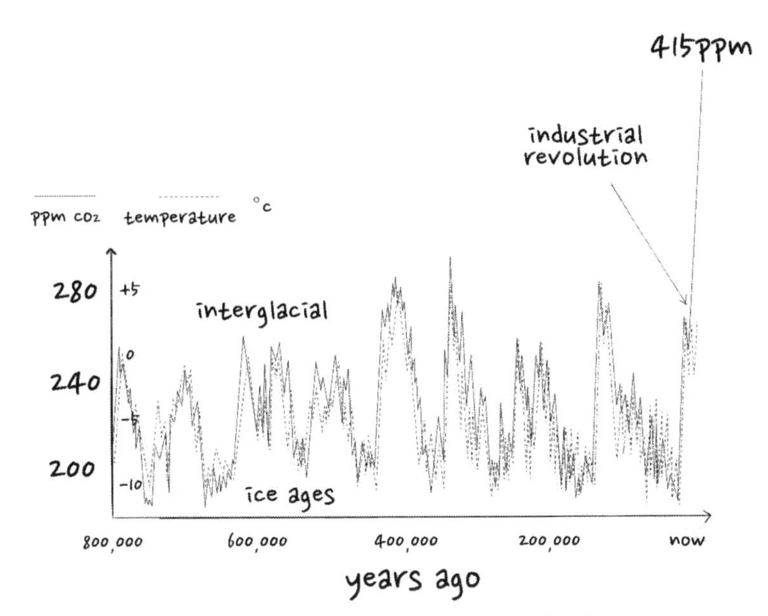

Atmospheric CO_2 concentration in ppm (labels left side of axis) and Antarctic temperature (relative to pre-industrial period from 1850-1900) for the last 800,000 years (labels right side of axis), data from NOAA Paleoclimatology data EPICA Dome C Ice Core Temperature estimates (from deuterium) and revised Antarctica composite ice core atmospheric CO_2 data.[25]

The Human Epoch: Destabilising Earth's Systems and Changing the Climate

Not long ago, humans were a tribe of subsistence co-operative farmers tending to the land. But fossil fuels have turbo-charged productivity and turned humanity into a collective superpower with planet changing consequences. We have reduced the number of trees on Earth from six trillion to just three trillion to make way for crops and livestock. We now move more rock, soil, and nitrogen every year than all the natural processes put together. If you weighed all large mammals on Earth, just 3% of the mass would be wild animals, 30% is human flesh, and the remaining 67% is domesticated livestock bred to sustain our energy hungry bodies. We account for just 0.01% of life yet have destroyed 80% of wild animals and 50% of plants.[22,26]

We have altered the rules of evolution through selective breeding, genetic engineering, health, welfare reform, and conservation. Global trade has reconnected the supercontinent Pangea and mixed once discrete species and diseases. We have learnt how to release, but not yet harness, the energy from atomic fusion with world-changing consequences.

We have also come to understand that humans are tampering with the global thermostat. Once thought to be the innocuous and invisible by-product of the industrial revolution, carbon dioxide has become the very definition of anthropic change to the planet.

The idea of the Anthropocene has been proposed and popularised by atmospheric chemist Paul Crutzen as the next geological epoch, one characterised by the human impact on Earth's geology and ecosystems.[27] Given the changes we have already made to the planet, many experts believe the Anthropocene has already arrived.

Coal, Oil or Natural Gas + Oxygen = Carbon Dioxide + Water + Energy

Our fossil-fuel-powered endeavours over the last 200 years have pumped over 1,500 billion tonnes of additional CO_2 into the atmosphere, increasing CO_2 concentrations from a stable 250-280 ppm to over 410 ppm.[28] As humans turn up the thermostat, the Earth is now moving away from the stable temperature[ix] of 14-15°C that we have enjoyed for the last 12,000 years.

To better understand how our activities have tipped the natural balance we must understand the flows of the carbon cycle and solar intensity today and into the future:

The Fast Carbon Cycle: Every year land biomass will exchange 120 billion tonnes of carbon with the atmosphere and oceans exchange around 90 billion tonnes of carbon.

Anthropic Emissions: We now release 9 billion tonnes of carbon (33 billion tonnes CO_2) per year from the combustion of fossil fuels and another 5.5 billion tonnes of carbon equivalent emissions (20 billion tonnes CO_2e)[x] from burning forests and other greenhouse gases. Human emissions may be just one tenth of land or ocean flows, but they have no natural counterbalance.

The Slow Carbon Cycle: The slow carbon cycle exchanges less than one billion tonnes of CO_2 emissions between atmosphere and rock each year. This exchange will move to counterbalance the higher CO_2 concentrations from human activity because higher temperatures speed up rock weathering, but this will take hundreds of thousands of years: the rate at which rock grows.

---

ix The Earth's average temperature is across all regions, both day and night, and through all seasons.

x CO_2e is shorthand for CO_2 equivalents which includes not just CO_2 but also the other greenhouse gases.

Earth's Orbit: Over the next 30,000 years the intensity of sunlight will reach its cyclical minimum at a level similar to the last ice age where global average temperatures were 3-4°C below today. A cooling rate of 0.01°C per century: too slow to offset anthropic warming in the near term.

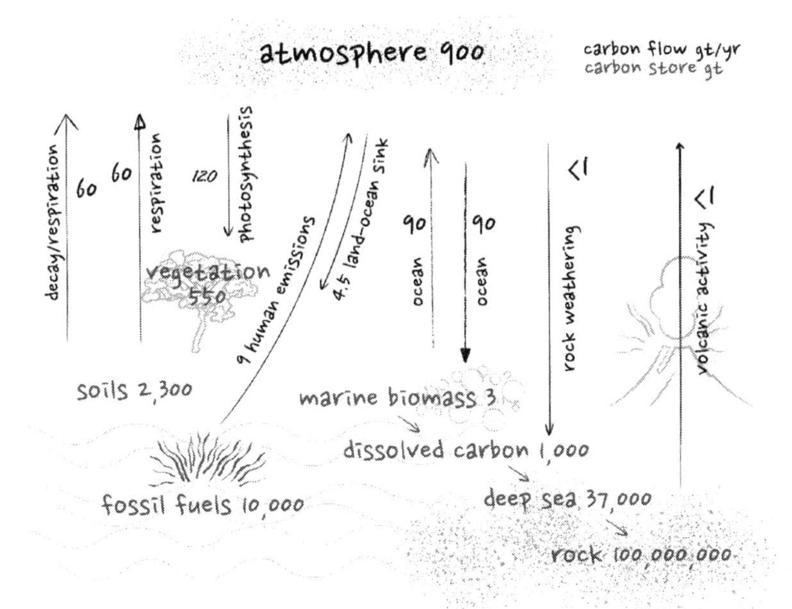

Simplified depiction of the Carbon Cycle. Grey text shows carbon stores in billion tonnes (Gt). Black text shows annual carbon flows in billion tonnes (Gt) (multiply by 3.7 for CO_2). Natural flows balance each other, human emissions into atmosphere are only half offset by the enhanced uptake of the land-ocean sink processes. Adapted from NASA Earth Observatory and NOAA carbon cycle data.[29,30]

Prior to the industrial revolution, the amount of CO_2 leaving the atmosphere was balanced by emissions from volcanoes, oceans, and land. The detailed reconstructions of the last 2,000 years show CO_2 concentrations between 270 and 280 ppm. Temperatures were below today, with small fluctuations during the medieval warming when temperatures increased by 0.5°C. Then the little ice or maunder minimum in the late seventeenth century cooled the planet by 0.5°C as sunspots became rare and solar intensity dropped. Even this small change saw the River Thames occasionally freeze over with ice thick-enough for Londoners to hold a winter festival on the water.

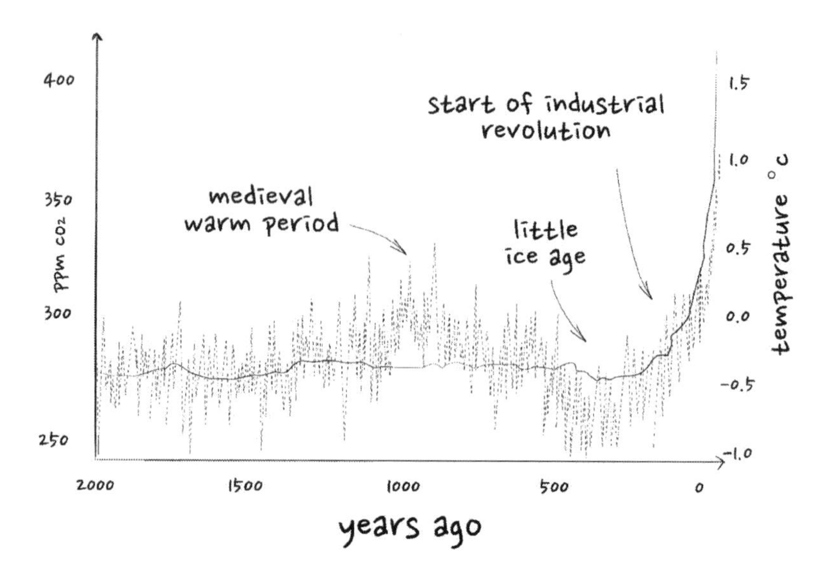

Reconstructed temperature (dashed line) and atmospheric CO_2 concentrations (solid line) for the last 2,000 years. Temperature data from NOAA and IPCC 5th report.[31] *CO_2 from Macfarling Meure et al. (CO_2 Law Dome Antarctica ice core).*[32] *Temperatures are relative to the pre-industrial average through the years 1850-1900.*

Global temperatures had started to respond to higher CO_2 concentrations by the 1920s increasing by about 0.2°C, first recognised by Guy Stewart Callendar. The temperature rise stalled from 1940-1970 as aerosol pollution in the atmosphere, due to the post-war industrial boom, reflected more sunlight away from the Earth, but pollution reduction and further CO_2 increases led a return to increasing temperatures from the years 1970 to 2000.

The temperature rises slowed in the 2000s as the oceans took on more of the heat energy for a period, before atmospheric temperature increases re-accelerated once again over the last decade. The planet is now over 1°C warmer than it was 100 years ago with another increase of nearly 0.5°C to come as temperatures play catch up with CO_2 concentrations already in the atmosphere.

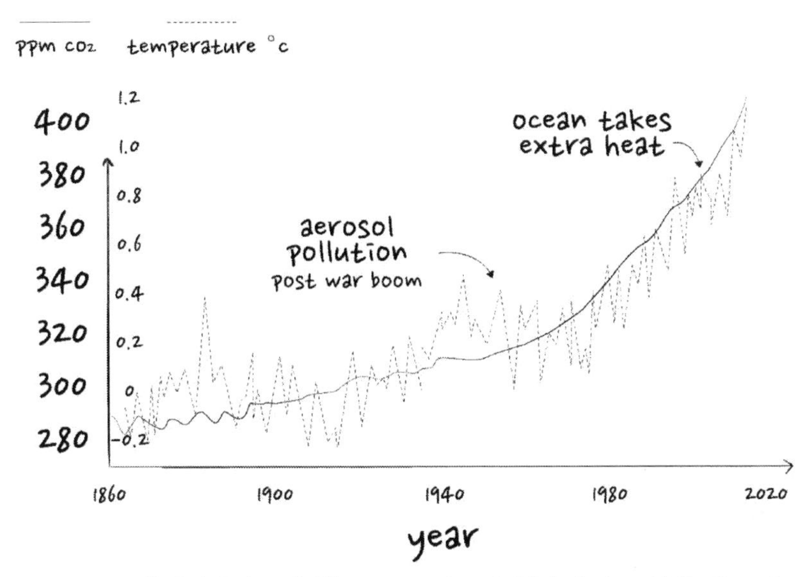

Temperature (dashed line) and CO_2 concentrations (solid line) since 1860. Data from NASA Land-Ocean Temperature Index and NOAA ESRL CO_2 data (Antarctic ice core data to 1959 Law Dome (D.M Etheridge et al.) and Mauna Loa, Hawaii direct measurements thereafter).[33, 34] *Temperatures are relative to the pre-industrial average between 1851-1880.*

So, whilst annual human emissions may still be ten times smaller than the flow of carbon from land or sea, they have no balancing flow in the short term. The land and the oceans have done their best to accommodate some of the additional CO_2;[xi] the ocean has turned from a net carbon emitter to a net absorber. This change has removed 50-60% of the additional atmospheric carbon at the expense of an increasingly acidic ocean. Humans have emitted nearly 2,000 billion tonnes of CO_2 since the year 1750 and raised atmospheric concentrations by 130 ppm; that's a 1 ppm increase for every 15 billion tonnes.

This rate of change is too fast for the land and oceans to accommodate. Carbon is entering the atmosphere faster than any other event over the last 66 million years, according to work published in *Nature* by Richard Zeebe and his colleagues.[35] The closest historical period of change was 56 million years ago during the Paleocene-Eocene Thermal Maximum (PETM) when carbon dioxide was pumped into the atmosphere at a rate of up to 4 billion tonnes per year for 4,000 years and temperatures increased by about six degrees. The total CO_2 emitted over this period

---

xi The land-ocean sink refers to the ability of oceans and land to capture CO_2 from the atmosphere.

was roughly the equivalent of burning half our remaining fossil fuel resources. Climate scientists consider this event a good comparison for the changes taking place today, except that we are emitting greenhouse gases more than ten times faster, at over 50 billion tonnes per year.

Temperatures and the Earth's systems have started to respond to the unprecedented CO_2 increases. With the planet already over 1ºC warmer through the last 100 years, the rate of temperature increase is accelerating towards the fastest rate of the last million years. If no change is made to the existing fossil fuel system then CO_2 and temperature will further climb, natural dampening mechanisms such as ocean uptake will saturate, and amplifying feedback mechanisms become increasingly dominant.

Reviewing the Evidence: The Fingerprints of Human Influence on Climate Change

Records throughout the last 150 years, ice cores for the last 800,000 years, and geological proxies for the last 500 million years, all show the same relationship between carbon dioxide and temperature. As one increases so the other will follow and vice versa. Simple statistical analysis shows the CO_2-temperature relationship is highly significant and a change in one explains nearly all the change in the other.[xii]

xii Our hypothesis is that temperature and CO_2 influence each other. The null hypothesis is that they have no influence on each other. The p-value is a statistical analysis which gives the probability that the null hypothesis is true. So, a smaller p-value and our hypothesis is more statistically significant and not just a fluke – for CO_2 and temperature it is close to 0%, generally below 5% is the cut off. The r-square value is another statistical test which gives the percentage of change (variance) which can be explained by the relationship. This time a higher percentage means a stronger relationship, CO_2 and temperature is 80-90%.

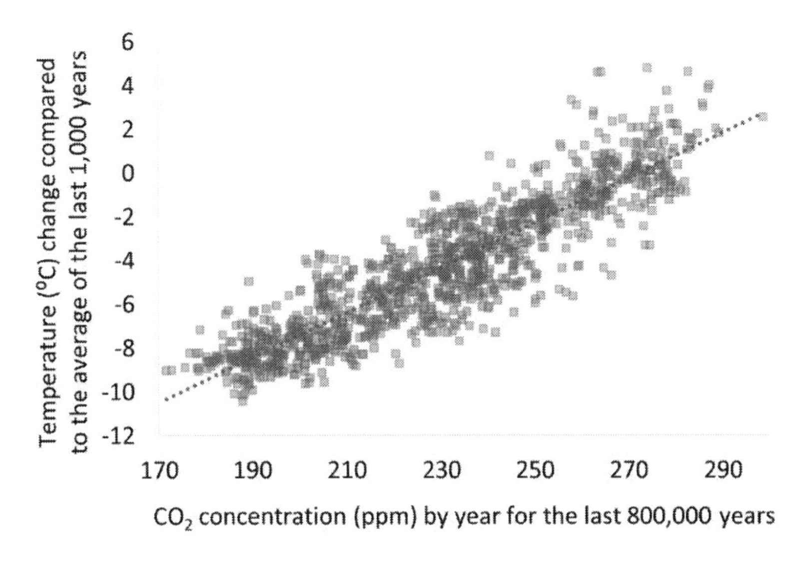

Atmospheric CO_2 concentrations going back 800,000 years plotted against the temperature difference in the Arctic at each CO_2 concentration – the tight fit of the data around a rising trend line yields a statistically significant relationship and a high dependence of one on the other.

Since the start of the industrial revolution cumulative emissions of CO_2 and other greenhouse gases from human activity have increased to nearly 2,500 billion tonnes and surface temperatures have responded. Small deviations from the trend are evident, and to be expected, as the warming or cooling impact of solar variance, El Nino, volcanic activity, or aerosol pollution creates short term variation. But the underlying trend between rising CO_2 and rising temperature is clear. Analysis by Philip Kokic et al. shows that there is a 99.999% probability that the accumulation of greenhouse gases in the atmosphere is responsible for the recent temperature rises.[36] The IPCC[xiii] is "virtually certain" human influence has warmed the climate system.

---

xiii[ii] The IPCC stands for the Intergovernmental panel on climate change, a United Nations body for assessing the science on climate change. The IPCC does not carry out original research but is rather a collection of around 200 experts nominated by governments all over the world which review and bring together the scientific literature. The IPCC release major Assessment Reports (AR) every 5-8 years which are considered the gold standard in climate change knowledge. The accompanying summary report for policy makers has summary statements which are negotiated and signed off by global governments.

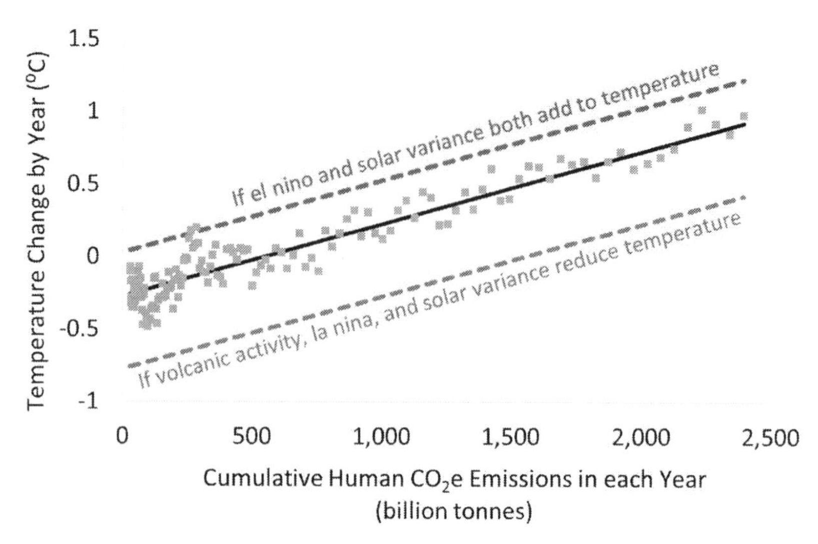

Cumulative CO_2e emissions emitted by humans since 1750 plotted against the global land-ocean temperature change at each level of CO_2e increase. The solid black line shows the underlying warming trend due to CO_2; the top and bottom dashed lines show the range of possible variability from the trend. The upper dashed line shows a possible combination of peak solar forcing (+0.1°C) and El Nino (+0.2°C), the bottom dashed line shows the combined cooling impact of volcanic eruption (-0.2°C), La Nina (-0.2°C) and solar minimum (-0.1°C). Importantly this variability is short lived whereas greenhouse gases persist for centuries in the atmosphere. Data Sources [25,33,34]

Alongside rising temperatures, other fingerprints of change are all too clear. Ice is melting, ocean oxygen is in decline, and sea levels are rising as waters warm. The oceans are becoming more acidic as they take on more carbon dioxide with mid-latitudes becoming saltier and the tropics and poles fresher as rainfall patterns change. The stratosphere (the atmosphere above the troposphere) is cooling because the atmosphere below is emitting less energy upwards, demonstrating the greenhouse effect as the root cause of surface warming. Oxygen concentrations in the atmosphere have declined by 100 ppm, used up as we turn fossilised carbon to carbon dioxide. That fossilised carbon contained in oil, gas, and coal has been buried for millions of years and no longer contains radioactive carbon-14, so concentrations measured in atmospheric CO_2 are also in decline.

Just one or two of these changes alone and maybe we could question the human impact on climate, but take all of these fingerprints combined and the evidence is irrefutable. We have already pushed Earth beyond its stable range of the last 12,000 years and we are destabilising the land, ocean, and atmosphere as fast as some of the most volatile periods in the history of rock and ice.

Climate is What You Expect, Weather is What You Get: Untangling the Impacts of Global Warming

The Earth is already over 1°C warmer and sea levels (which lag temperature) are 20 cm higher than before the industrial revolution. The effects of global warming are undoubtedly starting to impact the planet, but are they doing any damage?

Over the last 20 years the world has experienced 19 of the hottest years in two millennia. Over this period, we have witnessed some of the strongest natural catastrophes documented. Large-scale heatwaves killed 50,000 Europeans in 2003, 56,000 Russians in 2010, and 25,000 Indians in 2015.[20] The Russian heatwave crippled agricultural production and led to President Putin banning grain exports and global food prices doubled.

North America has been battered by hurricanes Wilma, Rita, and Katrina in 2005, Felix and Dean in 2007, Irma and Maria in 2017, and Dorian in 2019. That's nearly half of all the strongest Atlantic hurricanes on record hitting the US in the last 15 years. In 2012, extreme heat and drought in India forced widespread failure of the electrical grid; 600 million people lost power creating the largest blackout in history. In 2018 and 2020, California wildfires wiped out over two million acres and in 2019 Australian wildfires were the largest on record burning through 25 million acres of land and killing an estimated one billion animals.[37] Severe flooding in Venice now happens nine times per year compared to once every three years at the start of the twentieth century.[38]

Whilst it certainly feels like climate related natural catastrophes are on the rise, we must be careful not to mix up weather with climate. Weather encompasses short term changes to the atmosphere which can culminate in a natural disaster. Climate is the trend of weather events over longer periods of time. As British Geographer and Oxford Professor Andrew John Herbertson put it, "Climate is what you expect, weather is what you get".

We must be able to see past short-term volatility caused by natural variations or cycles like El Nino. We cannot claim that climate change was responsible for a particular weather event, but we can understand how it changed the probability of that event occurring. We need to understand if there is a statistical trend in climate-change-led impacts, just as there is a statistical trend in temperature.

The IPCC fifth assessment report states we are already observing increased drought in the Mediterranean, West Africa, East Asia, and Southern Australia whilst Central US and North Australia are getting wetter. Extreme precipitation events have increased everywhere. The world is experiencing 20-30 fewer cold days per year, 20-30 greater warm days, and more heatwaves in Europe, Australia, and the US. Glaciers are retreating all over the world, permafrost is melting in Siberia, and shrubs are slowly creeping into the once frozen soils of the northern tundra. Wildfires are increasing in North America, Greece, Portugal, and Australia. Sea creatures are heading for the poles.[39]

Another good source of trend data comes from the insurance industry. The ability to predict the impact of extreme weather, flooding, and fires is key to managing the premiums that the industry charges on the ships, buildings, and businesses that they insure. Charge too much and you win no business, charge too little and you are quickly going bust. Munich Re is a $40 billion German re-insurance company (they insure other insurance companies) and has an extensive database of natural catastrophes dating back to 1980 when global warming started to accelerate following the post-war hiatus.[40]

Over the last 40 years, the world has warmed by almost 0.7°C and the average number of climate related natural catastrophes – such as wildfires, heatwaves, storms, flooding, and droughts – has more than tripled from 200 to 650 major events per year. Associated annual damages have risen six times from an average of $25 billion to $150 billion per year (in today's money).

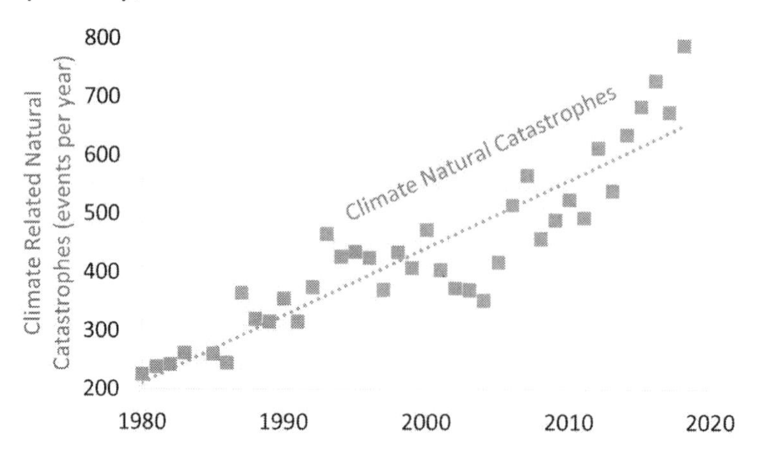

The number of significant events per year from climate related natural catastrophes (wildfires, heatwaves, storms, flooding, and droughts). Significant events threshold defined by number of deaths and total damages incurred. Data from the Munich Re NatCat Service.

Importantly, through the same period, geophysical events such as earthquakes, tsunamis, and volcanic activity, which are not influenced by global warming, have increased by just 40% from 30 to 40 events. Related damages have risen four times from $10 billion to $40 billion per year.

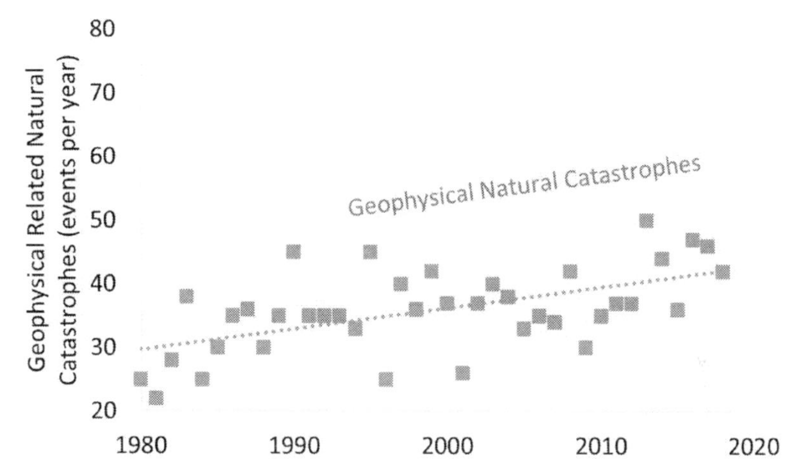

The number of significant natural catastrophes per year from (non-climate related) geophysical events (earthquakes, volcanic eruptions and tsunamis). Significant events threshold defined by number of deaths and total damages incurred. Data from the Munich Re NatCat Service.

We would expect the number of significant natural catastrophe (nat cat) events and the total damages to increase over the last 40 years because there are more people and more wealth in harm's way of any storm, flood, or wildfire, but this does not explain why climate related destruction is growing so much faster than geophysical. Both should increase at the same rate. If we assume the difference is rising temperature, then climate change has been responsible for about two thirds of the increase. Assuming this trend is linear, every one-degree C warming creates an additional 500 significant climate related natural catastrophes per year.

Beyond the human toll of (not so) natural catastrophes, the natural world is also starting to yield to higher temperatures. Since 1980, the extent of Arctic sea ice at the end of summer has shrunk by 40% and the great ice sheets covering Greenland and West Antarctica are now losing up to 500 billion tonnes of ice per year. The Northern Hemisphere growing season is starting 2-6 weeks earlier, bird migration is changing, and marine life is moving 35 km closer to the poles every year in search of cooler waters.[41]

Human activity and the burning of fossil fuels has made further change inevitable. Tropics will expand, the poles will retreat, and the weather patterns we are used to will alter. Yet Mother Earth will remain indifferent to the change taking place on the upper 0.5% of her spherical being. For some humans, creatures, or vegetation living in already hostile, cold environments, the change may be welcome. But for the most part human civilisation and the natural world have adapted to the stable climate and relatively predictable weather patterns of the last 12,000 years. The unprecedented rate of change in temperature, sea level, and ocean acidity will increasingly throw the planet out of equilibrium and, if not managed, will materially impact human quality of life and the natural world.

As Antonio Guterres, the secretary general of the United Nations, puts it, "Our planet will not be destroyed by climate change. Our ability to live on the planet will be destroyed". This is not planet Earth's problem; this is our problem.

It turns out the cost of escaping a life of destitution is a warming planet. And as Simon L. Lewis and Mark A. Maslin write in *The Human Planet*, "Once we recognize ourselves as a force of nature, we will need to address who directs this immense power, and to what ends".[42]

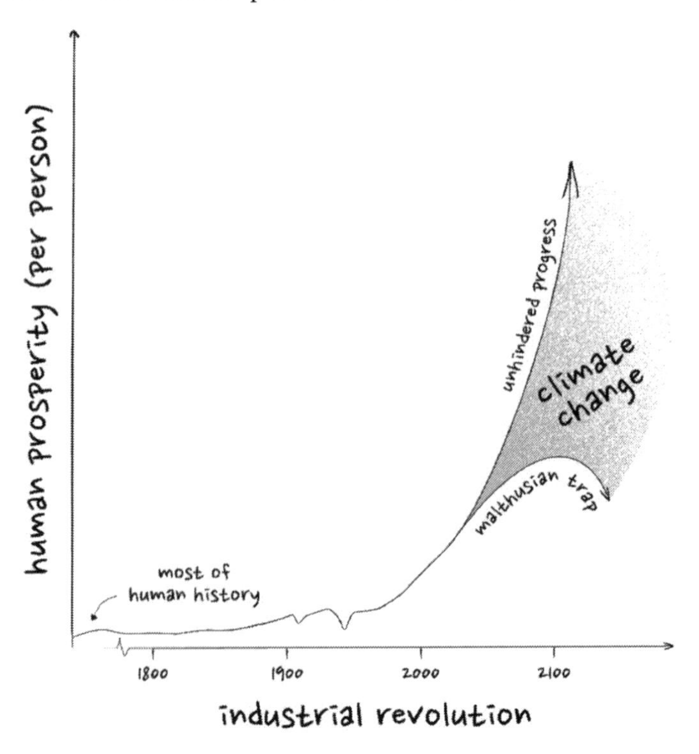

We now understand the clear relationship between CO_2 and temperature, scientists have built and tested a credible theory of global warming, and we have irrefutable evidence that human activity is already warming the planet and will continue to do so. The Anthropocene is upon us, and we can no longer rely on Earth to provide a stable habitat and abundant resources. Through our own actions, humanity has backed itself into a corner, pushing ourselves into the role of global caretaker. We must regulate Earth's temperature, ration her resources, and safeguard her natural order or we risk falling back into the Malthusian Trap.

But how quickly should we act? How fast will the planet warm? And how will the climate change?

In the next chapter we will take a look under the hood of the climate models that scientists use to predict future warming and changes to the Earth's systems.

Chapter 3: Forecasting Change

The Science of Warming

In this chapter we will learn how the early scientific pioneers pieced together the data to predict future warming, and how modern climate models can simulate the Earth's land, oceans, ice, and atmosphere hundreds of years into the future. We will meet our first dial of doom and boom: the science-based *climate sensitivity* which governs the rate of temperature increase with rising CO_2. And we will build our own approximation of a climate model to predict the temperature rise in a peak population, peak consumption world which still burns fossil fuels.

Calibrating the Thermostat: Calculating the Temperature Response to Increasing CO_2

In the late nineteenth century, scientists were piecing together a growing understanding of the Earth's past ice ages and the intertwining role of the carbon cycle and the sun's energy. Svente Arrhenius was the first to publish a mathematical basis for the greenhouse effect in his 1896 paper: "On the influence of carbonic acid {CO_2} in the air upon the temperature of the ground".[1]

Arrhenius approached the problem in three stages. First, he calculated the amount of energy absorbed by carbon dioxide and water vapour using the best experimental evidence of the time. Next, he split the Earth into 13 sections by latitude and estimated the average reflection of sunlight from cloud, snow, ocean, and land in each slice for every month of the year. Finally, he assigned an initial amount of atmospheric water vapour to each cross section and built a formula to allow the absorption from water vapour to change with temperature.

Now all he needed to do was insert the energy inflow from the sun and the different concentrations of CO_2 and he could calculate the planetary energy budget and temperature change. Armed with only paper and pencil it took Arrhenius one year to crunch the numbers over six different CO_2 scenarios; an arduous task which he only undertook as a distraction from his ongoing divorce from his former pupil, Sofia Rudbeck. Once the year of distraction was up, he estimated that if the concentration of CO_2 doubled, the Earth would warm by 5-6ºC.

Whilst Arrhenius' maths was sound, he was let down by the inaccuracy of the experimental absorption data and he missed some key physical changes such as the cooling impact of water evaporation from a warmer surface. This over-inflated his temperature estimates.[2]

Forty years later, British steam engineer Guy Stewart Callendar would rebuild Arrhenius' mathematical model with the benefit of more accurate CO_2 absorption recordings and a growing database of temperature and CO_2 measurements. He started by estimating that burning of fossil fuels were responsible for a 6% increase to CO_2 between 1880 and 1930 and that temperatures had risen by 0.3ºC through the same period. He built a similar mathematical representation of the climate to Arrhenius with fixed land, sea, ice, and clouds. He asserted that the increase in CO_2 was responsible for half of the 0.3ºC warming through the period. This was the first time CO_2 emissions from human activity had been linked to a recorded temperature increase of Earth. His model predicted that doubling CO_2 in the atmosphere would increase temperatures by 1.6ºC.[3]

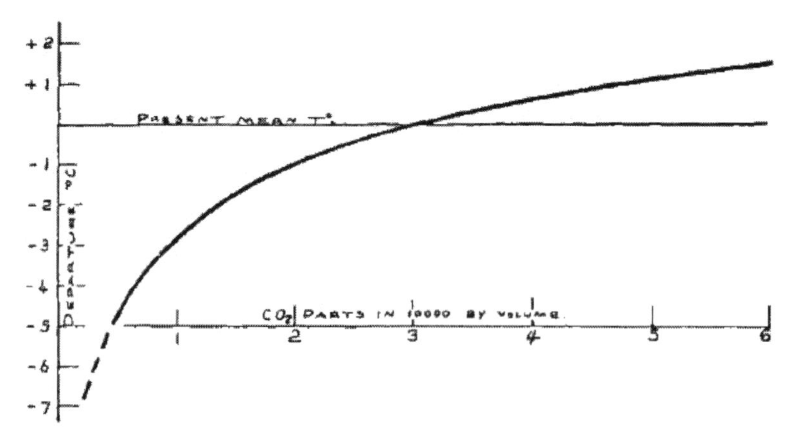

Original image from Guy Stewart Callendar's 1938 paper "The Artificial Production of Carbon Dioxide and its Influence on Temperature". Horizontal axis shows CO_2 in parts per 10,000 (multiply by 100 for ppm). Vertical axis shows his predicted temperature change in ºC. The increasingly shallow curve shows a logarithmic relationship.

Where Arrhenius was let down by the accuracy of experimental data, Callendar decided not to model water vapour change with temperature increase. He captured the thermostat but missed the boost button and underestimated the overall temperature increase.

Both models suffered from large oversimplifications, missing key feedbacks such as the amplifying effect of water vapour, the cooling impact of evaporation and changes to ice, clouds, and circulation. Nonetheless, by luck or intuition, the two estimates are broadly consistent with the upper and lower range of the best predictions today: between 1.5 and 4.5ºC temperature increase for double the atmospheric CO_2 (IPCC AR5).[4]

Arrhenius and Callendar had not only laid the foundations for understanding radiative forcing,[i] but also the relationship between CO_2 and temperature. A relationship which is non-linear. They follow what mathematicians call a logarithmic relationship – the increasingly shallow curve of Callendar's graph.

The logarithmic relationship means that higher CO_2 concentrations will always drive higher temperatures, but the increase in temperature for every extra unit of CO_2 becomes smaller with increasing concentrations. Doubling CO_2 from 300 ppm to 600 ppm will give you roughly the same absolute temperature change as doubling CO_2 from 600 to 1,200ppm.[5] Why? Well, infrared absorption in the atmospheric window increases most as CO_2 concentrations first rise – these are the easy pickings where CO_2 can absorb in frequencies of the spectrum not already blocked by water vapour. But as concentrations of CO_2 further increase, the infrared absorption also becomes increasingly saturated and only small further broadenings of the absorption band can trap extra energy.[6]

Think of global warming like putting on weight; the more you eat the fatter you get, but as you get heavier it takes more and more calories to maintain your larger body so the weight gains eventually slow.

Upgrading to Smart Heating: Developing Climate Models

In 1967, Manabe and Wetherald published what is regarded as the cornerstone of contemporary climate modelling.[7] They created the first computer simulation to represent the major radiative forcings and key feedback elements of Earth's climate such as water vapour, evaporation, and clouds, in a one dimensional slice of the atmosphere. They estimated doubling CO_2 would bring 2.4ºC temperature increases – this was the first scientifically convincing basis for the greenhouse effect.[8]

Since Manabe and Wetherald published that seminal paper, computing

---

i Radiative physics measures the flow of energy into and out of the Earth's climate. Radiative forcing arises when the energy flow is out of balance.

power has grown exponentially and models have moved from one-dimensional representations into a grid of three-dimensional cubes of atmosphere, land, sea ice, and ocean which interact with one another. These complex models are called General Circulation Models (GCM) and can be used to predict the weather or the future climate.

Our weather is controlled by the flow of energy through the Earth system. The bulk of the sun's energy enters Earth at the tropics and flows towards the poles, carried by wind, water currents, and influenced by the spin of the planet, finally escaping the atmosphere as infrared radiation. To predict the weather, simulations start at an exact state and must move to another exact state over an exact period of time. This limits meaningful weather forecasts to only ten days out.

However, climate is the probability of weather. Climate models simulate land, oceans, atmosphere, and ice, using scientific principles established 100 to 300 years ago and used in all aspect of modern life. Maxwell's equations of light describe how the sun's radiation enters the Earth and how infrared radiation leaves the atmosphere. Thermodynamics models how energy interacts with the physical world, the same equations used to calculate the power output of a combustion engine. Napier-stokes equations model the dynamic movement of air and water flowing across the planet just as they are used to predict how an aeroplane will fly. These equations are combined in the grids of Gerneral Circulation Models and used to simulate the statistical average temperatures, velocities, and pressures of the Earth system far into the future. Moden climate models have one million lines of code which is relatively compact compared to Google Chrome which contains just over six million. The difference is that Google answers just one question at a time using one computor processor whereas climate models must cycle through 65,000 cubes of code every minute for hundreds of years. This requires super-computers with millions of processers using upto 5,000 kW power – the same as a small town.[9]

Climate models are continually being refined and more feedback elements incorporated. Aerosols were added in the IPCC's second assessment, the carbon cycle and dynamic vegetation by the third assessment and land ice, the ocean pump, and atmospheric chemistry through the third and forth assessments. The next generation of models are called Earth Systems Models (ESM) which will add further elements such as melting tundra, wildfires, wetlands, and land ice sheets.

The basics of state-of-the-art climate modelling can be broken down into radiative forcing, fast feedback mechanisms, and biosphere response.

Building Climate Models: Radiative Forcing, Fast Feedback, and Biosphere Response

The models start with the underlying assumptions on how the sun's cycle, greenhouse gases, and air pollution will change the amount of solar energy entering and leaving the atmosphere to calculate the change in radiative forcing (the extra trapped energy in watts per square metre) and estimate the initial temperature change.[10]

Initial Change

Solar Forcing: The sun has an 11 year sunspot cycle which changes the output by ±0.1%, impacting temperatures by ±0.1°C.

Greenhouse Gas Emissions: The initial concentration increases of carbon dioxide (CO_2) and other greenhouse gases depend on predictions of future human emissions. Doubling CO_2 concentrations (from pre-industrial times) to 560 ppm will increase the added radiative forcing by 3.7 watts per square metre.[11,12]

Aerosols, dust, smoke and soot: Small particles suspended in the atmosphere are released from human activities such as burning fossil fuels or industry, alongside natural processes such as volcanic eruptions. Sulphur particles from volcanic eruptions reflect the sun and cool the atmosphere. Black carbon or soot from industrial activity is a strong warming influence as it absorbs the sun's energy, helps to form clouds and covers reflective surfaces. However, unlike greenhouse gases, these don't persist very long in the atmosphere. If we stopped emitting industrial sulphates this cooling effect would be gone within one to two years. The net impact of aersols today is to lower radiative forcing by 0.9 watts per square metre.[11,12]

Fast Ongoing Feedbacks

Next is the ongoing fast feedback mechanisms which can either amplify or dampen the initial radiative forcing from sun, CO_2, and aerosols. In turn this will amplify or dampen the initial temperature change. Fast ongoing feedbacks are some of the key elements missing in Arrhenius' and Callendar's calculations.

Planck Feedback: The response of the Earth to emit more radiation as the planet warms. This is what brings greenhouse-gas-led energy imbalances back into check. Planck feedback offsets radiative forcing by 3.3 watts per square metre per 1°C warming.

If the planck feedback was the only response to greenhouse warming then the climate sensitivity, the temperature increase for double the CO_2 concentration, would be just 1.1°C. [ii,13] *However, we already know there are further changes which will either amplify or dampen the initial temperature change. We can represent the overall impact using the following formula.* [11,12]

Final Temperature = Initial Temperature Change ÷ (1 - Feedback Strength)

Water Vapour Feedback (Amplifies Radiative Change): Water vapour increases by 7% in the atmosphere for every 1°C increase in temperature and amplifies an initial CO_2 led warming. Water vapour has the largest amplifying feedback strength of +0.6.[11,12]

Lapse Rate Feedback (Dampens Radiative Change): The lapse rate is the rate at which the temperature of the atmosphere gets cooler as you travel higher.[iii] As we saw in chapter 2, without this temperature gradient the greenhouse effect would not occur and Earth would be -18°C. Lapse rate *feedback* is the response of the lapse rate to an initial warming. The extra heat energy from the initial warming evapourates more water from the Earth's surface which moves up into the atmosphere by convection (latent heat). Once in the dry and cold atmosphere, the water turns back to liquid and releases the heat energy. This creates a warmer energy emission level than otherwise expected and so more radiation than expected is emitted to space. The extra transfer of heat from the Earth's surface to the colder atmosphere acts to reduce the temperature gradient

---

ii Based on the extra radiative forcing from a doubling of CO_2 (3.7 watts per square metre) divided by the Planck response (3.3 watts per square metre per 1°C) = 1.1°C

iii If the Earth's atmosphere were dry it would cool at a rate of 10°C per km due solely to the expansion air. However, the actual lapse rate is closer to 6°C per km because of water in the atmosphere.

of the atmosphere (the lapse rate) and so weakens the greenhouse effect. Lapse rate feedback acts to dampen the initial temperature increase with a strength of -0.3.[iv,11,12]

Cloud Feedbacks (Amplifies Radiative Change): Increasing temperature leads to more water in the atmosphere, increasing the number of clouds. Depending on their type and altitude, they can have either amplifying or dampening effects. Cloud tops higher up in the troposphere tend to warm the Earth as the cold water vapour traps more infrared energy. Low lying clouds tend to reflect more of the sun's energy (due to warmer water vapour). Overall, clouds have a small amplifying impact but are one of the most uncertain elements of climate modelling. Difficulties in modelling clouds arise because they are smaller than the gridded cubes of atmosphere simulated in climate models. It requires too much computing power to increase resolution so climate modellers approximate the net impact. Cloud feedback is estimated at +0.25.[11,12]

Sea Ice-Albedo Feedback (Amplifies Radiative Change): Increasing temperatures melt ice and reveal more sea. Ice reflects 75% of the energy which strikes its surface. The ocean reflects just 5%. Receding ice means less reflection and more absorption of the sun's energy amplifying the initial temperature change. Arctic sea ice covers ~10 million square km of surface and since 1979 average annual sea ice cover has declined by over 2 million square km. This is part of the reason the Arctic is warming twice as fast as the rest of the planet. At this rate the Arctic sea could be nearly ice free later this century. A study by Kristina Pistone et al., suggests that ice-albedo since 1979 has already increased solar energy input by 6.4 watts per square metre in the Arctic which is an average of 0.21 watts per square metre spread over the planet.[14] This equates to 0.1°C warming already and a further 0.5°C if all Arctic sea ice melts. Ice is an amplifying feedback with strength +0.1.[11,12]

So already summing up these feedbacks means an initial 1.1°C temperature increase from doubling CO_2 will be amplifed to about 3°C final warming. This is before we consider further response from the biosphere.[11,12]

$$\mathbf{3^{o}C = 1.1^{o}C_{\text{planck}} \div (1 - (+0.6_{\text{water vapour}} - 0.3_{\text{lapse rate}} + 0.25_{\text{clouds}} + 0.1_{\text{ice}}))}$$

iv The lapse rate, or cooling through the atmosphere, serves to keep water on and around the planet. If the temperature and pressure profile of the atmosphere were more like that of our close neighbour Venus, we would have lost our water to space long ago. Venus used to have water on the planet, but now has a dense CO_2 filled atmosphere, and runaway warming, with surface temperatures of nearly 500°C.

Biosphere or Tipping Point Feedback

Once the fast feedbacks to the initial temperature change have been captured, Earth System Models attempt to capture the slower changes or abrupt changes to once stable parts of the Earth system:

Changes to the Land and Ocean Sink (Dampens CO_2 change): The **ocean** has turned from a net carbon emitter to a net carbon absorber over the last 200 years, as the physical and biological ocean pumps responded to increasing CO_2 concentrations in the atmosphere and attempt to rebalance the equilibrium (Le Chatelier's principle). Vegetation on **land** has also increased the rate of growth and carbon uptake in response to higher CO_2 concentrations (greening of Earth). The net result of these initial changes has been to remove an average of 55% of human CO_2 emissions from the atmosphere. However, according to work by M. R. Raupach et al, the Atmospheric Sink is now starting to weaken.[15] In 1959, 60% of human emissions were absorbed by land and oceans. By 2012, with temperatures nearly 0.8°C hotter, the land-ocean sink removed just 53% of excess human CO_2 emissions. Warmer oceans hold less carbon, drought slows plant growth, and warmer soils increase microbial decay and respiration of carbon from soils. In other words, the fast carbon cycle (oceans and land) has a dampening response to CO_2 increase, but rising temperatures are now starting to weaken the efficiency of this offset. *A back-of-the-envelope calculation suggests if land and ocean uptake continues to weaken by ~10% for every 1°C initial warming (as it has done over the last 50 years) this will leave an extra 100 billion tonnes or 5 ppm more CO_2 in the atmosphere (compared to the fraction of CO_2 left in the atmosphere of the last 50 years).*

Forest Wildfires and Peat Fires (Amplifying CO_2 Change): Increasing temperatures dry out wood and peat which increases the risk of fires. ***Forest wildfires*** burn an average of 3.5 million square km per year (95% are caused by humans, 5% are due to lightning) compared to 0.15 square km deliberately burnt by humans for agriculture.[16] However, natural forests grow back and recapture the carbon, farmland doesn't. Since 1979 wildfires have increased by 20% or 0.5 million square km as the planet has warmed by 0.7°C.[17] If the burnt area continues to expand faster than it can regrow then the carbon remains in the atmosphere. Forests hold 250 billion tonnes of carbon (900 billion tonnes CO_2) over 40 million square km of land.[18] *A back-of-the-envelope calculation suggests that every 1°C initial warming will add up to 3%, 30 billion tonnes or 2 ppm of CO_2 into the atmosphere.* ***Peat fires and drying wetlands*** *may be an even bigger problem as peat holds 550 billion tonnes of carbon (2,000* *billion tonnes CO2) and may lose up to 10% of carbon, 200 billion tonnes or 10 ppm of CO_2 per 1°C initial warming.*[19]

Rapid Release of Greenhouse Gases (Amplifies CO_2e Concentrations): Tundra or permafrost is frozen soil which covers 19 million square km of the northern hemisphere. The Tundra holds 1.7 trillion tonnes of frozen carbon with more than 800 billion tonnes of carbon in the top three metres. If this frozen soil melts, the trapped gas is released into the atmosphere. The IPCC estimates up to 37-81% decline in the top 3.5 metres of permafrost at high northern latitudes this century unless we act on climate change.[20] Work by Kevin Schaefer et al. estimates that a moderate warming scenario could add another 100 billion tonnes of carbon (or 370 billion tonnes of CO_2) into the atmosphere by 2100.[21] Furthermore, a small fraction may be methane (3%) which could double the total warming potential of the release because methane is a 28 times more potent greenhouse gas than CO_2 over 100 years. *A quick back-of-the-envelope calculation based on this trend could mean an additional 400 billion tonnes or 20 ppm CO_2 in the atmosphere per 1°C initial warming.* The tundra could prove to be one of the biggest carbon feedback mechanisms.

Ice Sheet Albedo (Amplies Radiative Change): Glaciers around the world are already rapidly shrinking and will be more than half gone by the end of this century. The complete melting of all glaciers will add 0.4 metres to global sea level. The land based ice sheets of Greenland and West Antarctica are also melting but are much thicker and will take centuries to dissapear. Total melting of the Greenland Ice Sheet (GIS) and Western Antarctic Ice Sheet (WAIS) would each add six metres to global sea levels – this change may prove irreversible at temperature increases above 2-4°C.[10] The remainder of Antarctic ice could raise sea levels by another 60 metres but will take higher temperatures and millenia.

Amazon Rainforest Dieback (Amplifies CO_2 concentrations): The Amazon rainforest generates around half of its heavy rainfall through evapotranspiration from the forest itself. Evaporation from warm damp soils and transpiration from leaves drives water into the atmosphere, creating clouds and more rainfall. However, rising temperatures may reduce rainfall, rising CO_2 reduces transpiration and deforestation shrinks the forest. Combined, this process could trigger a sudden dieback (above 3°C), and rainforests would turn to more sparse savanah releasing CO_2 in to the atmosphere from the rotting forest vegetation and soils.[22,23]

Atlantic Meridional Overturning Circulation (Changing weather patterns): Melting Arctic sea ice and Greenland ice sheets release warmer, fresh water into the Arctic sea, reducing the saltiness and warming the upper ocean at high latitudes. This could weaken or divert ocean circulation currents such as the Gulf Stream which are driven by warmer, less dense tropical waters mixing with colder, denser Arctic waters.[24] This would cool Western Europe and severely change global weather patterns. The IPCC has found no conclusive evidence this is happening yet but asserts it is highly likely to weaken by 10-35% over this century and could completely collapse under high warming scenarios. Around 13,000 years ago, just before the climate settled into the current stable range, the Younger Dryas event cooled Greenland by 4-10ºC but warmed the southern hemisphere in a matter of decades. Given the speed and variability climate scientists believe this was caused by a weakening of the Atlantic Meridional Overturning Circulation (AMOC).

Thanks to the pioneering work by Arrhenius, Callendar, Manabe and many more, scientists can now capture the basic radiative physics, the logarithmic relationship of CO_2 and radiative forcing, and the key fast feedback mechanisms of water vapour, evaporation, clouds, ice, and aerosols in General Circulation Models. Work continues to fully integrate the remaining biosphere feedback mechanisms such as the carbon cycle response, fires, and melting permafrost in Earth System Models.

Forecasting the Temperature Rise: How to Approximate Complex Climate Models

Despite the significant advances in three-dimensional simulations, simple models are still used to give quick temperature forecasts without the need for spending months of calculation on a super computer. Simple radiative models give rough estimates for radiative forcing change which can be multiplied by a warming factor to yield a temperature forecast. The warming factor captures the net impact of water vapour, ice, clouds, and lapse rate feedback to approximate the output of General Circulation Models in simple form.

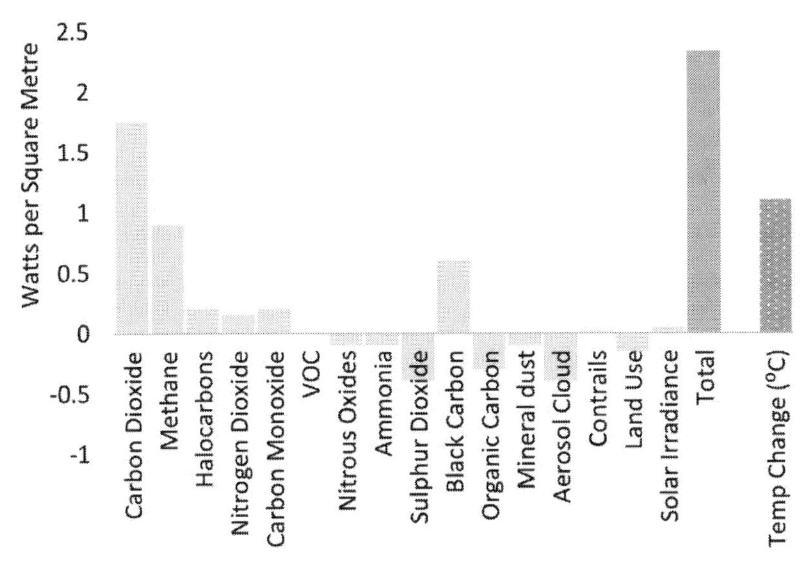

Estimated radiative forcing changes between 1750 and 2012 due to atmospheric component changes and solar irradiance changes.[25]

Since 1750, radiative forcing change has been predominantly driven by increasing concentrations of CO_2 and other greenhouse gases. The net impact of other changes to the atmosphere have broadly offset each other. So, we can estimate temperature change based on the change in CO_2 and CO_2 equivalents.

$$\text{Temperature Change} = 5.35 \times \text{natural log of} \left(\frac{CO_2 \text{ ppm now}}{CO_2 \text{ ppm past}} \right) \times 0.8$$

Simplified equation for temperature change, for a given change in CO_2 concentration in the atmosphere. Radiative forcing is the first part of the equation based on a simplified approximation of radiative transfer models by Mhyre et al. The 0.8x warming factor assumes a linear relationship between radiative forcing change and temperature change. From IPCC Assessment Report 3, 2001, Mhyre et al 1998.[26]

Assuming an increase in CO_2 from 280 ppm to 560 ppm (doubling) gives 3.7 watt per square metre increase in radiative forcing, this is the equivalent of a low power LED bulb shining on every square metre of Earth.[v] Multiplying by 0.8, the warming factor, gives 3ºC temperature increase.

We have now calculated the equilibrium climate sensitivity, which is the final temperature increase if CO_2 concentrations are doubled from pre-industrial times to 560 ppm. The IPCC range is 1.5-4.5ºC with 3ºC the consensus number. Recent work will likely narrow this range to 2.6-3.9ºC.[27,28,29]

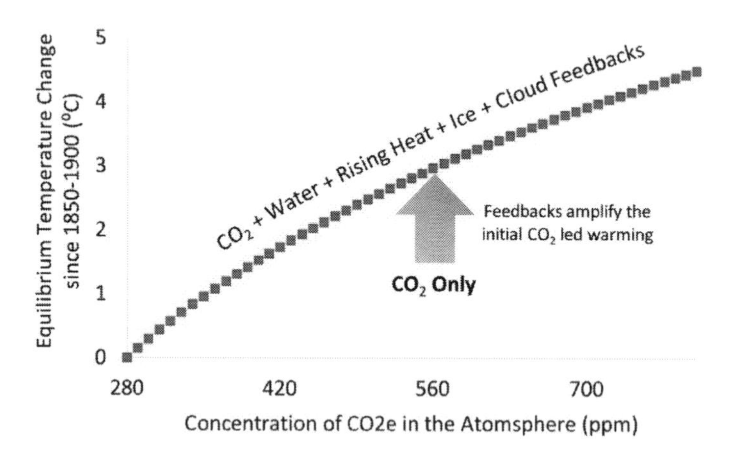

The simplified relationship between atmospheric CO_2 concentration and temperature increase (since pre-industrial average 1850-1900). Approximation based on simplified radiative forcing relationship of 5.35 x ln (CO_2 change) from Mhyre et al and using 0.8x Equilibrium Warming Factor to arrive at the mid-point of IPCC estimates.

We have calibrated the thermostat, and understood the relationship between CO_2, radiative forcing, and temperature. So now we can use our simple model to forecast temperature change based on future human emissions.

v Radiative forcing is measured in watt per square metre. 3.7 watt is enough to power a small LED bulb or charge a mobile phone.

Predicting Future Human Emissions: Defining a Baseline of Inaction

The dangers of climate change began to reverberate in the mainstream after oceanographer Roger Revelle and atmospheric scientist Charles Keeling started documenting the relentless rise of atmospheric CO_2[vi] in 1956.[30] Manabe and Wetherald published their definitive scientific basis for CO_2 and global warming in 1967. Then, in 1988, climatologist James Hansen famously testified in front of US congress stating that his team were 99% sure that human activities were already warming the Earth and creating extreme weather events. In 1989, the International Panel on Climate Change (IPCC) was established.

Despite the widespread concerns over the climate crisis, annual human CO_2 emissions continue to increase. Since the formation of the IPCC 30 years ago, we have released more CO_2 emissions into the atmosphere than the rest of human history combined. Thus far, humanity has blankly ignored the requirements and consequences of its own promotion to global caretaker, shirking the responsibilities of the Anthropocene, and is unwittingly, or knowingly, plotting a course for ever higher temperature change.

There are some small signs of promise with population and carbon emissions per capita showing indications of peaking in developed economies, but for most of the developing world population will continue to rise and increasing standards of living will demand ever more energy and resources.

Throughout this book we use one **baseline of inaction** scenario. This scenario is not a forecast, it's not based on current market trends or where the world is heading, it simply assumes that the future is a bigger version of the present. Given the ongoing efforts to shift away from fossil fuels, this is unrealistic, of course, but it serves as a benchmark for where the world would be if we decided to make no further effort to change our energy or agricultural systems. A baseline of inaction in a growing world.

Our baseline of inaction assumes the global population increases by 30% from 7.8 billion to a peak of ten billion people by the year 2060 (broadly in line with United Nations projections[31]). We also assume the average level of physical consumption increases to developed market standards throughout the world over the same period. A world where everyone has a top 10% standard of living. Where the 0.8 billion people with no

---

vi Atmospheric CO_2 is measured from the top of Mount Loa volcano in Hawaii where the air is well mixed and free from the influence of localised pollution or vegetation.

electricity access, 2.5 billion people with no clean cooking solutions or sanitation, and the 4 billion people with no running water, are brought up to modern standards.

In the ***transport*** sector, miles travelled by passengers double on land and triple by air and freight miles increase by 50%, taking everyone up to developed market averages. Across ***industry*** we budget for 12 tonnes of industrial production per person per year (from 8 tonnes today), bringing average goods consumption up to US or UK levels. We will assume ***basic amenities*** are provided for everyone on the planet, including penetration of space heating or cooling to all regions which require it, electricity access, clean cooking, and hot water for all. This is enough energy to deliver a top 10% level of consumption in sanitation, light, appliances, and cooking for each person on the planet. We assume ***agricultural*** output grows with population, production yields increase at a similar ongoing rate, and no change is made to diet or agricultural practices.

To fuel this increased consumption (assuming no change to the makeup of the existing energy system), the average final power demand will need to rise from 1.5 kW per person today to 2.3 kW per person by 2060. Total annual final energy demand doubles to 200,000 billion kWh[vii] and CO_2e emissions rise from 50 to 100 billion tonnes per year by 2060. Beyond this, let's assume the population, consumption, and carbon intensity[viii] remains the same and emissions hold at 100 billion tonnes per year.

In our baseline inaction scenario – peak population, peak consumption, peak emissions – the world would emit 7,000 billion tonnes of CO_2e over the next 80 years. These simple assumptions are broadly in line with OECD[ix] and IPCC baseline projections.[32]

vii **Peak Final Energy** = 10 billion people x 2.3 kW per person x 24 hours per day x 365 days per year = 200,000 billion kWh per year

viii **Carbon Intensity** = 52 billion tonnes of CO_2e per year ÷ 100,000 kWh final energy per year = 520 grams CO_2e per kWh final energy.

ix The OECD is the Organisation for Economic Co-operation and Development – an international economic organisation with 37 member countries founded in 1961 to stimulate economic progress and world trade.

Today	Peak
8 billion people	10 billion people
50b tonnes of CO_2 per year	100b tonnes of CO_2 per year
100,000 billion kwh energy per year	200,000 billion kwh energy per year
6,000 passenger miles per person per year	13,000 passenger miles per person per year
33% land used for agriculture	42% land used for agriculture
6kwh heating & cooling per person - every day	9kwh heating & cooling per person - every day
7 tonnes of stuff per person, per year	12 tonnes of stuff per person, per year

Our main assumptions used to forecast a baseline of inaction, peak population, peak consumption scenario. Our baseline of inaction is most closely aligned with the IPCC RCP6.0 CO_2 emission scenario, or somewhere between the SSP2 and SSP5 scenarios, under the new shared socio-economic pathways, which include population, wealth, land, and energy change assumptions.[33] For details on various pathways see Glen Peters' work.[34]

The CO_2 we release each year already has accumulated in the atmosphere and this will continue, steadily building up over time. For each slug of CO_2 added to the atmosphere, it takes 20 years for the ocean and land sinks to remove one third of the extra emissions. Over another few centuries another third is removed by soils and deep ocean mixing, until waters saturate and can hold no more. The increasing acidity of the ocean triggers a reaction between the dissolved carbon and calcium carbonate helping the ocean absorb a further sixth of the CO_2 over another 10,000 years. The final sixth will remain in the atmosphere for hundreds of thousands of years waiting to turn to rock.

Climate activist Bill McKibben puts this irreversibility in rather more colourful terms: "we're not going to get the planet back we used to have. We're like the guy who ate steak for dinner every night and let his cholesterol top 300 and had a heart attack. Now he dines on Lipitor and walks on the treadmill, but half his heart is dead tissue. We're like the guy who smoked 40 a day for 40 years and then had a stroke. He doesn't smoke anymore but the left side of his body doesn't work either... no one is going to refreeze the Arctic".[35] Emissions are irreversible over human timescales unless we artificially draw down the CO_2.

Predicting Future Temperature Rise Under Our Baseline of Inaction

In our baseline of inaction pathway, CO_2 concentrations will have doubled from pre-industrial times by mid-century (280 ppm to 560 ppm) and will reach nearly 750 ppm by 2100. CO_2e concentrations, including other greenhouse gases, are already over 500 ppm and will reach 1,000 ppm by 2100.[36] Based on our equilibrium climate sensitivity, this will lock in an eventual equilibrium temperature rise of more than 3.5ºC by mid-century and 5ºC by the end of this century.

Fortunately, it will take many decades or centuries to reach the full equilibrium temperature increase as the deeper ocean initially takes on more energy before giving it back to the atmosphere.[x] The ***transient*** surface temperature increase will be around half to two thirds of the eventual ***equilibrium*** temperature increase by year.[37,38] A transient warming of 1.8ºC in 2050 and over 3.3ºC in 2100. Once greenhouse gas concentrations stop increasing (or slow) the transient warming gradually catches up to the implied equilibrium temperature.

---

x It takes the same amount of energy to warm the top 10 metres of the ocean as all the atmosphere. Oceans have absorbed 93% of the extra energy over the last 50 years with just 1% going into heating the atmosphere.

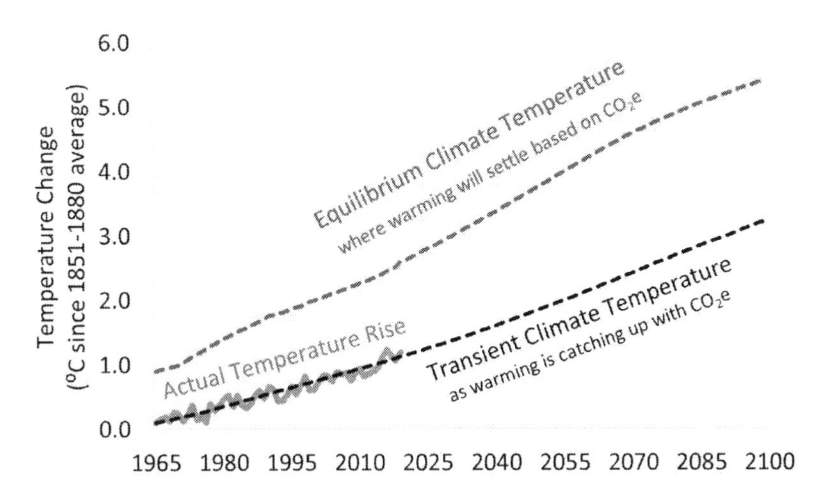

Our simplified temperature increase forecasts based on peak population and peak consumption by 2060. Approximations use simplified radiative forcing relationship of 5.35 x ln (CO_2 change) from Mhyre et al. CO_2e concentrations based on linear rise to peak emissions of 105 billion tonnes per year in 2060 then flat. CO_2e concentrations calculated at gross 0.13 ppm per billion tonnes CO_2e and drawn down by the carbon cycle in 5 sinks over the differing rates leaving between 0.07 and 0.05 ppm CO_2e in the atmosphere per billion tonnes emitted by humans. Equilibrium Climate Sensitivity multiplies radiative forcing by 0.8. Transient surface temperature catches up to equilibrium at ~2% per year minus ocean temperature increase. Ocean temperature increase catches up to atmosphere at ~1% per year.

Next, we can turn our approximation of a General Circulation Model into an approximation of an Earth System Model. To accomplish this, we need to include the impact of melting tundra, peat, forest fires, and a weakening land and ocean sink which are not all included in the last round of IPCC models. As increases in temperature trigger additional releases of carbon and more CO_2 is left in the atmosphere, our back-of-the-envelope calculations showed these biosphere feedbacks may increase atmospheric CO_2 by 20 ppm, 12 ppm, and 5 ppm for every 1°C of initial heating, respectively. Including these feedback mechanisms increases the forecast CO_2e to over 1,150 ppm by 2100 (1,000 ppm without biosphere feedback) and implies transient temperatures are over 3.6°C higher by 2100 (an extra 0.3°C).

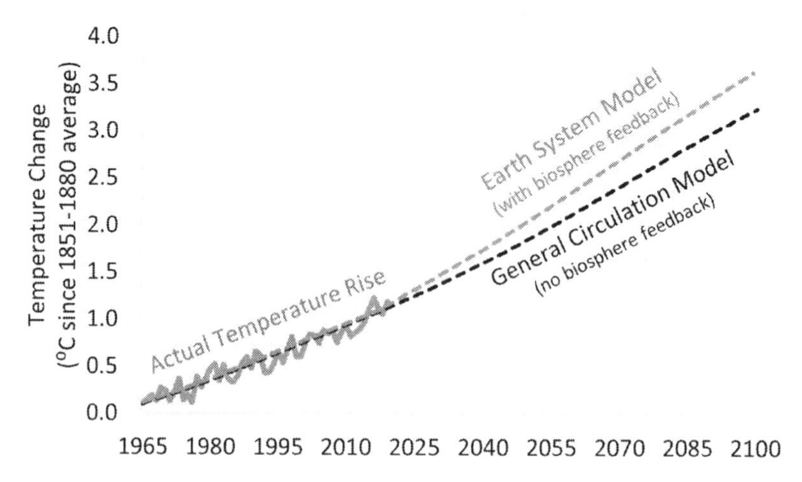

Transient temperature increase forecasts based on peak population and peak consumption by 2060. Approximations use simplified radiative forcing relationship of 5.35 x ln (CO_2 change) from Mhyre et al., bottom dashed line shows basic transient temperature increase calculation. Top dashed line includes extra CO_2 in the atmosphere from melting tundra, peat and forest fires and a weakening land-ocean sink as the planet warms (sensitivities referenced in preceding text).

Finally, let's revisit Enrico Fermi's play book and use a back-of-the envelope estimation to make sure our science-based forecasts seem sensible. First, let's look at the trends in recorded data. We can plot temperature change, CO_2 concentration, and sea level rise over the last 150 years. All three show a tight correlation and an increasing trend. What happens if we simply project these trends into the future based on our peak population, peak consumption, emissions timeline?

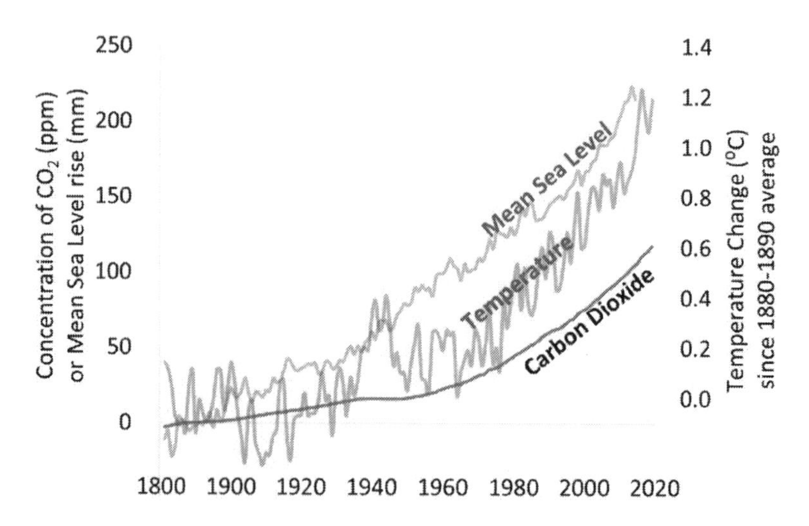

Direct Climate Measurements showing change in CO_2 concentrations, temperature, and mean sea levels since the average of the years 1880-1890.[39,40]

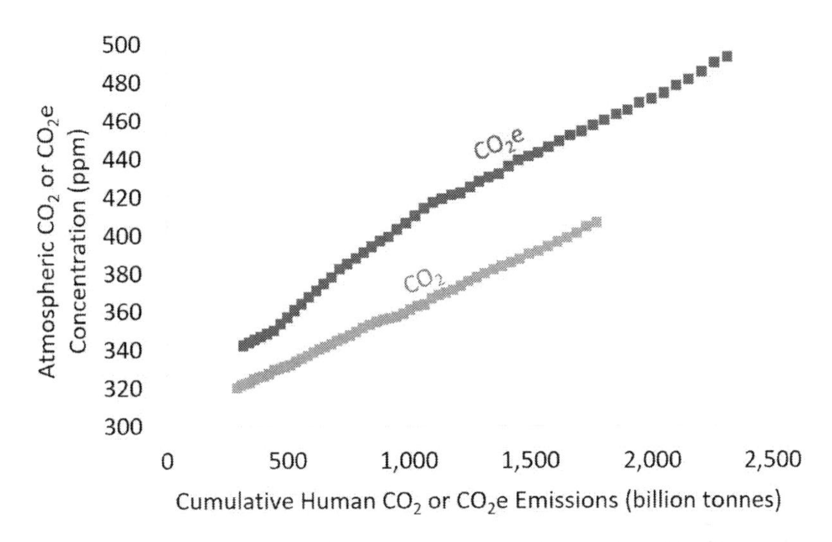

Scatter plot of total human CO_2/$CO_{2}e$ emissions from fossil fuels and land use versus atmospheric concentrations of CO_2/$CO_{2}e$ (1965 to 2017), top is $CO_{2}e$, bottom is CO_2.[41,42]

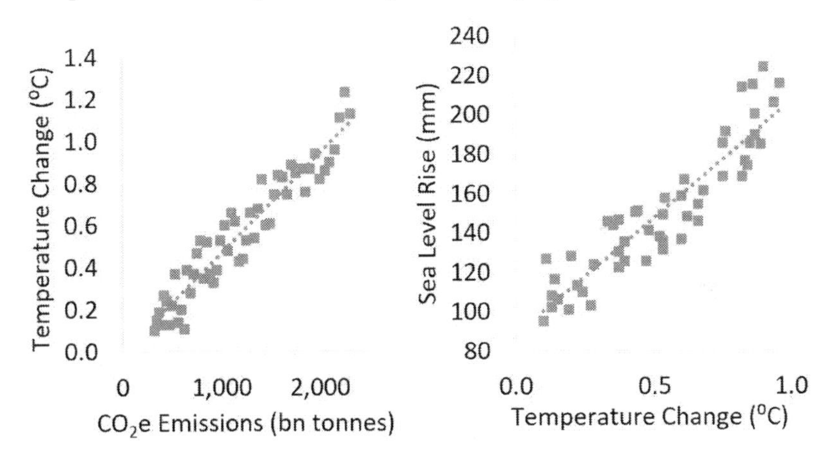

Left: Scatter plot of total human $CO_{2}e$ emissions (including other greenhouse gases) versus temperature rise (1965 to 2017). Right: Scatter plot of temperature increase versus mean sea level rise (1965 to 2017). [39,40,41]

Using a regression analysis[xi] on total human emissions versus temperature rise, and fitting a linear trendline, gives a very rough version of a climate model using nothing more than the historical data. Extrapolating this basic relationship for peak emissions implies a temperature increase of 4.5°C by 2100. This is higher than the mid-range of IPCC models

xi Regression analysis examines the statistical relationship between two or more variables.

because it assumes a linear (straight line) rather than logarithmic (Callendar's downward curve) response – the only way this happens is if biosphere feedback amplifies human CO_2e emissions beyond our back-of-the-envelope assumptions.

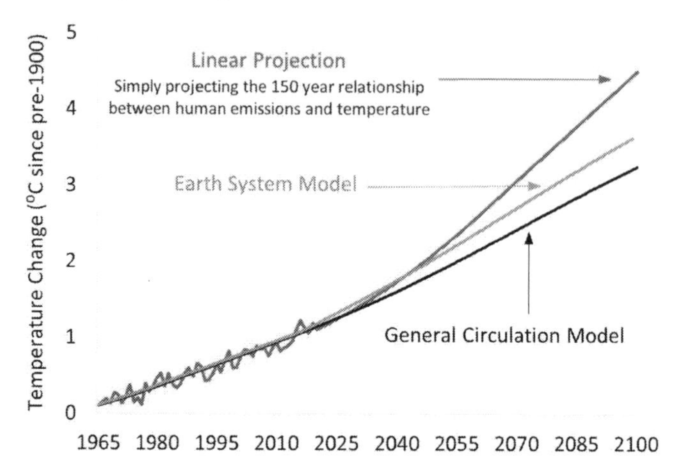

Temperature increase by year based on our simplified modelling and baseline of inaction scenario.

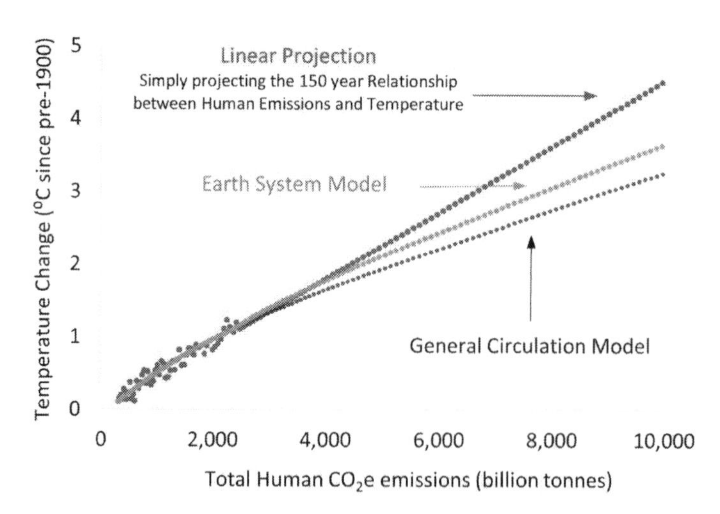

Temperature increase in the 21st century in relation to total cumulative human emissions of CO_2 equivalents – based on our simplified modelling. Cumulative human CO_2 equivalent emissions stand at nearly 2,500 billion tonnes today (2020) and will reach 10,000 billion tonnes by 2100 under our scenario for peak population, peak consumption by 2060 (100% of current fossil fuel reserves, 15% of total fossil fuel resources used by 2100). This chart shows how biosphere response amplifies warming, by amplifying human emissions, and taking the human-CO_2 and temperature relationship from logarithmic to towards linear.

Comparing End of the Century Climate Change Scenarios

We can compare our approximate projections to the IPCC high end greenhouse gas emissions scenarios called RCP6 and RCP8.5.[xii] RCP8.5 is a scenario based on 3 times increases to energy consumption this century, which is predominantly served by fossil fuels, including 10 times increases to coal use and coal-to-liquid fuel technologies.[43] Given the changes to policy and coal economics over the last ten years, the high end RCP8.5 now appears an extremely unlikely pathway. These levels of atmospheric CO_2e concentrations would likely only become possible if the world makes no further decarbonisation efforts and the biosphere response (added CO_2 from melting tundra, fires, and a weakening land-ocean sink) is twice as large as the best scientific estimates anticipate.

Whilst RCP8.5 is now seemingly unlikely, RCP6 has a simlar emissions profile to our baseline of inaction scenario and serves as a good comparison for a future where little further systems change is made and population and consumption grows to peak. Our simple calculations show a temperature rise of 3°C by the end of the century (2080-2100 average) which is slightly above the RCP6 warming range due to the inclusion of biosphere feedback.[xiii]

---

xii RCP stands for Representation Concentration Pathway. The number represents the radiative forcing change by the year 2100.

xiii RCP pathways were used in the last IPCC report in 2014 to map out future scenarios of greenhouse gas concentrations in the atmosphere. Since 2014 the world has shown signs of shifting away from heavy emissions technology like coal power so RCP8.5 now looks like an extreme, worst case scenario this century, rather than 'business as usual'. Our model assumes no further emissions efforts are made from 2020, this is our baseline. The IPCC is introducing Shared Socioeconomic Pathways for the next report which better reflect possible policy routes and socio-economic responses with a more realistic human narrative.

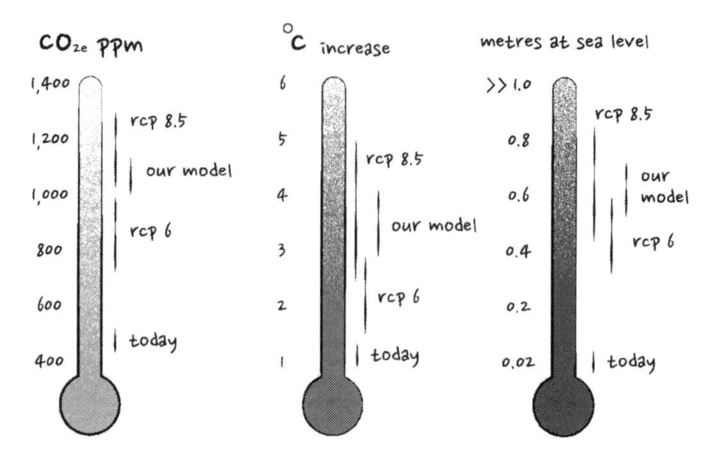

Comparison of forecast atmospheric CO_2e concentration (ppm), temperature increase (⁰C) and mean sea level rise (metres) based on IPCC RCP8.5 scenarios (2080-2100 averages), IPCC RCP6 scenarios (2080-2100 averages), and our three simplified models based on peak CO_2 by 2060. End of the century defined as 2080-2100 averages.

So, we have a feel for the basic physical processes of the greenhouse effect, an insight into how scientists forecast global warming using both simple radiative forcing models or more complex General Circulation Models, and a glimpse into why the development of Earth System Models mean temperature projections may need to increase. The Grantham Institute briefing paper "Biosphere feedbacks and climate change" points out that only a fraction of models include biosphere feedback, such as changing land-ocean sinks, and no model in the current AR5 IPCC temperature forecasts (from 2014) includes the release of carbon from tundra, potentially one of the most important feedback mechanisms.[44] As the next set of models start to include land-ocean sink feedback, nutrient availability, wildfires, and permafrost, some scientists are concerned about the idea of Hothouse Earth, where even 2⁰C initial human warming is enough to trigger Earth system tipping points which could push temperatures to over 4⁰C even if we eliminated further emissions.[45,46]

The Climate Sensitivity: Our First Dial of Doom and Boom

Whilst modelling and predicting climate is a highly complex feat, the underlying science is robust. We know there is a statistically significant relationship between CO_2 and temperature that extends deep into geological time. Climate scientists can explain this relationship based on the fundamental physics, chemistry, and biology of the Earth system. This scientific basis can be used to predict future temperature changes and even the earliest of these models from the 1970s have proven accurate so far.

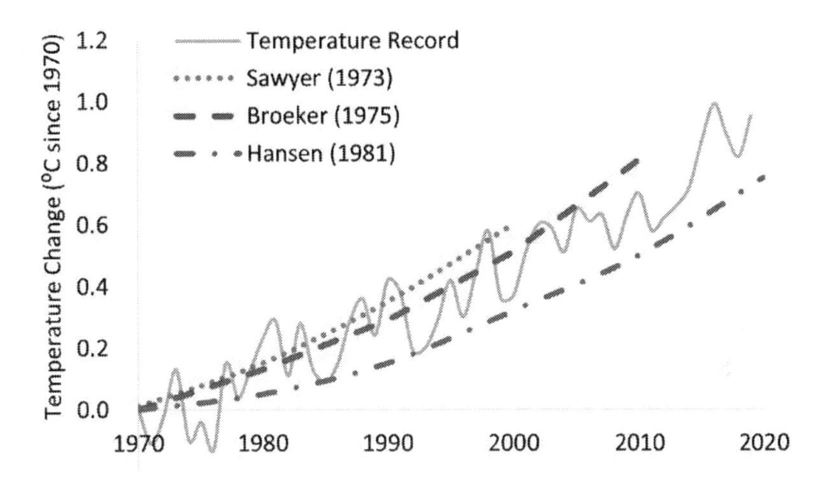

Comparison of temperature forecasts, made by scientists between 1973 and 1981, with the actual temperature rise since the 1970s. These predictions represent the first scientific basis for global warming built into predictive models. [47, 48, 49]

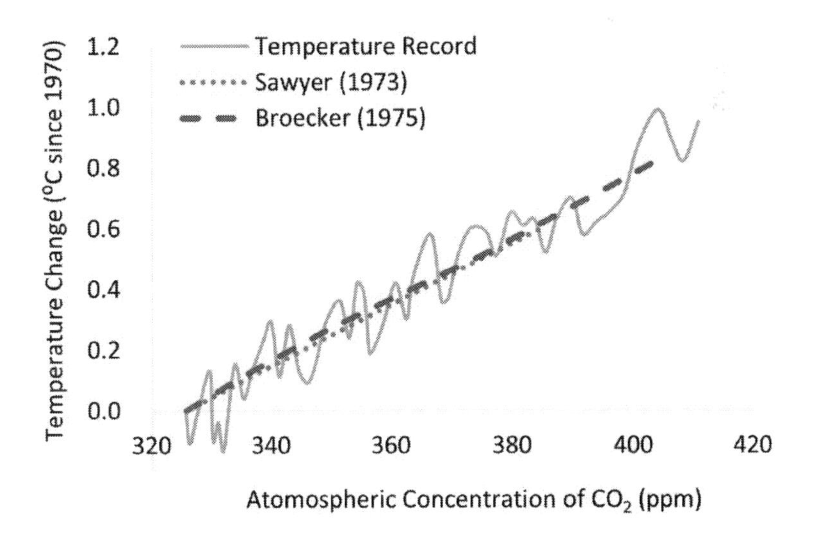

The same temperature predictions by Sawyer and Broecker but in relation to CO_2 ppm concentrations rather than by year.

It is now undisputed that human activities are increasing CO_2 in the atmosphere and it is clear that this has already pushed temperatures over 1ºC higher. Remaining on the current trajectory, human activity will very likely warm the planet another 2ºC this century and, as the biosphere responds and the deep ocean equilibrates, it seems the risk is increasingly skewed towards higher average temperatures still.

Beyond the year 2100, temperatures and sea levels will continue to rise as the Earth's system equilibrates. Our peak population and peak consumption scenario will have used up all of today's extractable fossil fuel *reserves* (1,300 billion tonnes carbon[50]) by 2100. However, six times more fossil fuel *resources* are yet to be uncovered or cannot be reached with today's technology.[51] Burning the full 10,000 billion tonnes of fossil fuel carbon over the next few centuries would take atmospheric CO_2 concentrations towards the highest levels in 500 million years of planet Earth, temperatures to over 10ºC warmer, and climate conditions last encountered over 50 million years ago.

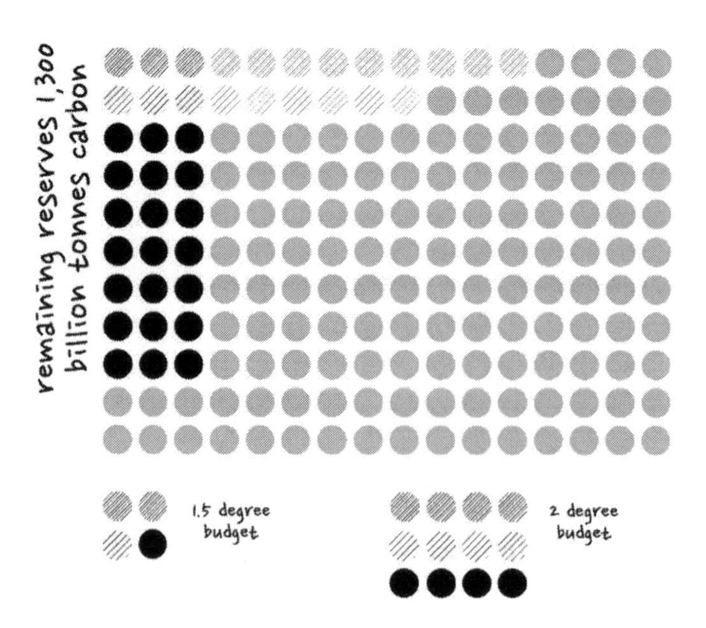

Estimated fossil fuel resources left on Earth are all the coal, natural gas, and oil deposits. Remaining reserves can be extracted using today's technology at reasonable cost. Each bubble represents 50 billion tonnes of carbon (multiply by 3.7 for CO_2*). Fine Stripe = Natural Gas, Medium Stripe = Oil, Solid = Coal. Faded bubbles are resources yet discovered or hard to extract. Carbon budget shows the amount of carbon which can still be burnt before 1.5ºC or 2ºC warming is breached (based on our simplified warming model). Data from BP statistical review (2019). World Energy Resources (2016).*

The last time living organisms had such a drastic impact on the climate was two billion years ago when the Great Oxidation event altered the chemistry of the atmosphere and fundamentaly re-wrote the evolution of life on Earth. And unless we leave the majority of fossil fuel resources buried underground we are set to create a planet better suited for cold blooded reptiles than for human inhabitants.

From the painstaking calculations of the early pioneers to the groundbreaking work of modern climate scientists, the world has slowly come to understand and quantify the impact of loading the atmosphere with greenhouse gases. The direction of change is known and it is only the precise regional impacts and precise speed at which these changes take place that remains most uncertain.

The ***climate sensitivity*** is our first dial of doom and boom, the science-based value which defines how quickly the climate will change as we load the atmosphere with CO_2. Dial it down and perhaps warming is slow enough to manage, turn it up to full and catastrophic change could await.

So, we have the ability to predict future temperature change and sea level rise should we not act to combat climate change. But how big a problem is a few degrees of warming? What does this mean for the environment? What damage will be done to human systems? And might there be some benefits?

In the next chapter we will look at the physical changes that a warmer world could inflict upon the environment, the economy, and on human social systems.

Chapter 4: Fearing Change

The Physical Impacts from a Warming Planet

In this chapter we will use our temperature and sea level forecasts to better understand how the changes to climate will drive physical changes in the frequency or magnitude of storms, flooding, heat, fire, droughts, and natural catastrophes should we make no change to our use of fossil fuels. We will also quantify the toll of increasing air pollution, water stress, and ocean acidification, the potential declines in agricultural yield and labour productivity, and the impacts these changes will inflict upon society and the economy.

Risk Aversion and a Peak Consumption World

If we move towards a peak consumption world where we continue to power the economy with fossil fuels, climate models forecast over 3°C higher temperatures by the end of this century, with a risk to further increases. A 3°C average temperature increase may not sound huge but remember averages don't tell the full story. Average land-ocean surface temperatures may increase 3°C, but big cities and the Arctic will warm over 2-3 times as fast, mid-latitudes 2 times faster, and land 1.5 times faster, with all of these increases balanced by an ocean which warms at two thirds of the average rate.[1] The extra trapped heat means nights will warm more than days, winter more than summer, and the chance of extreme hot weather will rise exponentially as the chance of extreme cold weather fades. But how will these changes impact humanity? Do they warrant urgent action?

Humans have evolved to be naturally risk averse. We feel the pain of loss acutely more than the pleasure of gain. Offer someone £10 straight up or £20 if they win a 50:50 coin toss and the majority will settle for the former.[2] Statistically, each option is worth the same, but while most people ***like*** to win, they ***hate*** to lose. This is also why climate change is such a divisive issue. Those who campaign for immediate action on global warming fear the changes brought about by a warmer planet.

Those who drive for the continued use of fossil fuels fear the overreach of government or the social and economic cost of changing our energy system. To figure out who is right we first need to understand the impact of making no further changes to our existing energy mix and agricultural system and letting it grow with population and consumption. We can estimate the physical change to the planet as temperatures rise and benchmark the social and economic impact of continuing to power the economy with fossil fuels. This serves a baseline of inaction.

How will these climate changes impact the weather we experience every day? How urgently is action needed? Will the change prove large enough to drag us back into the Malthusian Trap?

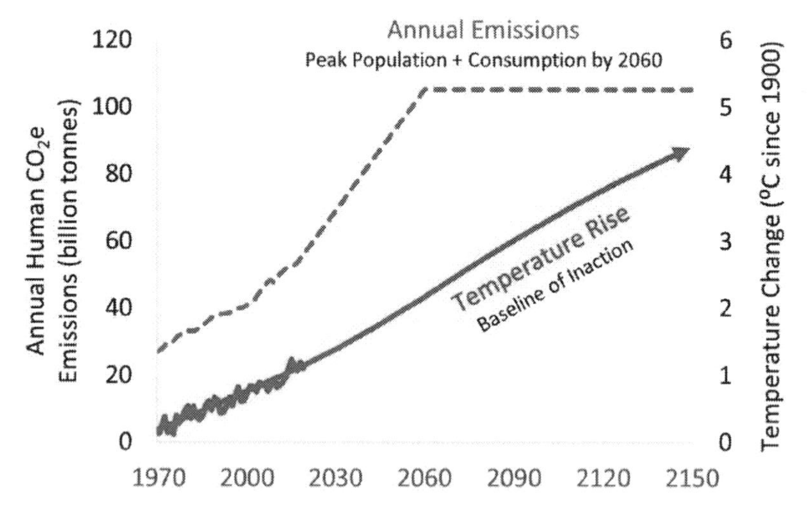

Our basic peak population, peak consumption scenario assumes global population grows to 10 billion people by 2060 then holds flat. We assume consumption per person increases to developed market levels bringing total annual CO_2 equivalent emissions to over 100 billion tonnes per year.

Sea Level Rise

The combination of melting ice sheets and warmer water results in a rising Mean Sea Level (MSL). The oceans are already 0.2 metres higher than 100 years ago driven by melting ice (0.14 m) and warmer expanding water (0.06 m). Every degree of warming is estimated to increase the sea level by 2.3 metres over 2,000 years.[i,3]

---

i Every 360 billion tonnes of ice melting from land will raise sea levels by 1 mm. Since 1970 glaciers have lost 5,000 billion tonnes of ice, Greenland 2,500 billion tonnes and Antarctica 2,000 billion tonnes of ice contributing nearly 3 cm to mean sea level rise. The 2.3 metres rise is 0.4 metres from expanding warmer water and the rest from melting ice.

By the end of this century more than half of all glaciers will have disappeared and the great ice sheets could be irreversibly melting for centuries to come. The IPCC estimates between 0.3-0.85 metres of sea level rise this century under a peak emissions scenario. But scientists continue to find evidence that ice is melting faster than anticipated and estimates may need to be revised higher.

The last time Earth had 400 ppm CO_2 was 2-4 million years ago; temperatures were 4ºC hotter and sea levels were 25m higher with little Arctic ice and no West Antarctic Ice Sheet. The faster temperatures rise, the greater the risk of reaching dangerous tipping points which can accelerate melting. One of the most significant of these would be if large areas of the West Antarctic Ice Sheet broke off, sliding into the ocean as warmer waters melt it from below.[4]

Globally, 382 million people live in areas which are less than five metres below sea level.[5] The Maldives, Bahamas, Bahrain, and Suriname have 80-100% of their combined 1.4 million population in low lying coastal regions. China, India, Bangladesh, and Indonesia have a combined 300 million people in low lying areas (~20% of their population).[6] Based on this data, in our baseline of inaction scenario by the end of the century up to 90 million people might have been displaced (1% of the global population) by rising seas.

Beyond the year 2100, unless CO_2 and temperatures are reduced, sea level will continue to rise over centuries to more than ten metres higher as waters expand and the Greenland Ice Sheet disappears, drowning Miami, Manhattan, London, Shanghai, Bangkok, and Mumbai[ii,7] and displacing one in ten people across the world.

Burn the remainder of all fossil fuel resources on the planet and we set a course for >10ºC higher temperatures and tens of metres of sea level rise over the coming centuries. Half of the global population would eventually need to move.[8]

This compares to the Internal Displacement Monitoring Centre estimate that 50 million could be displaced by the year 2100. The last IPCC assessment puts the number between 72 and 187 million displaced due to submergence and flooding by 2100 if no action is taken.[9]

---

ii Elevation above sea level in metres: Miami (2 m), Manhattan (5 m), London (11 m), Shanghai (4 m), Bangkok (1.5 m), and Mumbai (14 m).

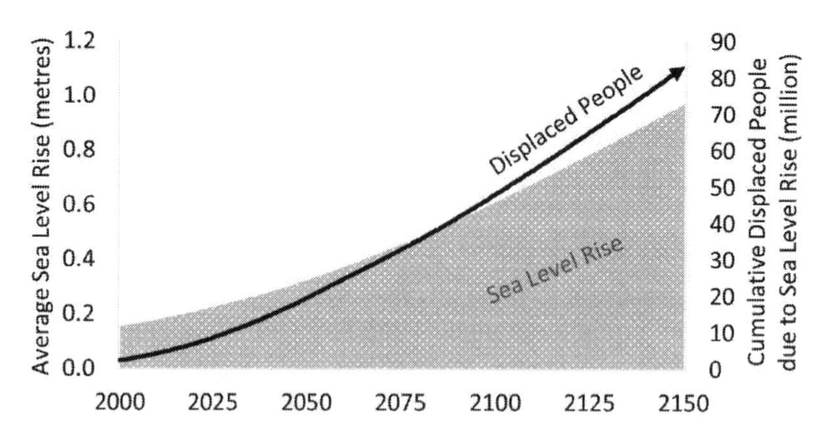

Black Line: Total cumulative number of people displaced due to rising sea levels based on our peak consumption, inaction scenario (3.3ºC warming by 2100, 4.4ºC by 2150), calculated based on population living below mean sea level as water rises. Shaded Area: Increase in mean sea level in metres projected by our model (in line with IPCC).

Storms and Flooding

Every 1ºC temperature increase adds 7% more water vapour into the atmosphere, increases precipitation by 2% and extreme rain by 4.5%.[10] It also warms the surface waters of the ocean. When ocean waters reach above 26ºC (down to 100m) this creates optimal conditions for hurricanes, cyclones, and typhoons to form and helps them travel further, potentially onto land.[11,12] Higher energy means higher wind speeds and the destructive power of these storms are proportional to the wind speed to the power of 2 or 3.[13] The damages are further amplified by rising sea levels increasing the risk of storm surge flooding. Category 5 hurricanes now account for 35% of all hurricanes, this was just 25% thirty years ago.

Data from Munich Re shows the annual number of global storms and floods has steadily increased from an average of 220 in the early 1980s to 650 per year. Some of this change can be attributed to greater coastal populations and wealth, but the data and most research agrees that global warming is already impacting the intensity of this class of natural disaster.

The last IPCC report (AR5) states flooding has already ramped up and it is virtually certain intense tropical cyclone activity has increased in the North Atlantic, though hurricane data is less conclusive elsewhere. With rising temperatures, the frequency of tropical cyclones may flatten or even decline, but most research suggests the intensity of storms will increase. Every degree of warming is expected to add 3-5% wind speeds or up to 16% to the destructive power. More intense tropical storms will continue to batter the Gulf of Mexico, south and east coast US, Central America, India, Bangladesh and South East Asia. The reach of tropical cyclones may extend further as warmer waters expand Poleward.

More water vapour in the atmosphere means mid-high latitude regions of the northern hemisphere are already experiencing 5% more rain and snow.[14] This will continue to rise at a rate of 5-10% per 1ºC warming and continue to increase the risk of flooding across South East Asia, Tropical Africa, Northern South America, North East Brazil, South Africa, the UK, and many other parts of the planet.

In the last 20 years we have experienced both the costliest weather-related event and one of the deadliest weather events on record. Hurricane Katrina in 2005 devastated New Orleans and much of Louisiana and Florida with nearly $200 billion of damages and lost income, killing over 1,700 people.[15] The deadliest weather-related event of the last 40 years was Cyclone Nargis which struck Myanmar killing 140,000 people (the fifth worst on record).[16] Nargis had winds of over 200 km per hour (category 4) and made landfall on Friday, 2 May 2008, sending a storm surge 40 km up the densely populated Irrawaddy Delta.[17] The Labutta Township alone was reported to have 80,000 dead and over 1 million people became homeless after the event. Rising temperatures increase the likelihood of more high energy destructive events.

Long-Term Intolerable Heat

Rising oceans aren't the only change which will force human migration. Rising heat will probably prove the bigger factor over the next century. As the IPCC states: "If the body temperature rises above 38ºC (heat exhaustion) physical and cognitive functions are impaired: above 40.6ºC (heat stroke) risks of organ damage, loss of consciousness, and death increase sharply". Humans don't like being too hot for too long.[iii]

Death Valley, Inyo County in the US, is one of the hottest places on Earth. The average high temperature in July is 47ºC with the average low 27ºC at night.[18] The record high reached 54ºC in summer 2020.[19] These temperatures are intolerable for most humans; the population density of Death Valley is just 0.06 brave people per square kilometre compared to a global average of 60.

Rising global temperatures will create more Death-Valley-like temperatures around the world. Another degree on land and 7 million

---

iii Wet-bulb temperature at 100% humidity is the best measure of intolerable heat. Healthy humans struggle to survive above 35ºC heat and 100% humidity.

people in Kuwait and Qatar will be experiencing average summer highs of 47ºC.[20] Add another degree and a further ten million people in the United Arab Emirates reach Death Valley status. One more degree and things become ever more serious as big populations such as Algeria, Niger, Iraq, and Mali also reach Death Valley temperatures bringing intolerable heat to over 150 million people. Beyond this the Middle East, North Africa, and Pakistan will suffer from similar levels of searing summer heat.

Assuming migration levels step up as Death Valley temperatures are breached, then around 400 million people may have attempted to move by 2100 according to our baseline of inaction scenario. Combine heat migration with sea level rise and 13 million people may be forced to abandon their homes every year. With today's average global property valued at $36,000 per person (and rising with GDP growth), this adds up to nearly $100 trillion of real estate losses by 2100 or 3.5% of accumulated future capital (land, real estate, machinery, infrastructure).

These numbers compare to official estimates from The International Organisation for Migration which estimates between 225 million and 1 billion people may be unable to live in their areas of origin by the end of the century, based on the IPCC high emissions pathways (RCP6/RCP8.5).[21]

Extreme Temperature, Fire, and Drought Events

A warmer lower atmosphere means higher average temperatures and the number of extreme hot days increases as the number of extreme cold days declines. This boosts the risk of deadly heatwaves in areas ill-equipped to deal with extreme heat. Despite more water vapour in the atmosphere and greater average rainfall over the planet, dry areas will become drier as wet areas become wetter. Rain will fall heavier but less often and the evaporation of water from soil and vegetation will increase surface run off and reduce water retention. Dry forests increase the risk of wildfires and the extent to which they spread.[iv]

Data from Munich Re shows the number of global heat waves, fires and droughts have steadily increased from an average of ten episodes per year in the early 1980s to over 40 per year in the last five years. As the planet continues to warm, we can expect more deadly heatwaves across much of the world. Research suggests every 1ºC warming could increase the amount of land experiencing extreme summer temperatures (3-5%

iv Over the last 200 years the annual total burned-area of land on Earth is down from 5 to 3.7 million square km because deliberate burning by humans for agriculture has declined from 2 to 0.15 million square km. However non-deliberate wildfires in areas like Australia, California and Russia have increased with rising temperatures, despite less forest cover and better management.

hotter than normal) by 10%, leading to a 50-fold increase in the most dangerous heatwaves.[22] India and Pakistan will be the first major nations to feel the full impact of more frequent lethal heatwaves where even healthy individuals will be unable to cope with the elevated temperatures without cooling shelters or air conditioning – a luxury most cannot afford today.

Central United States or northwest Australia may experience fewer droughts but already high-risk drought-prone areas such as southern US, the Mediterranean, and Africa will increasingly turn to desert. Access to fresh water will become even more difficult in India, Southern Africa, Australia, central South America, the Mediterranean, and parts of the US. Every 1°C warming could increase the risk of drought by 5-10%.[23] Wildfires will burn harder and more frequently in regions such as the Americas, Russia, Canada, Southern Europe, and Australia.[24,25] Research estimates every 1°C warming could increase lightning strikes by 10%, the wildfire season by more than 20%, and the potential burn area of each fire may double.[26,27]

The heatwaves across Europe in 2003 and Russia in 2010 were the third and fourth deadliest climate events of the last 40 years and had the highest death rates for extreme heat on record, killing 68,000 and 56,000 people, respectively. France was hit hardest in 2003 with eight consecutive days of more than 40°C heat even in northerly regions. The lack of air conditioning and lack of respite at night meant that nearly 15,000 died and undertakers had to set up shop in industrial refrigeration units on the outskirts of Paris to handle the sheer volume of dead bodies.

Natural Catastrophes

Rising CO_2 concentrations trap more energy in Earth's climate system. More energy means either more powerful or a greater frequency of natural catastrophes. Combine this destructive energy with stresses from rising water and heat and you have a recipe for increasing disaster.

Since 1980, temperatures have risen by nearly 0.7°C and climate related storms, flooding, wildfires, heatwaves, and drought have tripled from 200 to 650 major events per year. Rising temperatures, warmer oceans, and more water vapour will increase the intensity of tropical storms – rainy areas will become rainier, dry areas drier – and as the number of unusually hot days grows this will extend wildfire seasons and boost the chances of deadly heatwaves. One positive will be a reduction in extreme cold days and fewer cold related deaths in winter. However, on balance, the overall impact is one of rising destruction and a rising death toll.

The average natural catastrophe kills 50 people and incurs over $200 million in damages.[28] If the correlation between rising temperature and natural catastrophes continues, then under our inaction scenario the frequency of storms, floods, droughts, and wildfires may increase by another 1,000 events per year before the end of the century. If the average damages rise with our GDP growth (~2% per year) then total annual losses will increase from $200 billion to nearly $3 trillion per year. A cumulative impact of $65 trillion over the next 80 years which wipes out nearly 2.5% of accumulated future capital. The death toll will rise from 20,000 towards 100,000 people per year by 2100 (0.1% of deaths).[v]

For comparison, the Cambridge Climate Change Business risk index estimates that total natural catastrophe damages could increase by $100 billion per year by 2040 – a similar trajectory to ours.[29]

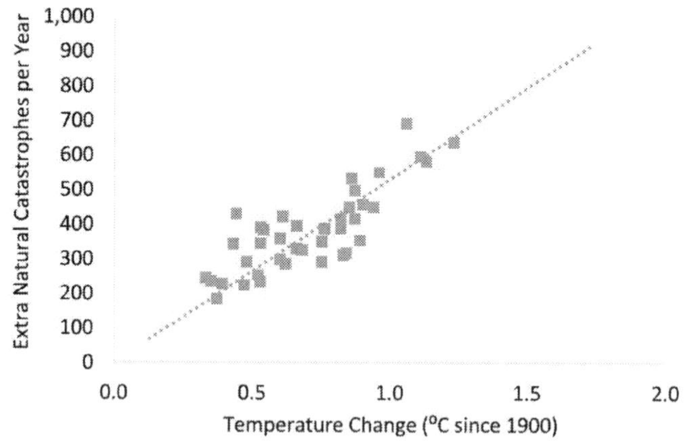

Regression of annual additional natural catastrophe events per year as global temperatures have increased with a linear trend line shown (the underlying trend from population growth and wealth growth have been removed from the data using the trend in non-climate natural catastrophe events). Historical data Munich Re NatCat database, temperature change from NASA GISTEMP data.

Warmer Winters

A warmer lower atmosphere means less extreme cold events and milder winters. This is at least some good news for many living in northerly reaches of the world. In winter death rates increase by 10% above the

v Munich Re record Natural Catastrophe events which cause greater than one death and/or cause damages in excess of $100k-$3mn (inflation adjusted) depending on the wealth of the country affected.

average and in summer and autumn death rates are 5% below average.[30] Two thirds of the winter increase is due to cold weather and one third due to the spread of influenza. The balance between more heat deaths and fewer cold deaths remains unresolved by climatologists, epidemiologists, and statisticians.[31] More people die in winter, but the death rate is less sensitive to changes in temperature, whereas fewer die in summer but deaths rise very quickly with abnormally high temperatures. We are better at dealing with the unusually cold rather than the unusually hot.[32] By the middle of this century the IPCC estimates that heat related deaths will outweigh the fewer cold deaths.[33]

Air Pollution

Where rising water, heat, and extreme weather will launch an all-out frontal assault, air pollution is the environmental Trojan horse: desecration of the air we breathe is often overlooked in relation to climate change, but the combustion of fossil fuels is also responsible for more than 90% of all pollutants[vi]. Burning coal, oil, gas, and wood in our homes, cars, industrial processes, and power plants not only releases CO_2 but a whole host of atmospheric pollutants such as fine particulate matter, nitrous oxides, sulphur oxides, and ozone. These invisible gases linger in the lower atmosphere, infiltrating our every breath, traversing through our lungs, and into our bloodstream. The effect is to agitate the airways, reacting in our bodies and heightening the risk of respiratory diseases such as asthma, pulmonary conditions, chronic bronchitis, and lung cancer. The strain on the body stops blood vessels relaxing, natural anti-clotting mechanisms are reduced, and fine particles help block arteries and thicken the blood leading to degenerative heart disease and heart attacks.[vii]

Here are some of the offending pollutants:[34]

Particulate Matter is one of the most dangerous. PM are small particles of carbon not converted to CO_2 during combustion. Transport accounts for half of these emissions, electricity generation and industry the other half. Large PM10 (10 thousandths of a millimetre wide or less) give smoke its hazy appearance and can penetrate deep into the lungs. Smaller PM2.5 particles, more commonly produced in modern combustion processes, are invisible and readily pass through our lungs and into the

---

vi Sources of air pollution: 80% Particulate matter, 10% other human emissions, 10% natural.

vii Deaths from air pollution: 21% pneumonia, 20% stroke, 34% heart disease, 19% COPD, 7% lung cancer.

bloodstream. Particulate matter is emitted by all combustion processes but is particularly high in log burners, coal fired power plants, industrial coal heating, and cars (particularly diesel).

Nitrous Oxides are another class of pollutant responsible for degrading the air we breathe. Nitrous Oxide or NOx represents NO and NO_2 which are formed from the nitrogen and oxygen in the air during high energy combustion processes taking place in car and plane engines (80%) and in gas boilers (20%). NOx agitates the lungs and acidifies soil and water.

Ammonia or NH_3 is a common fertiliser which, when used excessively, releases ammonia into the air. It is also released from the manure of livestock reared on protein intensive diets. Ammonia, like NOx, agitates the lungs, eyes and acidifies the environment.

Ozone is formed when NO_2 and VOCs (Volatile Organic Compounds from the fumes of petrol or released naturally from forests) meet in the presence of intense sunlight. Ozone (O_3) is very useful at higher levels of the atmosphere, shielding us from harmful ultraviolet radiation and breaking down other atmospheric pollutants, but at lower levels, ozone is dangerous. It is a highly reactive species which inflames lung tissues, triggers asthma attacks, and is toxic to plants and animals.

Sulphur Dioxide is formed from the oxidation of sulphur impurities found in so-called sour crude oil or coal. SO_2 is the key ingredient of acid rain which kills aquatic life, erodes buildings, and releases aluminium from soil, killing trees and plant life.

The World Health Organisation (WHO) and *Lancet* research estimate that over four million people die prematurely each year due to outdoor air pollution.[35,36,37] More recent research suggests that the number of deaths from outdoor fine particle air pollution could be double these estimates, closer to nine million premature deaths each year – responsible for 30% of all deaths in East Asia, 17% in Europe and 13% in the US.[38]

Another four million die from indoor biomass (wood or dung) burning on open stoves in developing countries.[39] More than 600,000 deaths are of children under the age of 15 and air pollution accounts for 1 in every 10 deaths of children under five years old.[40]

Add the ongoing but non-fatal health burden and you multiply this suffering multiple fold. With an average price tag of around $3,000 per treatment the annual cost on the healthcare system is around $50 billion globally and that's before you factor indirect costs from sick days and lower productivity which add another $200 billion lost economic income (0.3% GDP).

Air pollution reduces total life expectancy of all humans by 150-200 million years or an average of 1.5-2 years per person making it one of the leading risk factors for death.[41] Not everyone will succumb to air pollution, but for those eight million people that do their lives are cut short by nearly 20 years.[viii]

Air pollution ranks as the number one risk factor leading to death in China and India with over 2 million deaths per year or about 100 deaths per 100,000 people. Developed countries fare better, but air pollution still ranks as a leading risk factor, responsible for nearly 100,000 premature deaths per year in the US or nearly 20 deaths per 100,000 of the population.[42]

Progress has been made to clean up the air by switching away from wood or coal burning stoves, moving power plants out of cities in the 1950s, adding catalytic converters to cars in the 1970s, and taking sulphur out of fuel in the 1980s. Death rates have declined in developed countries, mostly driven by lower indoor air pollution. but less progress has been made in developing regions. Despite these efforts, eight million people continue to die each year. That's 14% of all deaths worldwide.

Regulation such as the UK's clean air act in 1956, and the US's in 1963, have cut much of the visible smog from major cities like London and New York. But, as Tim Smedley illustrates in his book *Clearing the Air*, many invisible dangers still lurk. More than 80% of those living in urban areas are breathing air with pollutants exceeding recommended limits. The air in many cities across the developed world is rife with invisible PM2.5, nitrous oxide, and dangerous ozone. Most cities across the developing world still lack basic air pollution regulations at all.

My wife's parents grew up in the east end of London in the 1960s and 70s. They still remember when the smog regularly enveloped the city. The "pea soup" as they called it, could get so thick they would often get lost on their way home from school. London's air may look cleaner today, but it still regularly breaches the recommended limit of 25 micrograms of PM2.5 per cubic metre creating risks for those with existing health problems, children, and the elderly.[43] Beijing can reach hundreds of micrograms of particulate matter for weeks at a time – the equivalent

---

viii Researchers at the University of Chicago highlight that fine particulate matter is now one of the biggest threats to global health, with sustained exposure reducing life expectancy by 1 year for every 10 micrograms per m^3.

of stepping into a busy smoking lounge every time you venture outside. The International School of Beijing now has a pressurised dome for its children to play "outside".[44] Delhi has some of the most toxic air on the planet which can breach one thousand micrograms; one in three adults and two in every three children have respiratory problems.

As the global population continues to grow and energy consumption rises per person, so does the air pollution. Total deaths will climb from eight million today to 14 million per year by 2060 if no change is made. Annual direct healthcare costs[45,46] could rise to $100 billion per year and productivity losses will move into the trillions (0.4% of GDP).

Agriculture

Over the last 200 years, mechanisation, selective breeding, and genetic modification have increased the average crop yield from 1 tonne to 20 tonnes per acre per year. The future evolution of farm productivity hangs in the balance with rising CO_2 and a warming atmosphere.

The rising concentration of CO_2 provides plants with more of their key feedstock for photosynthesis and reduces the length of time they must open their leaves to feed which helps limit water loss. Combined, these benefits drive accelerated growth. Experiments show that every 50% increase in CO_2 boosts growth by 8%. A 2016 study led by 32 international experts using high-res NASA satellite data showed the Earth's land is already 10% greener than it was 35 years ago.[47] This is a key dampening mechanism, helping to remove human CO_2 emissions from the atmosphere and it also helps improve agricultural yields.

However, not all change is positive. Rising heat will bring more drought to many key agricultural hubs including the US, Latin America, and Africa. Rising sea levels and extreme weather will bring more flooding to areas such as India, Bangladesh, and Asia. Add the increasing probability of damage from pests, a lower number of pollinators, risk of fire and storm damage and the mounting burden will eventually cripple agricultural yields despite the CO_2 boost.

Experts are torn between small net agricultural benefits up to about 1.5-2ºC whilst others believe we are already on the downward slope. Most agree that beyond 1-3ºC, whilst some areas like Canada and Russia will benefit, overall agricultural productivity will decline on existing global farmland. Rising temperatures will force farmers to relocate towards the poles. Progress will be hindered by the availability of land and quality of soils which can take centuries to form. As L. Hunter Lovins points out in *A Finer Future*: "A Handful of healthy soil has more living organisms in it than there are people on Earth... Treating soil like dirt threatens all life on the planet".[48]

The last IPCC report highlighted corn and wheat are already negatively impacted globally and soy and rice not far off.[49] Frances Moore and team at the University of California Davis estimate productivity will hold broadly flat to 2ºC and will then decline quickly towards 25% losses at 4ºC warming, based on the trade-off between CO_2 and temperature.[50] If we assume that beyond 2ºC agricultural productivity declines at a rate of 10% per 1ºC warming, under our inaction scenario production yields will be 10% lower by 2100 (3.3ºC warmer) and 20% lower by 2150 (4.4ºC warming). Coupled with 30% increases to the global population this will place major pressure on food availability around the world.

The current agricultural system provides enough crops, meat, and dairy to provide 2,500 calories[ix] per person per day for each human on the planet. The average person consumes 2,000 to 2,500 calories per day. Surprisingly, farmers produce more than enough calories to feed everyone on the planet. Yet there are an estimated 800 million undernourished and 1.9 billion overweight adults. Equal distribution of calories is our current problem, shortage of calories might be our future problem.

As warming continues, agricultural productivity declines and the population grows bigger, so feeding the world becomes ever more difficult. By the middle of this century the world will start to become short of food if no agricultural changes are made. Climate induced hunger (people forced to halve their calorie intake) may grow to 2.5 billion people by 2100 adding nearly 2 million premature deaths per year.

Our estimates are close to those of The World Bank where work by Tubiello et al. estimated 1.3 billion more people at risk of undernourishment by 2080 if there is no further gain from CO_2 fertilisation.[51]

---

ix Calories or kcal are a measure of energy available in food. One small calorie has enough energy to increase one gram of water by 1 ºC. One kcal is 1000 calories or 0.0012 kWh.

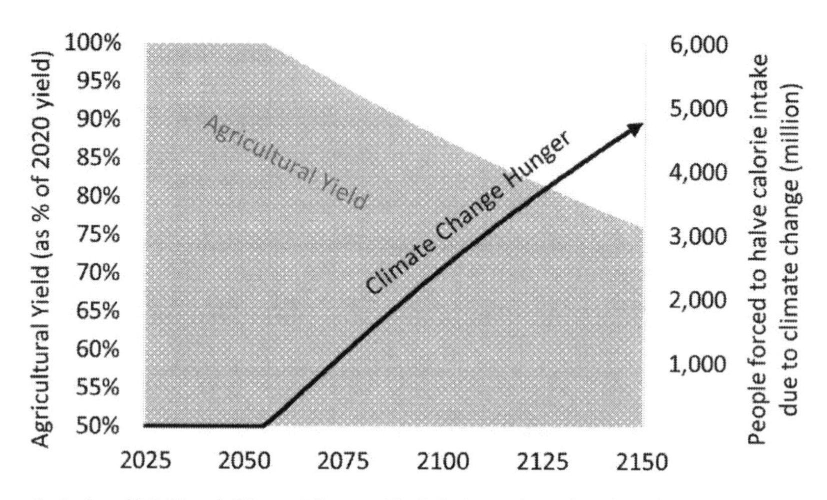

Agricultural Yield and Climate Hunger. Shaded Area: Agricultural productivity relative to today (every 1°C temperature rise above 2°C reduces yield by 10%). Solid Line: Number of people required to consume half the number of calories compared to today due to climate change led reduction in agricultural output. Based on our damage modelling and sensitivities references in preceding text.

Fresh Water

Higher temperatures and changes in weather patterns are predicted to affect the availability and distribution of rainfall, snowmelt, river flows, and groundwater. The world will experience more rain, but it will fall in less regular, intense bursts creating more run off over dry, hardened soils. Rising sea levels and extreme flooding further threaten the destruction of water points, sanitation facilities and the contamination of water sources.[52] According to the United Nations, water availability is already becoming less predictable.[53]

There are 1.4 billion km^3 of water on planet Earth which is 350,000 times more than we use each year. Unfortunately, 97% is salt water and 2% is frozen at the poles. That leaves just 10 million km^3 of potential useable water in the ground and 100,000 km^3 in lakes and rivers.[54]

Accessible water stores are replenished at a rate of just 15,000 km^3 per year. Today the world uses around 4,000 km^3 of fresh water each year or about one third of the sustainable supply for agriculture[x] (70%), industry (18%) and households (12%).[55] Another 700 km^3 is used to cool fossil

x On average each tonne of global crop product uses 300 tonnes of freshwater irrigation. Every tonne of meat requires 8 tonnes of crop feed and therefore 2,500 tonnes of water.

fuel and nuclear power plants but reintroduced into the waterways.[56]

Yet because of the unequal distribution of water on the planet, nearly 800 million people still lack access to clean drinking water, half the world has no basic sanitation, and one third of countries have medium to high levels of water stress.[57] Climate change will place further burden on these statistics. Globally, every 1°C warming could severely reduce water availability by 5-10%.[58] *Hafsa Munia from the Water and Development Research Group estimates that under a high emissions scenario 380 million more people will suffer from water stress by 2050.*[59]

We expect the global population to expand by around 30% so if no changes are made to diets or our agricultural system, we will need to increase crop production from 9 billion tonnes to 12 billion tonnes per year. The need for higher yields or more agricultural land coupled with an increasing frequency of droughts will require the amount of irrigated cropland to triple from 20% to 60% of all acres (irrigated land averages double the yield of non-irrigated land[60]). Combine this with a greater demand for water in households, industry, and power generation and by 2100 we may well breach the sustainable supply of fresh water globally. This would place most areas in high water stress unless greater storage, water management, or desalination can be effectively deployed.

The impacts will be felt earliest and hardest in the Mediterranean, Africa, and in countries such as Egypt, Iran, Mexico, and Turkey. Lack of water will cause disruption to drinking water supplies, thermal power plant management, services, and agriculture.[61,62] Competition over water resources is also tied to conflict and areas such as the Sahel in Northern Africa have ten countries sharing the same water course. Ethiopia, Egypt, and Sudan are already locking horns over potential water shortages following the ongoing construction of the Grand Ethiopian Renaissance Dam.

Fresh Water per Year (km³ / billion tonnes)	Today	2100 (Inaction)
Domestic use	475	608
Industry use	721	2,532
Agriculture use	2,767	8,997
Total Water Withdrawals	**3,963**	**12,138**
Thermal pollution from power	719	1,413
Total Water Used	**4,682**	**13,552**
Warming	1	3
Renewable fresh water supply	53,688	53,688
Additional capture from dams	3,500	3,500
Flood run-off	-40,000	-40,000
Climate change decline	-417	-2,396
Inaccessible water	-2,000	-2,000
Net Water Available	**14,771**	**12,792**
Water Surplus (Shortage)	**10,090**	**-760**

Table showing fresh water use and renewable supply based on data from The World's Water Vol 8 (2014), Limits to Growth 30-year Update, and USDA and FAO Water and Food Security. 2100 projections based on our baseline of inaction (3.3C warming by 2100, 7% decline in water availability per 1C, peak population, peak consumption by 2060, agricultural irrigation triples to increase yield/ available cropland based on no change to diets).

Ocean Acidification

The ocean has absorbed more than one quarter of all the CO_2 emitted by humans over the last 200 years. This has helped dampen potential warming, but oceans are already nearly 30% more acidic. Surface waters in polar seas and up-swelling regions risk becoming under-saturated with calcium carbonate. If atmospheric concentrations of CO_2 breach 560ppm waters become so acidic they will dissolve the shells and skeletons of sea life not protected by an organic layer.[63] Sea molluscs and the start of the aquatic food chain face significant impact. More acidic oceans may also reduce fertilisation rates, hinder larval growth, and slow protein formation in larger sea creatures.[64] Some plankton might benefit from higher CO_2 availability, but the lack of other essential nutrients as deep ocean mixing slows will offset the positives. These changes may substantially reduce fishery catch potential, affecting the quantity, quality, and predictability of future harvests. McKinsey highlight that with seas already 0.7°C warmer and more acidic there has already been a 35% reduction in fish yield across the North Atlantic.[65] Cheung et al estimate that every 1°C of warming will reduce the global catch of fisheries by 3 million tonnes from a total of 90 million tonnes per year at present.

The well documented decline in coral reefs bears testament to the effects of warmer and more acidic oceans. Corals consist of hundreds of thousands of tiny soft bodied creatures called polyps which secrete a calcium carbonate outer skeleton to attach themselves to rocks. These tiny animals exist in symbiosis with microalgae living in their tissues. The microalgae feed off the corals' waste and the coral use the oxygen and energy products from microalgae photosynthesis. Living on shallow ocean shelves, coral reefs are provided with plentiful nutrients so when waters warm the microalgae grow faster, disrupting the symbiotic relationship and forcing the coral to expel their colourful partners. The coral turns from vibrant colour to ghostly white – a so-called bleaching event. This first starves the coral of its energy source and then more acidic waters start to dissolve the calcium carbonate exoskeleton leaving the coral even more vulnerable. At least 20% of coral reefs are already gone and the IPCC estimates that 70-90% of coral will be gone at 1.5°C and nearly 100% by 2°C warming.[66,67]

Productivity

It has been the improvement in productivity that has driven the increasing wealth and quality of life over the last 200 years. But climate change is now threatening to slow or even reverse these gains. Extreme heat, hunger, water shortages, and poor health will impair our ability to work.

If we estimate the productivity losses from our baseline of inaction, annual economic output is reduced by $2 trillion dollars in 2050, $30 trillion in 2100 and $150 trillion by 2150 – this equates to around 1%, 4% and 7% of GDP. Lower labour productivity is likely the largest market-economic burden of climate change.

Hunger: Nearly 2.5 billion people will be forced to consume less than half the current calorie intake by 2100 unless our current agricultural system is changed. Malnutrition impacts productivity by 10-20%[68] meaning at least $18 trillion of lost income in 2100 (2.4% of GDP).

Water Availability: With at least 5% of the population impacted by water availability for every 1ºC of warming, a minimum of 1.5 billion people will struggle with water supply disruption by 2100. Assuming a 2% decline in productivity for each person struggling with water shortages and the economic impact adds up to at least $3 trillion per year by the end of the century (0.4% GDP) and more if we are forced to increase irrigation of crops.

Heat: Nearly one third of the working population work outdoors and research shows that abnormally high temperatures lower productivity by 2%[xi] for every 1ºC.[69,70] By 2100 nearly 4% of all working hours will be impacted by extreme heat, reducing income by over $3 trillion by 2100 (0.4% of GDP). Areas such as India will be hit hardest where outdoor labour accounts for half of GDP and three quarters of employment.[71] *(Consultancy group McKinsey forecast up to a 1.5% of GDP lost to heat related productivity declines by 2050.*[72]

Respiratory Disease: By the year 2100 there will be nearly 30 million new cases of respiratory illness per year which typically reduce time at work and productivity by 30%[73] – a further $3 trillion hit to global income (0.4% of GDP). Areas such as China, India and Eastern Europe will be worst affected.

Migration: By 2100 nearly 500 million people may have been displaced by rising seas and intolerable heat and will be migrating at a rate of 13 million people per year. Assuming this means no job for at least one year and it adds another $1 trillion of lost income (0.1% GDP).

Forced Industrial Downtime: Fossil fuel and nuclear power plants lose 1-2% efficiency per degree increase in temperature and extreme heat and water shortages create added downtime. Industry will become increasingly impacted by weather disruption, infrastructure breakdown and water shortages.

The list of possible (or probable) impacts could go on to include the spread of infectious disease such as malaria, dengue fever, Zika or Lyme disease, plagues of pests, and destruction of much of the world's cultural heritage.

---

xi Based on current trends and projected future warming, heatwaves will increase from 40 to over 100 events per year by 2100. Each heat wave impacts ~3% of the global population for about 1 working week, with temperatures over 5ºC warmer than usual.

Global warming will take its toll on natural eco-systems faster than human systems because they simply don't know what's coming and cannot adapt or evolve fast enough. The WWF[xii] point out that 85% of wetlands have already been lost, 32% of the world's forest has been cut down, 33% of fish stocks are overfished, and over the last 50 years we have recorded 83% declines to freshwater species, 60% declines to vertebrate species, and 41% declines to known insect species.[74]

Although a few degrees warming may not sound like much, we have run through the profound physical changes that this increase could inflict upon society, the economy, and the environment. As we pump more greenhouse gases into the atmosphere, we lock-in ever greater physical change and we will move further away from the stable climate that human civilisation has enjoyed for the last 12,000 years. The net impact of this change is certainly negative with increased death, suffering, and economic toll, but how big a problem is climate change? How do we compare climate damages in the future with other problems today? And what proportion of the world's limited resources should we use to combat climate change?

The next chapter will sum up the damages and look at how we compare climate change with other pressing global problems.

---

xii World Wide Fund for nature is an international non-governmental organization that works in the field of wilderness preservation and the reduction of human impact on the environment.

Chapter 5: Damage Assessment

Quantifying the Social and Economic Costs of Climate Change

In this chapter we will quantify the total death, suffering, and economic toll from climate change using our baseline of inaction and the physical impacts of warming. We will explore how we can compare these future climate damages with other global problems today. We will investigate whether we could soften the blow using adaptation, examine the risk of dangerous tipping points which could accelerate change, and define the limitations of damage forecasting. We will also meet our second dial of doom and boom: the discount rate, which is the economic sensitivity that defines how much value we place on future generations and the magnitude of future climate damages in comparison with other problems today.

Adding up the Damage to Society

Our basic projections of physical, social, and economic damages from the last chapter give us an idea of how climate change will begin to disrupt our everyday lives. Whilst it's typically the headline grabbing catastrophes that we associate with the worst impacts of a warming planet; the biggest damages are actually rather more subtle. For every migrant forced to abandon their home due to rising seas or extreme heat, there will be ten people forced into malnutrition, lacking fresh water, or struggling with respiratory disease. For every recorded death from a hurricane, flood, or wildfire there will be nearly 200 people who die a slow death from hunger or air pollution.

By 2050, as today's children are thinking about bringing their own children into the world, there may be over one billion more people on this planet who lack proper access to clean water and 300 million more people may have died from diseases linked to air pollution. Twenty million people could have been displaced due to sea level rise and one million more may have died due to the rising intensity of not-so-natural catastrophes.

By 2100, as our children should be looking to retire, more than two billion or 20% of the population may lack access to sufficient clean water and nutrition. One billion more people may have had their lives cut short by respiratory disease and heart failure. Over 500 million people may have been forced to migrate and at least one third of the world will have been severely impacted by climate change.

If the economy continues to be powered by fossil fuels into the next century, hunger, water shortages, migration, disease, and disaster may impact everyone on the planet in some form. Once the numbers get as big as this, predictive models break down, chaotic effects take over, and the risks of socio-economic breakdown, pandemic disease, and war become significantly elevated.

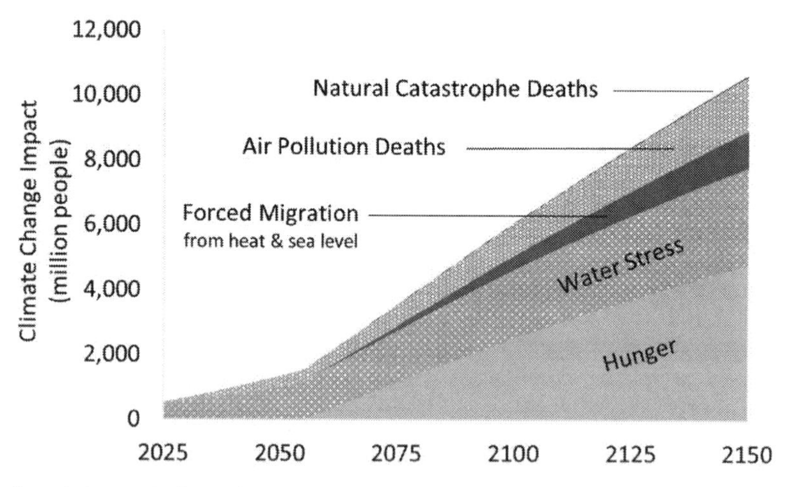

Cumulative total of people impacted by climate change in millions based on our peak consumption, inaction scenario (3.3°C warming by 2100, 4.4°C by 2150). Basic sensitivities to temperature, CO_2, and sea level change referenced in preceding text. Same people may experience multiple impacts.

Adding up the Damage to the Economy

Just as the hidden factors do the most social damage, so the subtle changes most negatively impact the economy. Mention the cost of climate change and it's the damage to homes, industry, and real estate from heat, fire, floods, and storms that spring to mind. But for every dollar lost to disaster, five dollars will be lost to productivity. Heat exhaustion, malnutrition, and lack of water reduce economic output and will create a serious dent in future income and wealth. To better understand the damage that climate change may inflict upon the economy let's sit back for a moment and reflect on what drives income and wealth.

Income is generated in two basic ways: from our labour and from our accumulated capital. Think about when you first leave school and start a job and, unless you are lucky enough to have inherited a trust fund, all you have is your ability to work and bring in money through your labour. But over time you become more skilled, you start to earn a higher wage, your income grows faster than your spending and you can save or invest money into financial assets like shares or bonds in your pension or into non-financial assets like a house or machinery to start a business. These investment assets are called capital and they provide an income stream for you in the future.

Global GDP is also measure of income: it is the monetary value of all the goods and services in the economy and when divided by the global population gives (near enough) the average income per person. Global GDP is around $90 trillion dollars per year today, which is about $12,000 per person or about $18,000 if measured using a comparable price[i] of goods.[1] Just as you generate personal income from a mix of labour and capital, so global GDP is the combined output of labour and capital of each person on the planet. The value of global capital adds up to around $400 trillion today and generates about one quarter of GDP whilst the labour of the 8 billion people on Earth accounts for the other three quarters.[2,3,4]

Classical economics use this basic relationship to forecast the potential for long run GDP growth as a function of labour and accumulated capital multiplied by a productivity factor which represents the level of technological progress, education, and productivity. These functions are usually a variant of the Cobbs-Douglas model.

Total Output = Factor Productivity x Labour$^{(\beta)}$ x Capital$^{(\alpha)}$

Note: Labour and Capital are raised to the power of β and α. These represent output elasticities, if α were 0.25 then every 1% increase in capital will increase output by 0.25%. If β and α add up to 1 this represents a constant return to scale where doubling labour and capital will double output if factor productivity remains unchanged.

So, GDP can grow through either growing the population, growing capital, growing the productivity factor, or a combination of all three. In our baseline scenario we assume factor productivity increases at 1.75% per year, the population peaks at 10 billion people by 2060 and 25% of

i Purchase Price Parity uses the comparable price of goods in different countries to quantify money in terms of its ability to purchase goods and services, rather than the absolute dollar value: USD$1 can buy a lot more in Africa than it can in the US. So global GDP is $90 trillion in market exchange rate US dollars, but more like $140 trillion in PPP.

GDP is reinvested into capital to offset wear and tear and support future growth. Our forecast GDP per person grows at 2.4% per year for the rest of this century if unhindered by climate change.[ii] So how will climate change and air pollution impact future GDP?

It will diminish our factors of global output. The productivity of human labour is hit by rising hunger and heat. The productivity of capital is reduced with heat and water availability disrupting industry. Labour hours are diminished as heatwaves limit outdoor work, ongoing respiratory illnesses increase sick days, and forced migration disrupts the labour force. Physical capital such as buildings, roads, and industrial equipment will be destroyed by natural catastrophes and rising seas. Natural capital such as the air, water, and soil will be desecrated by drought, flooding, and acidification. The ability to generate income declines, labour hours and capital is lost, and the opportunity to reinvest for future growth is diminished. GDP output is lowered, and we are forced to reduce either consumption spending, investment, or a bit of both.[iii]

By 2050, lost income will total nearly $3 trillion or 1% of GDP, driven two thirds by productivity declines and one third by lost investment. That's $300 lost annual income from an average wage of $23,000 per person.

By 2100, lost income will have grown faster than the economy to reach $40 trillion per year or 5% of GDP. The biggest factors are productivity losses from hunger and lost investment due to slower growth and climate induced damages to capital. That's $4,000 lost annual income from an average wage of $75,000 per person.

---

ii Our 2.4% GDP growth per capita is given in real terms so it excludes the impact of inflation (changes in the price of goods). For a large part of the twentieth century inflation added more than 5% per year on top of real growth but has settled at 1-2% per year in the twenty-first century. Real GDP represents how purchasing power changes through time. Economists may also adjust real GDP to reflect purchasing power across regions. This would take our 2.4% growth and slow it to about 1.5% because prices in different regions should converge as incomes converge over this century. Economists consider about 1.5% real GDP PPP as the fastest any country at the technological frontier has sustainably grown in the past.

iii The economy is made up of billions of two-way trades where one person's spending is another person's income. So, GDP is simultaneously a measure of income and a measure of spending. Economists break spending down into two components called consumption and investment. Consumption is money spent on goods and services which are used up in the short term, things like food, entertainment, or healthcare. Whereas investment is money spent on long lived assets like factory equipment, buildings, or education which adds to capital and allows for income in the future. Consumption provides satisfaction today and investment provides consumption and satisfaction in the future.

Beyond 2100 our lost income due to climate change scales towards 10-15% of GDP by the end of the next century. The risk of economic collapse or global conflict becomes significantly elevated as significant proportions of the global population are severely affected and global tensions rise. As environmental activist Mark Lynas puts it, "As social collapse accelerates, new political philosophies may emerge, philosophies which seek to lay blame where it truly belongs – on the rich countries which lit the fire that has now begun to consume the world."[5]

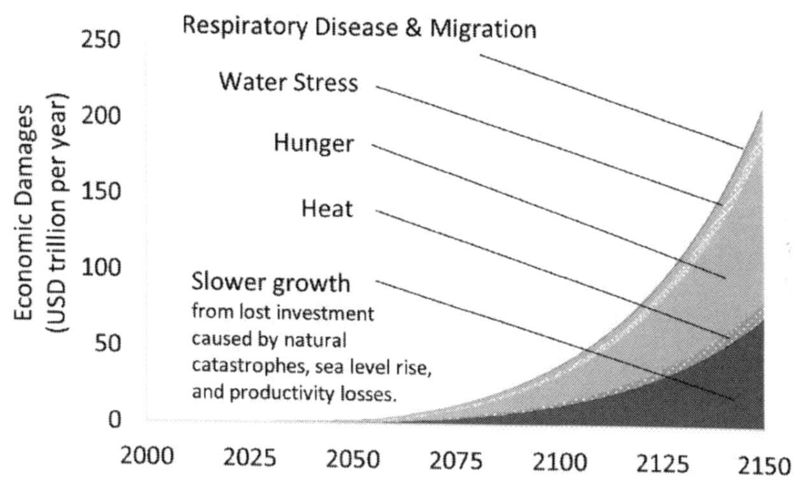

Annual lost income (excluding inflation) due to climate change related factors in our baseline inaction scenario of peak population and peak consumption by 2060 (3.3°C by 2100 and 4.4°C by 2150). Based on our damage modelling and sensitivities references in preceding text.

For comparison, the OECD baseline damages scenario, which has a similar emissions profile to our inaction scenario, estimates 1.5% lost GDP by 2050 and 6% lost by 2100 under their central projections.[6]

Utility and Discounting: Comparing Problems Tomorrow with Problems Today

In our baseline inaction scenario, and by the year 2100, we have estimated climate change and air pollution could kill more than one billion people, throw over two billion people into malnutrition, force hundreds of millions to abandon their homes and wipe out $40 trillion dollars or 5% of annual GDP. Does this mean the neo-Malthusians are correct? Must we limit population and reduce consumption and by how much and how fast? One of the difficulties of conceptualising the impacts of climate change is the timescales involved. Humans are used to making decisions seconds, days, or years into the future. So how do we compare the magnitude of damages decades or centuries ahead with problems today? How urgently do we address climate change?

Something called **utility** attempts to answer these questions. Jeremy Bentham, an eighteenth century English academic whose philosophy revolved around the greatest happiness for the greatest number of people, introduced the idea of utility to economics. Bentham's ideas were very progressive for the time, advocating individual and economic freedom, separation of church and state, equal rights for women, decriminalisation of homosexuality and the abolition of slavery.[7] And it's not just his legacy which was preserved; Bentham left specific instructions that he should be medically dissected before being embalmed and displayed "in the attitude in which I am sitting when engaged in thought".[8] His body remains on show in the foyer of University College London to this day and has even been wheeled out to attend special university council meetings (however, his mummified head is now kept under lock and key after students at rival university Kings College London kidnapped it and demanded a charitable ransom of £100 for its safe return).

Bentham's idea of utility is a measure of happiness, satisfaction, or usefulness of our actions, and modern welfare economists study how the allocation of resources can impact social utility today and in the future. Economists use consumption spending as a proxy for utility based on a logarithmic relationship; they assume happiness rises with increasing consumption, but the added happiness gets smaller and smaller, just like temperature and increasing CO_2.[iv] The decision becomes how far we should reduce consumption and utility today to slow climate change to increase utility for our future selves and future generations. Estimating the magnitude of this trade-off requires that we take our future dollar damages and convert them into the dollar equivalent utility today.

Economists use something called discounting. I have mentioned "today's money" several times already through the opening chapters, and it's time we looked at the concept of discounting and time value of money in more detail. Specifically, the idea that money today is worth more than the same amount of money in the future. That statement sounds wrong but imagine if I asked you whether you wanted £100 now or £100 in one year's time. If you took the money now you could deposit it in the bank, earn 3% interest, and in one year you would have £100 plus another £3. If you took the money in one year your missed opportunity

---

iv The saying goes that money can't buy happiness, so consumable income may seem like a crude way to gauge utility. But many studies have shown that happiness is indeed proportional to income up to a certain level. Beyond this point happiness tends to be more closely correlated to the relative magnitude of your income compared to the society in which you live. This is called the Easterlin Paradox.

cost was the accumulated interest.

That 3% difference may seem small in the grand scheme of things but imagine how 3% per year compounds on an 80-year time horizon. We ran through the tremendous power of exponential growth in chapter two. In fact, a quick sum tells us £100 invested for 80 years at 3% annual interest grows by more than 10x to £1,064. Put it another way, £1,064 of damages in 80 years' time are worth £100 today. The discount rate turns future climate change damages into the equivalent loss today and has a significant bearing on the required urgency of dealing with climate change.

Discounting Climate Change Damages: The Value We Place on Future Generations

So, what discount rate should we use? There are two schools of thought on this. The first approach is to use a purely market-based rate or opportunity cost. This theory points to the fact that spending money on climate change solutions is just like any other investment. You spend money today and you receive a benefit from lower damages in the future. Therefore, climate change spending must compete with other investment opportunities across the economy such as investing into machinery, infrastructure, buildings, companies, human capital (e.g. education or health), or real estate. A leading proponent of this descriptive approach is William Nordhaus, a Nobel winning climate change economist, who typically applies 4-5% discounting in his calculations.[9,10]

However, the problem with using a descriptive approach is that there are very few assets traded in the market which provide a return for longer than 30 years and so it doesn't fully reflect the time scale of the problem. There are frictions, uncertainties, and distortions in the market which means the rate is never true. We are also making investment decisions impacting future generations who do not trade on today's markets and have no influence on the market rate.

The second school of thought is to calculate the discount rate based on the value we place on future generational welfare and the rate at which wealth will increase. The leading proponent of this prescriptive approach is Nicholas Stern, a British economist who has written extensively on the financial impacts of climate change. The most common determinant of this social discount rate is the Ramsey equation.

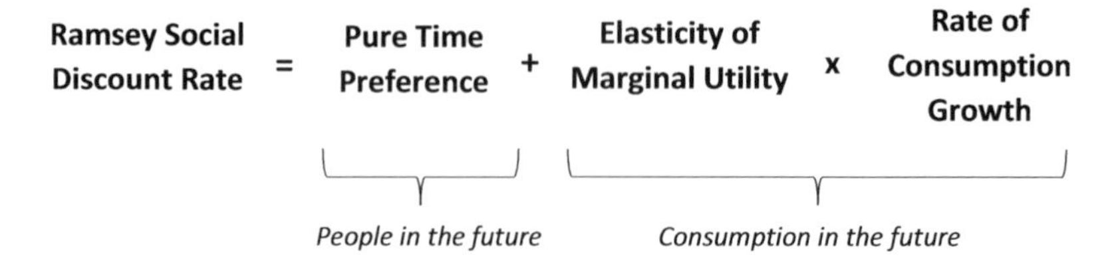

Let's start by first explaining the second part of the Ramsey equation which is the least controversial. This sets the discount rate equal to the growth rate, times something called the elasticity of marginal utility.

Setting the discount rate to the same level as GDP growth means that damages remain proportional to incomes. So, if future generations become wealthier and better able to pay for higher damages, we are still comparing like for like. In other words, damages at 5% of income in 2100 correspond to 5% of current average incomes. We can then compare the magnitude of the damages to other problems today.

The elasticity of marginal utility takes the concept one step further. Marginal utility states that as you have more of something, its value declines. Say you got a pay rise at work and could eat out twice a week instead of just once, that second night out will probably bring you nearly as much pleasure or utility as the first, but next imagine you went from eating out six days to seven days a week. You might still enjoy that seventh meal, but it would certainly not bring as much pleasure as adding the second. This is the logarithmic relationship between utility and consumption we mentioned earlier. The same principles apply to climate change damages. Losing 5% of your $75,000 annual income in 2100 will hurt a lot less than losing 5% of your average $12,000 income today because each percentage point of income brings less utility for your wealthier future self. The principle that providing for basic needs of food, shelter, and health creates far more utility than upgrading to a Lamborghini.

Adding the elasticity of marginal utility multiplier (typical value of 1 to 2)[11] acts to discount damages faster than consumption growth to account for the fact that we care less about losing money when we have more of it in the future.

So far so good, but now for the first part of the equation which is more controversial and more mind boggling. The pure time preference is an added element of the Ramsey discount rate to represent the fact that we value having something sooner rather than having it later, even if we make no financial gain in the process. This represents the impatience of humans. Essentially, those who argue for a higher pure time preference in relation to climate change, also argue that we should value the welfare of current generations higher than future generations.

Nordhaus makes the argument that we have less anxiety for our great grandchildren than our children because they are more remote and have our children and children's children to look after them. We can't worry about everyone into an infinite future so the pure time preference rate must be positive, and he uses a value equivalent to 1-2%. On the other hand, Lord Stern argues the only reason we should value future generations any less than current generations is the remote possibility the human race doesn't exist in the future. He uses a 0.1% pure time preference based on a worst case 10% chance humans don't last another 100 years. Many famous economists argue for and against the pure time preference rate.

Throughout this book, I will base all calculations on a total discount rate of 3% real (excluding inflation). Generally social discount rates range between 1-3% and market discount rates range between 3-6%, so our estimate is a compromise between the descriptive and prescriptive methods and a compromise between Stern (1.4%) and Nordhaus (4.5%).[12,13]

A 3% real rate of return also represents the weighted average return or interest paid by global bills, bonds, and shares over the last 100 years excluding price inflation.[14] This is the real return of a market investment, on a very long-term view. This is the long run opportunity cost or discount rate using a market based descriptive approach. It is also the current market rate of return for low-risk large-scale investment projects such as wind and solar farms in developed countries.

A 3% real rate of return also tallies with the prescriptive approach. If we assume 2.4% growth with 1.2 times elasticity of marginal utility and 0.1% pure time preference our discount rate comes to 3%. But why side with Stern on the pure time preference? I would argue if we are trying to find the best course for humanity then having something sooner rather than later becomes rather meaningless, because humanity is effectively infinite on our time horizons. Therefore, barring an extinction event, the time preference should be zero.[v]

---

v Compare this to an individual who knows life is finite and they are unlikely to live past another 100 years and of course you would prefer the benefits sooner. So, for a specific individual's project a positive pure time preference makes sense.

	NET-ZERO	Stern	Nordhaus
Pure Time Preference	0.1%	0.1%	1.5%
Elasticity of Marginal Utility	1.2	1.0	1.5
Consumption Growth	2.4%	1.3%	2.0%
Ramsey Social Discount Rate	**3%**	**1.4%**	**4.5%**

Ramsey Social Discount Rate for This Book, Stern, and Nordhaus. Stern given in Stern review. Nordhaus doesn't explicitly use the Ramsey method, but he approximated the elements in his review of the Stern report.[15]

So, what does this mean for our climate change damage estimates? Well, if we estimated damages at 5% of GDP by 2100 and 9% of GDP by the year 2150, this is the equivalent of around 4-5% of GDP in today's money after discounting at our 3% rate. In other words, losing 5% of GDP in 2100 or 9% of GDP in 2150 will feel like losing 4.5% of GDP today.

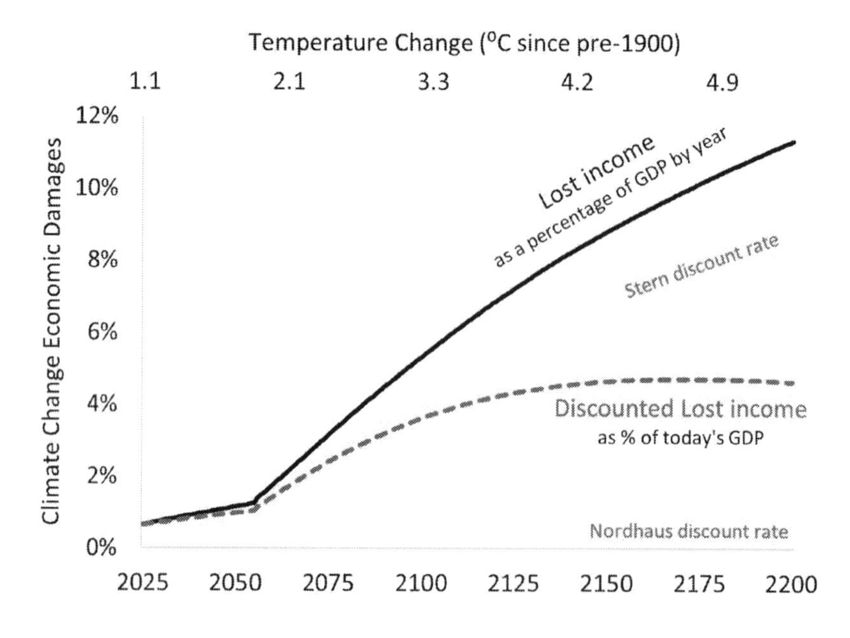

Chart showing how our projected climate change damages increase with rising temperature over time. Solid line shows forecast damages per capita as a percent of forecast GDP per capita. Dashed line shows future USD damages discounted into today's money as a percentage of today's GDP per capita – the equivalent utility or welfare. We use a GDP per capita growth rate of 2.4% and a discount rate of 3%. The chart also shows our discounted damages using the Stern discount rate (1.4%) and Nordhaus discount rate (4.5%).

Using our same damage projections, Lord Stern's discount rate would argue the losses are closer to 6.5% or more in today's equivalent and Nordhaus would argue they are equivalent to less than 0.5% of today's income. This gives you an idea of why climate change is such an intractable problem. The long time periods create such uncertainty that the smallest change in assumptions can completely alter the outcome. Are we warranted spending 6.5% of GDP or more to solve climate change and air pollution or should it be less than 0.5%?

For those of you whose eyes have yet to glaze over or brains freeze, the subject of discounting runs even deeper into the rabbit hole. There are many more arguments and counterarguments between economists for increasing or decreasing this all-important value.

The Precautionary Principle, which was first proposed in the 1970s for dealing with environmental risks, is a key pillar of the IPCC declaration, stating, "lack of full scientific certainty shall not be used as a reason for postponing cost-effective measures to prevent environmental degradation".[12] In other words, not having all the facts is not a reason for inaction on climate change. This idea influences the discount rate. We may not be able to forecast the exact future damages or the exact cost to prevent those damages, but we know enough to assert that damages run the risk of being essentially unlimited and effectively irreversible, whereas the cost of solutions are capped. The Precautionary Principle suggests that given we are getting wealthier and every dollar brings us less incremental happiness, we should be more willing to pay to avoid the risk of unlimited, irreparable damage.[16] A case for lowering the discount rate and justifying greater action to avoid climate change.[vi]

Equity and Utility Across the World: Comparing Damages Here with Damages There

Not only must we compare damages through time (intergenerational equity), but also across the world (regional equity). The rising heat,

vi The idea of uncertainty can be further extended to justify a decreasing discount rate through time. Remember £1,064 of damage in 80 years' time is the equivalent of £100 today if discounted back at 3% per year. But maybe we are unsure on the average growth rate and want to include the potential for acceleration of technological progress or economic shocks. We could justify a discount rate 2% higher or lower than our base case. Repeat the calculation for 1% and 5% discount rates and the damages become £480 or £21 in today's utility. The average of these two extreme outcomes is £225, and to arrive at this value today from £1,064 in 80 years in the future requires an annual discount rate of 1.8% (not the 3% average of 1% and 5% you would intuitively think). The average of uncertain outcomes will tend toward a lower discount rate with time, again justifying greater more immediate action.

water, and extreme weather will not be equally distributed across the planet. Whilst 3°C average warming may erode just a few percentage points of GDP from more northerly countries such as Canada or Russia, many of the already hot regions closer to the equator such as India, Bangladesh, Southern Europe, or Africa could suffer losses of 10% or more.[17] Many of the areas worst affected also have some of the lowest income economies on the planet. A triple whammy of more damage, less ability to cope, and greater loss of utility.

The historical correlation of warmer climate and weaker economies may be more than just coincidence. For much of human history, the warmer regions probably had some small economic advantage over the cold northern areas of the world. However, following the scientific revolution, the birth of global trade, and later the industrial revolution in the UK and Europe, fortunes had changed course. Progress unlocked great European wealth but with it came ambitions of global conquest.

Europeans first began to colonise the Americas through the early sixteenth century, arriving in a land 11 times larger than Europe with less than one third of the population. The indigenous peoples of the Americas had no vehicles, ships, or weapons and mustered little opposition to the more technologically advanced Europeans. However, it was disease that the invaders accidentally transmitted which tore a path of destruction through the continent. According to economic historian Angus Maddison, newly introduced diseases had killed two thirds of the indigenous population by the middle of the sixteenth century. Europeans settled and colonised the temperate northern lands, establishing governance, legal systems, new crops, technology, and infrastructure and they created new trading routes with Asia. America was set on course to become a new world power.

Around the same time as Europeans were colonising the Americas, they had also begun to establish trading posts in Africa, having failed in most attempts to develop colonies. Europeans had developed immunity to the animal borne diseases in Europe (smallpox, measles, flu and typhus) but settlers in Africa had no immunity to insect borne tropical diseases such as malaria which thrived in the warmer climate. One third of new European arrivals succumbed to disease and a lack of settlers meant no investment in education, technology, or infrastructure. Instead, the Netherlands, France, Portugal, and the UK began shipping slaves to the Americas to build the new world: put to work in the hot, malaria-ridden plantations and mines of the southern United States and South America. By the end of the nineteenth century more than 1.5 million Africans had died during the brutal journey across the Atlantic and the 11 million slaves that did make it were subject to hard labour, inhumane treatment, and premature death. Europeans didn't colonise Africa until the late nineteenth century when the discovery of quinine helped protect against malaria. The English, French, Spanish, Belgians, Germans, and Italians divided up the continent as they saw fit with no regard for ethnic or religious boundaries. The warmer climate had led to exploitation rather than knowledge transfer and investment.[18]

Fast forward to the middle of the twentieth century and, following two World Wars, new international organisations were established to promote peace, stability, and development such as the United Nations, The World Bank, and the International Monetary Fund. Expectations for a dramatic reduction in poverty and inequality by the end of the century were high. Financial support from developed countries was to target transformation through education, trade, scientific capabilities, stable governments, and finance. But insufficient aid, lack of technology transfer, Cold War spending, and corruption meant far less than expected was accomplished.

Globalisation since the 1980s has supported a transformation of China and India's economies, but much of Africa remains caught in a poverty trap. A historic lack of knowledge transfer, lack of investment, exploitation, and ruthless treatment has left Africa the poorest continent on the planet, with an average $4,000 of accumulated wealth (capital) per adult compared to the new world of North America with $400,000 per adult.[19]

Inequality remains rife across the planet with the 10% of the global population on the highest average incomes (over $16,000 per year) accounting for half of global GDP. The 10% of the global population with the lowest incomes (less than $1,000 per year) account for less than 1% of global GDP. This is where regional economic damage estimates can run into some serious problems. We know climate change disproportionately impacts already poor countries, so let's take this to the extreme and imagine for a moment that climate damages completely wiped out the livelihoods of the poorest 800 million people on Earth (10% of the global population). This impacts global GDP by less than 1%. Extreme inequality completely skews regional damage models.

Economists attempt to solve this problem by equity-weighting regional damages.[20,21] The economic losses from each region are adjusted to mirror utility and reflect the fact that a dollar is worth significantly more to the poor than to the rich. Once each region is reflecting like for like utility, the total damages can be summed up and converted back to a global monetary exchange rate. Equity-weighting regional damages can nearly double the global total.

Non-Market Value of Life and Nature: Comparing Damages Inside and Outside the Economic System

Once economists have settled on a suitable discount rate and equity weightings, the next problem is how to include non-market costs. These are the toll of human suffering and death plus the destruction of the natural world.[16] We have kept these damages as separate categories, but economists may attempt to monetise these damages into their forecasts by assigning a value to human life[vii] and nature.[22] The three basic concepts are:

- **Stated Preference:** Simply asking a person how much they would pay, for example, to add an extra year onto their lives or to save a coral reef.
- **Revealed Preference:** Trying to interpret what people have paid for non-market gains. For example, by analysing house prices next to a forest rather than a main road.
- **Willingness to Pay or Willingness to Accept Compensation:** Asking how much a person would contribute to growing a forest at the end of their road or conversely how much they would be willing to accept in compensation for the forest to be bulldozed. Compensation is generally higher than willingness to pay because we like to win but hate to lose.

None of these methods are anywhere near perfect and all suffer from huge inequality differences. For example, the average willingness of an American to preserve or prolong their life is 20 times greater than that of an Indian. Does this mean Americans are 20 times more valuable than Indians? Of course not. They just have more money and utility and therefore a greater capacity to pay. Including non-market damages must be done with caution.

How Does Climate Change Compare to Other Problems?

We have run through some very big sounding numbers, hundreds of millions of migrants, over one billion dead, two billion suffering from malnutrition, and more than 40 trillion dollars per year wiped off the global economy by 2100. But just how big are these numbers? We now

vii Value of Statistical Life is an estimate of the monetary value for each year of a human life. Usually based on surveys asking about willingness to pay to offset certain risks. The figure is highly dependent on the survey sample due to factors such as wealth, age, or cultural bias. Surveys have returned anything from $3 thousand to $20 million per life year.

understand how we can compare damages in the future with damages today and we can level up the regional damages depending on inequality across the world. So, let's take a step back and rationalise the impact of climate change in relation to other pressing global problems.

Death Toll

For comparison, the total number of deaths for the global population is nearly 60 million people per year. High blood pressure is the biggest risk factor leading to death but is closely followed by air pollution.

Indoor and outdoor air pollution from the combustion of fossil fuels, wood, and dung already accounts for nearly eight million deaths per year. This number will rise to over 14 million deaths if we let the current energy system grow with population and consumption. Add another two million deaths per year from malnutrition and hundreds of thousands more from natural disasters by the end of this century, and climate change and air pollution could become the number one risk factor leading to premature death.

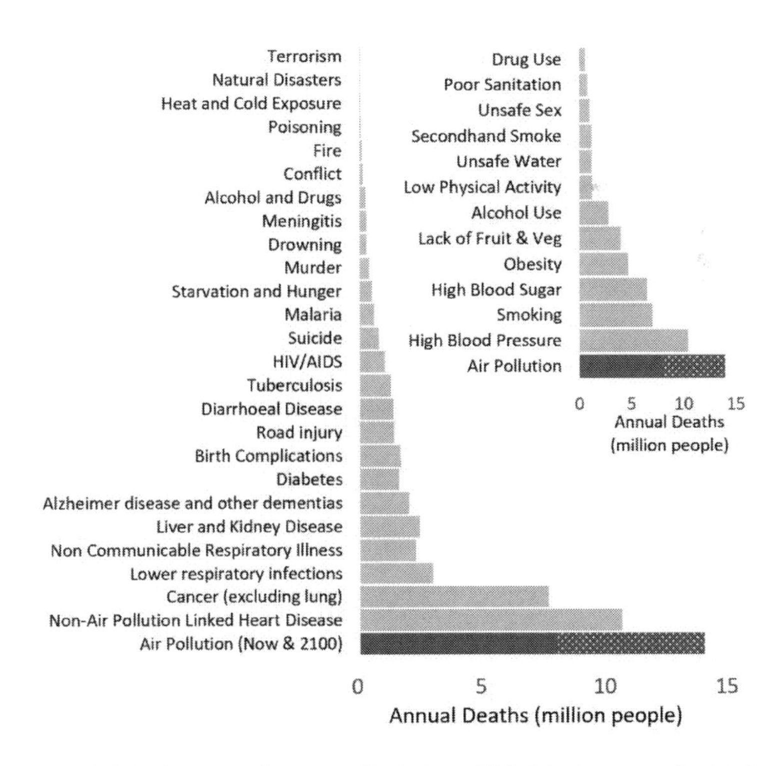

Left: Global deaths per year by category. Right inset: Global deaths per year by risk factor. Our estimate for air pollution, other data from sources [23,24,25,26,27]

Suffering Toll

There are already over 300 million people affected by climate change according to former UN secretary general Kofi Annan and his think tank, The Global Humanitarian Forum.[28] Based on our forecasts for hunger, water shortages and forced migration those suffering at the hands of climate change will rise to nearly three billion people by 2100 in our baseline inaction scenario.

Compare these numbers to people suffering from a lack of basic amenities, corruption, serious health problems, or extreme poverty and climate change already ranks as a top ten issue. By 2100, it will likely be the biggest cause of suffering globally.[viii]

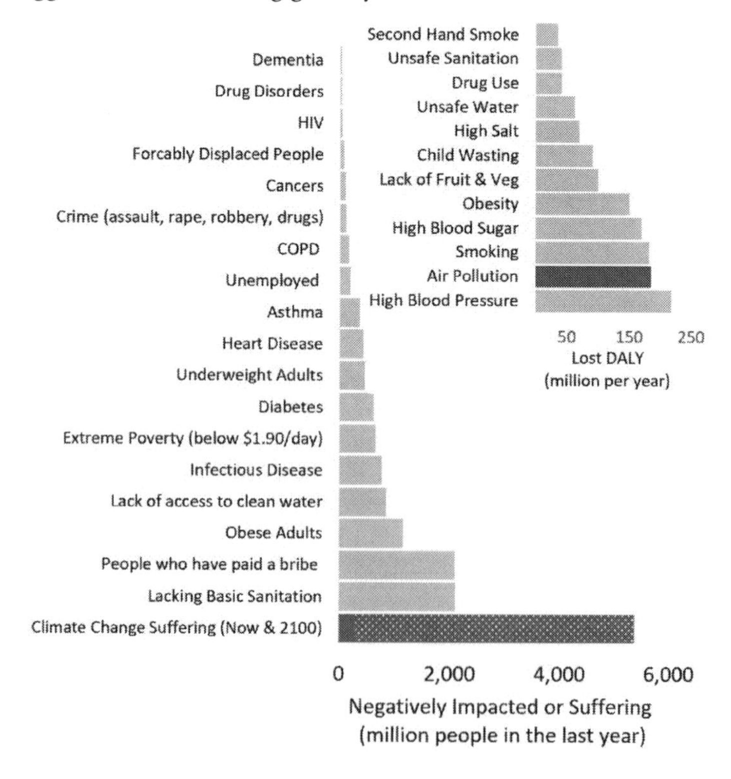

Left: Number of people suffering from lack of basic amenities, poverty, corruption, health and climate change impacts over last 12 months. Right: Disability Adjusted Life Years Lost every year due to risk factors. Our estimates for climate change, other data from sources.[29,30,31,32,33,34,35,36,37,38,39,40]

viii Another way to look at suffering is to estimate years of quality life lost to premature death and disability (DALYs). Air pollution again already ranks as the second biggest health risk factor for suffering.

Economic Toll

Annual economic losses from climate change due to lower productivity, healthcare costs, and destruction to property may rise to $40 trillion by 2100 under our inaction scenario. This is equivalent to average annual losses of just over $4 trillion in today's money (3% discount rate). Placing climate change losses firmly in the top three most costly global spending areas.

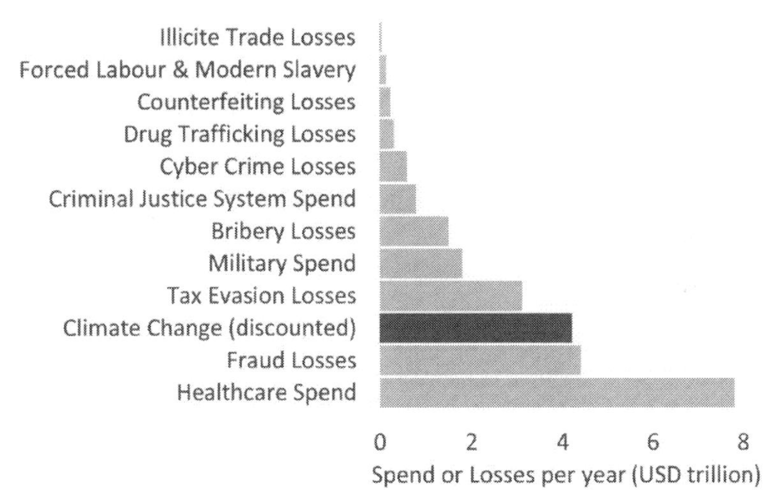

Annual spending or annual losses to Global GDP. Climate change impact is the annual lost GDP in 2100 discounted into today's money at 3%. Our estimate for climate change, other data from sources [41,42,43,44,45,46]

Manageable Change and Adaptation: Spending Our Way Out of the Problem

Climate change qualifies as a top three global problem and through the last two chapters we have estimated that damages may rise towards 5% of GDP by 2100 and 9% of GDP by 2150, but growing wealth will make the damage feel more like 4.5% of GDP. Climate change will slow growth by 0.1% per year, but the average annual income per person will likely continue to rise, from $12,000 to $70,000 by 2100. The average human will become six times richer. Climate change will slow prosperity growth, but not stop it in its tracks.

Our damage estimates so far have simply assumed the future is a bigger, warmer version of the past and humanity has just watched idly whilst rising temperatures and rising seas destroy food supplies, homes, work, and health. In reality, growing wealth and technology could be used to adapt to a warmer climate, employed to clean the air we breathe, improve respiratory healthcare, and build sea walls, cooling shelters, and alert systems. We could develop new technologies to grow food using less land and less water, and to relocate humans, animals, and plants out of harm's way. Adaptation will of course come with added cost, but spent well adaptation could lessen the death toll, suffering, and lost income, softening the blow, so that losses may be just a fraction of our estimates.

Natural catastrophes are a good example of successful adaptation over the last century. We have already seen how the frequency of significant climate related natural disasters has tripled due to climate change. Yet compare the number of deaths from natural disasters over the last decade to a century ago and significantly fewer people die. How do these trends tally? It's because people are over six times wealthier than they were at the start of the twentieth century. We have used our growing wealth and knowledge to develop early alert systems, evacuation procedures, event forecasting, resilient buildings, and emergency services. We have been able to adapt and to reduce the number of fatalities.

Next, let's take agricultural production as an example of future adaptation. With 10% yield declines for every 1ºC warming (over 2ºC), we calculated that by the year 2100 nearly 2.5 billion people would be underfed. It follows that if hungry people are 10% less productive, this leads to $18 trillion or 2.5% of GDP losses: one of the biggest social and economic impacts of climate change. But does it really make sense that we can continue to accumulate knowledge, improve technology, and boost productivity, thereby driving average incomes up another six times, and yet one quarter of the world is going hungry? Maybe. In a world of extreme inequality this could certainly happen. Remember, we produce more than enough food for everyone on the planet today, and yet 10% of the population are undernourished. We aren't so good at sharing.

But of course, we could use our increased income to do something about the declining agricultural production. We would need to boost yields by 30% and offset the damage done by extreme weather and droughts. We already calculated this would require three times more irrigation and probably a combination of resistant crop varieties, more fertilisation, and improved farm management. We would likely need to de-salinate (remove salt from) as much sea water as we use fresh water today (5,000 km^3 per year). De-salination treatment uses ten times the energy of freshwater treatment (4kWh vs 0.4kWh) and would add 25% to global energy use with processing costs of at least $3 trillion per year. A fair price to pay to feed an extra 2.5 billion people and save $18 trillion in productivity losses, but still a major economic burden.

If climate change follows our baseline of inaction, worst case we lose 10-15% of GDP by next century. Depending on the success of adaptation measures, however, that number could be anywhere between 0.5% and 15%. Either way, average incomes grow, and the life of an average human being is still improving with rising wealth, driving longer life expectancy, lower rates of child mortality, more education, greater protection of human rights, and a higher overall life satisfaction. Maybe the neo-Malthusians got it wrong. Maybe we need to listen to the cornucopians. Should we just let adaptation, free markets, and technology solve the problem?[47] It seems like average prosperity will continue to improve, so why worry?[48]

Enter unmanageable change and tipping points.

Unmanageable Physical Change and Tipping Points: The Problems Money Can't Fix

So far, we assumed change is linear and manageable. Every 1ºC temperature increase will also increase disasters, hunger, water shortages, and productivity by a set amount. But what if change accelerates as temperatures rise? Our version of the future can't possibly account for tail risk, tipping points, and non-linear change. The climate, the economy, and the structure of society are complex systems. With so many interconnections, dependencies, feedback loops, competing processes, relationships, and self-referencing dynamics, they are impossible to precisely model and predict. Despite the huge development in computer processing power, we simply cannot fully capture every interaction in these systems. Instead, we must simplify the system so we can make good short-term predictions and identify longer term trends. But at some point, our original system moves so far from its original state that it hits a tipping point and becomes unpredictable.

Thomas Shelling, a Harvard economist, was the pioneer in modelling tipping points. His 1971 work on racial segregation of neighbourhoods suggested that when neighbours of a different colour, settling in a predominantly white neighbourhood, reached a certain threshold, this created a "tipping" or acceleration towards total segregation as whites began rapidly to move out of the neighbourhood.[49] He described the tipping point as a small change in some variable which results in a large system change.

The idea of tipping points or accelerated change is common across many systems from banking runs to political endorsements, or rapid changes in ecosystems. But the more complex the system, the more difficult it is to identify at what point the system will tip. Just as we struggle to predict economic crashes and election outcomes, so we are likely to miss many of the possible outcomes of climate change and here are a few:

Underestimating Tail Risk: Gaussian distribution functions or bell curves are often used to describe the probabilities and risks of a natural system. A good example is the frequency of temperature anomalies in the Northern Hemisphere. Unusually hot or cold days are distributed around an average temperature in the form of a bell-shaped curve. Over the last 40 years the whole curve has moved higher by 1ºC, or the average temperature increase, but the curve has also widened: the chance of extreme heat events has risen exponentially from 0.2% to over 10%. Get the exact shape of the bell curve wrong and we underestimate the tail risk of physical events such as natural catastrophes. The equilibrium climate sensitivity is in itself a fat tail risk because we know the lower limit of warming for double CO_2 is 1.1ºC but there is still small probability that the upper limit could be higher than 4.5ºC.[50]

Sub-System Change: As temperatures rise, this impacts all aspects of life. Management consultancy Mckinsey define the five key sub-systems for sustaining life as: liveability and workability, food systems, physical assets, infrastructure services, and natural capital. The change in these sub-systems is unlikely to prove linear and equal.[51] As temperatures and physical stresses increase, regional sub-systems will not simply deteriorate at the same rate, but they will at some point completely collapse. Global average forecasts become meaningless if individual regions break down.

Systemic Breakdown: Rising physical change, exponential tail risk, and sub-system collapse will at some point push the entire economic and social system past a point at which we risk either accelerated physical change, economic crisis, pandemic, or global conflict. Even a combination of all four. This means our estimated death toll and economic burden skyrocket. It is the risk of this unprecedented, unpredictable change which skews our damage predictions into two binary outcomes – manageable change or unmanageable change. This could be the difference between human civilisation continuing to drive increasing prosperity or spiralling back into the Malthusian trap.

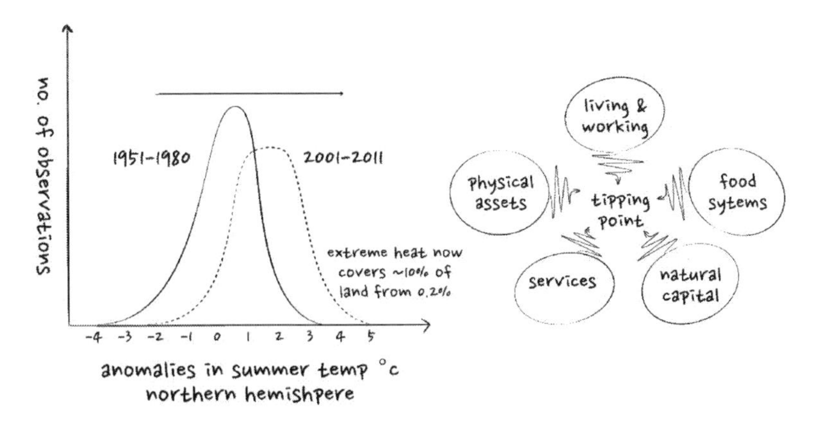

Left Chart: Depiction of temperature anomaly distribution data for 1951-1980 and for 2001-2011. Temperatures have increased by ~1°C but the curve is shorter and broader. Extreme heat events have increased exponentially.[52]

The consensus amongst scientists is that tipping points for many physical climate systems lie somewhere beyond 2-3°C warming.[53] This is where the predictable rate of ice sheet melt, agricultural productivity decline, ocean circulation slow-down, forest cover loss, and weakening land-ocean sink may suddenly become unpredictable and the rate of change rapidly accelerates.[54] How far beyond is unknown.

The depletion of the ozone is a good example of underestimating the rate of change.[55] Interest in damage to the ozone first began in the 1970s as the airline industry was developing the first supersonic passenger planes which would fly higher than standard aircraft through the stratosphere. Scientists had concerns that emissions would damage the ozone layer and although the supersonic passenger industry never reached more than 20 Concordes, the early work led to an understanding of the damage done by another class of pollutants called CFCs.

Scientists including James Lovelock, Paul Crutzen, Sherwood Rowland, and Mario Molina figured out that the CFCs released into the atmosphere from aerosols, fridges, and air conditioning units had no natural sinks in the lower atmosphere and would make their way up to the stratosphere, break down, and destroy the protective ozone layer leaving the Earth's surface vulnerable to carcinogenic high energy UV radiation from the sun. James Anderson would go on to make measurements proving that reactive chlorine compounds were indeed present in the stratosphere. Then, in 1985, the British Antarctic Survey announced they had found a hole in the ozone above the Arctic. This shocked both the world and the scientific community because no model or theory had predicted the damage could happen so fast. It turned out that NASA satellites had detected the ozone hole as early as 1979 but the processing software had been written to discount low readings which were thought impossible and the data never surfaced. Urgent Arctic expeditions and breakneck revisions to atmospheric models would reveal that a combination of polar vortex winds and clouds of ice crystals above the Arctic acted to liberate more reactive chlorine and destroy more ozone and block replenishment from other parts of the atmosphere. Global averages had underestimated the risk of regional sub-system breakdown and the world scrambled to cut CFC emissions.

The Carbon Crunch versus The Credit Crunch: Social and Economic Tipping Points

Whilst the risk of accelerating physical change is all too real, the risk of social and economic tipping points also looms large and could perhaps arrive even sooner. Financial markets don't operate on what happened in the past, but rather what is expected to happen in the future. If the physical effects of climate change arrive sooner than financial markets expect, this will very quickly become reflected in lower prices for real estate, shares, and commodities. If it happens too quickly, this can destabilise the whole system – a so-called Carbon Crunch.

The financial crisis or Credit Crunch in 2008 serves as a good analogy to the unknown economic risks of a changing climate. In the run up to the financial crisis, the global economy had been through a period of rapid expansion. Confidence was high. Consumers, business owners, and governments saw no end to the strong growth and continued to borrow more money to spend on houses, industrial equipment, and infrastructure. Investors, banks, and mortgage companies were also caught in the boom mentality and more than happy to extend more credit or lending in order not to miss out on ever greater returns. Incomes were rising, house prices were increasing, and share prices kept beating all-time highs. The economy was in a positive or amplifying feedback cycle – more growth, more confidence, more credit – a bull market. But with every debt cycle there comes a point where that extra borrowing returns little extra income. You already bought the new car to get to work faster, the new machine to builds widgets faster, or the new shipping terminal to export more goods. Soon the additional debt repayments start to outweigh the additional income and growth grinds to a halt – it turns to a negative cycle of slowing growth, reduced confidence, and credit dries up further thus shrinking the economy – a bear market. If this happens slowly and is well managed, most people will barely notice and the economy can reset. But the intersection of risk factors in 2008 meant that the economy hit a tipping point and that forced unmanageable change.

The first mistake was underestimating **tail risk** in the housing market. Mortgage companies had started lending money to high-risk borrowers without the proper checks to ensure they could meet their repayments: sub-prime mortgages. This debt was being offloaded to investment banks who would combine it with debt from cars or credit cards to create Collateralised Debt Obligations or CDOs. The idea is sound: different streams of repayments mean the chance of a large number of defaults in one CDO is lower.[56] Overall the debt is less risky. Rating agencies could rubber stamp CDOs as investment grade and the banks could sell them on for a fee. But the problem was not with the idea but in the maths. Derivative risk is calculated using Gaussian functions or bell curves – just like certain impacts of climate change – and the bankers underestimated the risk of default. CDOs with too much exposure to bad housing debt or CDOs made of bundles CDOs (CDO squared) skewed the maths and these investment products were way more dangerous than everyone thought.[57] Warren Buffet wrote in a 2002 shareholder letter, "derivatives[ix] are financial weapons of mass destruction".[58]

Next came the **sub-system collapse** with the downfall of Lehman Brothers, a US investment bank founded in 1850. Lehmans had been aggressively expanding and had built up a massive sub-prime and CDO business. When homeowners ran into trouble on their repayments and mortgages started defaulting, the financial world realised the value of those CDOs was significantly lower than originally thought. The price of CDOs, or any asset (houses, shares, bonds, gold, or currency), is simply the last price agreed between a buyer and a seller. So as the confidence in the investment return of CDOs fell apart, everyone rushed to sell at the same time. With no buyers in the market, prices crashed.[59] Lehmans had underestimated the risk; they weren't holding enough cash or liquid capital to weather the losses and the bank collapsed.

No private financial bail-out was found to save Lehmans and the government took the view the bank should be allowed to fail. What the authorities didn't realise was this event would prove to be the tipping point, the final straw, which would lead to **systemic breakdown**. The financial sub-systems lost trust in one another, transactions ground to a halt, liquidity dried up, no one was willing to extend credit to the economy, and the initial slowdown in growth turned into an economic collapse or Credit Crunch. Global GDP declined by 5%, millions of people lost their jobs, and the value of global stock markets halved.

---

ix Derivatives are a financial arrangement or product whose value is determined by an underlying asset such as the repayments from mortgages, car loans, and credit cards.

The total collateralised debt obligation market was worth $2 trillion dollars at peak and an estimated half of these products defaulted.[60] Returning to our climate change analogy we know there is tail risk, sub-system risk, and ultimately systemic risk of socio-economic collapse. But how does the value at risk compare if we turn our focus on climate?

Let's start with fossil fuels. If we discount the future revenues from coal, oil and gas reserves over the reminder of this century, that's a total of $100 trillion of sales in today's money.[x] Assume the fossil fuel companies can make a 10% profit and that's $10 trillion to the bottom line. Burning these reserves would release 4,000 billion tonnes of CO_2 into the atmosphere and raise temperatures by over 3°C. To keep warming below 1.5°C we could use at most half the oil and gas and just 10% of the coal.[61,62,63] That's $75 trillion of revenue and $7.5 trillion profit left stranded.

But fossil fuel write downs are just the beginning. Fossil fuel power stations and generation equipment add another $15 trillion of assets. There are $20 trillion worth of cars, trucks, and planes which run on oil and coal and gas-powered industrial equipment worth more than $20 trillion. Plus, another $5 trillion worth of gas boilers, cookers, and general equipment. A total of $60 trillion of assets which run on fossil fuels. How many will have to be decommissioned before the end of their useful life? Will there be even more stranded assets?

The Carbon Crunch is a scenario where inaction or rapidly accelerating physical change forces the world to switch away from fossil fuels in an extremely short time frame. The transition would be chaotic, with energy shortages, skilled labour shortages and businesses, governments, and citizens unable to adapt fast enough. The result: a large devaluation of that total $70 trillion pot.

Beyond this first wave of destruction, future damages (already locked in) become increasingly ominous. Our discounted damages to 2100 add up to more than $160 trillion dollars; does the world simply write-off the coastal property, agricultural lands, and human capital in the worst exposed areas? The Carbon Crunch could lead to mass job losses, bankrupted businesses, failed governments, and breakdown of financial systems.

---

x There are 1,000 billion tonnes of coal, 200 trillion m^3 of gas, and 1,700 billion barrels of oil reserves which have been discovered and can be easily extracted. Enough to last out the century based on our peak emissions profile and worth $90 trillion, $25 trillion, and $130 trillion of revenue, respectively.

Michael Bloomberg (co-founder of Bloomberg L.P and former mayor of New York), Henry Paulson (former secretary of the US treasury) and Thomas Steyer (hedge fund manager and philanthropist) stated in their Risky Business report in 2014: "The US economy faces significant risks from unabated climate change. Every year of inaction serves to broaden and deepen those risks". Mark Carney, the former head of the Bank of England and now the UN envoy for Climate Action, puts it simply: "Companies that don't adapt [to climate change] will go bankrupt without question. The longer the adjustment is delayed in the real economy, the greater the risk that there is a sharp adjustment".[64]

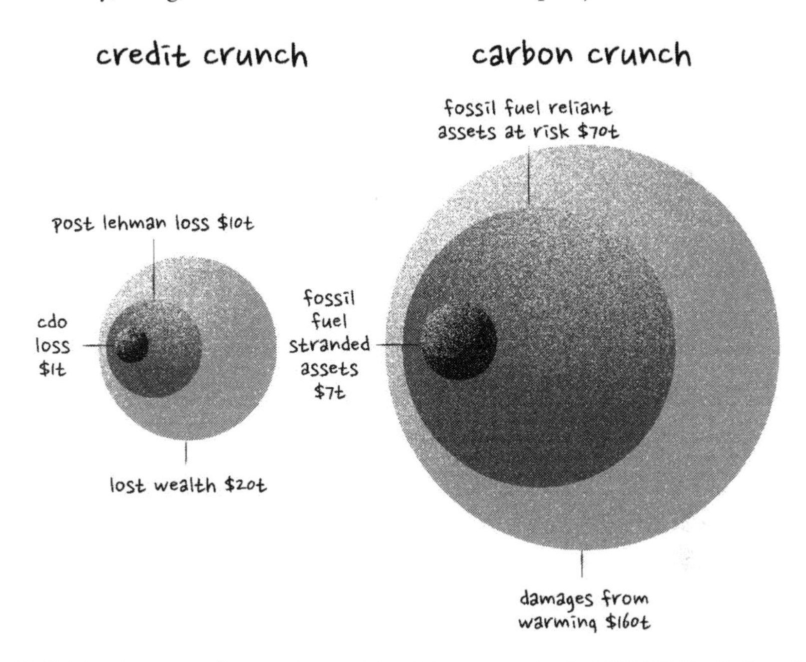

Tail risk, sub-system collapse and systemic breakdown events for the 2008/09 Credit Crunch and the potential elements of a Carbon Crunch. CDO = Collateralised Debt Obligation. Post-Lehman losses reflect market losses immediately after Lehman Brothers collapsed. Lost Wealth is the reduction in global household wealth between 2007 and 2008. Fossil fuel stranded assets are 70% of the future $10 trillion profits on fossil fuel reserves (in today's money). Fossil fuel reliant assets are the present value of power plants, vehicles, industrial equipment, and home/commercial appliances which rely on fossil fuels. Damages from warming represents the present value of future climate damages to 2100 under a baseline inaction scenario using a 3% discount rate.

Damages and the Discount Rate: Our Second Dial of Doom and Boom

Our damage model includes no adaptation, price dynamics, or regional break down, but it does provide a back-of-the-envelope take on the scale of possible damages from continued inaction on climate change. The total cost is 15 million deaths per year, ongoing suffering, and hardship for over one third of the global population, and 5-9% of global GDP wiped out by the start of the next century or the equivalent of 4.5% of today's GDP lost forever. Climate change and air pollution certainly qualify as a top three global problem.

Most Integrated Assessment Models (IAMs), which link physical change to damages, as we have done in this chapter, predict around 0.1-5% negative impact on GDP at about 2.5-3.0°C warming.[65] These models such as DICE and RICE developed by Nordhaus, FUND developed by Anthoff and Tol, or PAGE developed by Hope and used by Stern, use simplified mathematical functions to represent the damage from rising temperatures on energy, water, coastal areas, tourism, and health. The range of climate sensitivities, difficulties in estimating damages, the uncertainty of adaptation, and the problems of time discounting, equity-weighting, and non-market value judgements makes estimating the impact of climate change incredibly difficult. This may explain why there are just 27 global models which have attempted to do so and why the IPCC states "the aggregate damage functions used in many IAMs are generated from a remarkable paucity (low amount) of data and are thus of low reliability". Nonetheless, having something is better than nothing and these estimates should be used as a guide not as accurate predictions, and should be held alongside other possible futures containing dangerous tipping points or greater technological promise.

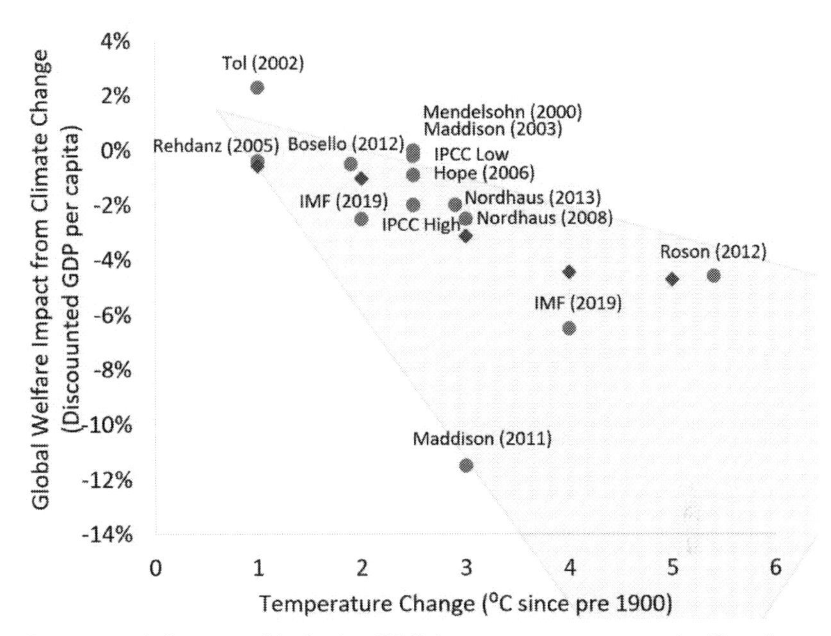

Comparison of damage models showing GDP impact versus temperature rise. Data from IMF working paper authored by Matthew Kahn et al. [66] *Published estimates labelled dots, our estimates from this chapter shown by diamonds.*

Our 4.5% losses sit at the higher end of the IAM range before even including the non-market damages from loss of life or destruction of the natural world. Nordhaus is one of the most established economists in the field and his latest DICE models estimate 1-2.5% GDP impact. The IPCC gives a range of 0.2-2% lost income by 2.5°C but with significant uncertainty and risk of a larger number. Lord Stern had the more severe predictions of 5% lost income from direct market impacts and this rises by a further 2-3% when the probability of dangerous tipping points is included, with a further 6% loss to include non-market damages and multiplied by an extra third to account for regional inequality. Stern could arrive at a total of 20% equivalent GDP losses now and forever.[13]

Apart from Lord Stern's numbers, most damage estimates (including ours) assume a manageable outcome. We know tipping points exist but it's difficult to assign a probability for these very large changes, so this uncertainty makes them near impossible to model. Late economist Martin Weizmann drew attention to the possibility of large tail risk in the climate sensitivity itself which makes extreme warming events more likely.[67] In turn, these outsized damages start to dominate the statistical output of model runs. He argued that you cannot ignore potentially catastrophic outcomes.

The further we push Earth's physical systems and humanity's socio-economic systems, the greater the chance of a tipping point. Beyond this limit, global warming would not just exert pressure on all other global problems, but would also spill over into social unrest, crime, pandemics, inequality, corruption, health, unemployment, migration, extremism, and financial instability. Just as water vapour amplifies the warming effect of CO_2, so the damages of climate change will amplify many global problems.

A linear rise in climate damages would slow but not hinder the exponential growth in human prosperity and cornucopian solutions tied to human ingenuity, free markets, and technology might be enough. But breach a known but unknowable tipping point and exponential damage could drag humanity back into destitution.

In an ideal world we would have unlimited resources to throw at each of our global problems. Unfortunately, we have limited time, money, and human intellect to go around. We must therefore find a way to allocate these precious assets for the best overall outcome. Lord Stern's damage estimates justify spending 5% or more of GDP on climate solutions, but Nordhaus' conclusions warrant significantly less.

The **discount rate** is our second dial of doom and boom, the economic sensitivity which defines how much value we place on future generations: dial it up and damages seem small, manageable and distant, dial it down and climate change threatens not only continued improvements in prosperity but, if combined with dangerous tipping points, could pull us back into the Malthusian Trap.

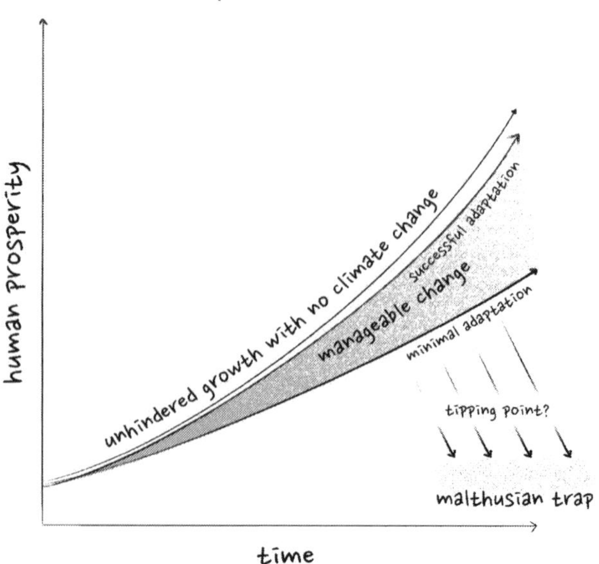

The Precautionary Principle states we cannot risk the accelerated physical or socio-economic impacts of climate change – we must move fast enough to avoid dangerous tipping points and keep to a least-cost but manageable path. But where do we start? What activities are responsible for greenhouse gas emissions? And what are the possible strategies which could address climate change?

The next chapter runs through where exactly emissions come from, what we can do about them, and how we select the best strategy.

Chapter 6: Opening for Planetary Caretaker

Building a Climate Change Strategy

For the first time in history, we have improved our quality of life and broken free from the Malthusian Trap. But for how long? Left unchecked and the best scenario is that physical change will incur added cost, reduce productivity, and slow prosperity increases. At worst, we will see increasing temperatures, rising sea levels, and unpredictable tipping points threaten to unravel the last 200 years of progress. The Precautionary Principle suggests action is required and we must now accept the unsolicited role of planetary caretaker. So, can we rise to the challenge? Or will today's prosperity become another short-lived anomaly in Thomas Malthus' predictions?

This chapter will run through the warming potential of other greenhouse gases compared to CO_2, we will identify emissions from different parts of the economy, and understand the link between wealth, energy use, and emissions. We will examine the pros and cons of the three main strategies to address climate change, namely: adaptation, climate engineering, and mitigation. And, we will show that reducing and eliminating greenhouse gases is the only low risk strategy, and that reaching net-zero is the only full proof plan.

It's time to accept our unintended position as planetary caretaker and get to work. As Spider-Man eloquently puts it: "With great power there must also come great responsibility."

Identifying the Greenhouse Gases: The Warming Potential of CO_2 and other GHGs

The thicker the blanket of CO_2 surrounding the planet, the hotter the surface temperature becomes to balance the incoming and outgoing energy. But it's not just CO_2 which increases the tog: other problem gases include methane, nitrous oxide, and a class of industrial chemicals called halocarbons.[1] One of the major jobs undertaken by climate scientists was to figure out the impact of each gas and understand where these emissions were coming from.

Methane: Chemical formula CH_4 is emitted from leaks in the oil and gas network (50%) and released from livestock enteric fermentation, or more commonly referred to as cow and sheep burps, (25%), paddy rice farming (5-10%), landfill (5-10%), and coal mining (5-10%).

Nitrous Oxide: Chemical formula N_20 is mostly released from the use of fertilisers, manure, and soil management in agriculture (>70%), and the remaining emissions are from internal combustion vehicles.

Halocarbons: Chemicals containing a mix of carbon, fluorine, chlorine, bromine, hydrogen, or oxygen. Halocarbons are used as the cooling or heating fluids (refrigerants) in air conditioning, heat pumps, and refrigeration (70%) or in industrial processing and semiconductor fabrication (30%).

Greenhouse gases all act to warm the planet, though many have a bigger impact than carbon dioxide. Methane, nitrous oxide, and halocarbons have a much lower concentration in the atmosphere so their ability to absorb specific infrared frequencies is less saturated than CO_2. Small concentration increases have an outsized impact on warming.

The Global Warming Potential (GWP) is gauged by the ability of each gas to increase radiative forcing and warm the planet for a given weight of emissions when compared to CO_2. Every one-tonne of methane is the equivalent of adding 28 tonnes of CO_2 into the atmosphere. Every one-tonne of nitrous oxide is the equivalent of 265 tonnes of CO_2.

But to complicate the matter further the warming impact must be compared over a given timescale. The warming potential of methane is 28 times more powerful than CO_2 on a 100-year basis, but it is 84 times more powerful over 20 years. This is because methane only lasts about ten years in the atmosphere before it reacts with oxygen to form CO_2.

Multiply the increase of each gas by the warming potential and scientists can estimate the relative contribution from each. Although CO_2 is the least powerful greenhouse gas per tonne, the sheer quantity we release from the combustion of fossil fuels makes it responsible for more than half of all human induced global warming over the last 200 years. Methane is responsible for one quarter. Nitrous oxide and halocarbons 10%, and the remainder is mostly due to soot or black carbon. CO_2 takes centre stage in climate analysis not only because it is the largest contributor to global warming, but also because CO_2 persists in the atmosphere for centuries and, without intervention, creates irreversible change over human timescales. To help simplify the annual stock take, greenhouse gas emissions are often quantified in carbon dioxide equivalents or CO_2e.

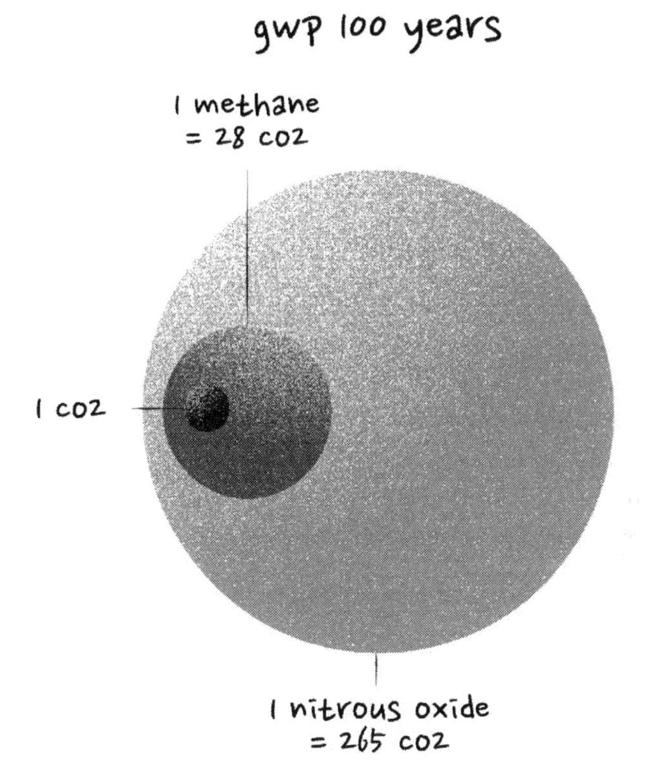

Global Warming Potential of methane and nitrous oxide compared to CO_2 over 100 years.[2]

The Greenhouse Gas Stock Take: Energy Use and Emissions from Human Activity

Energy drives the modern world, from the gas fed into our boilers, the electricity running through our meters, the gasoline pumped into our cars, or the embedded energy and emissions in the goods and food we buy. The combustion of fossil fuels for power plants, cars, boilers, cookers, industrial machinery and heating processes is responsible for 70% of all emissions. The remaining 30% come from agricultural sources or industrial reactions.[3]

The growing consumption of energy has been the biggest driver of increasing CO_2 and over the last 50 years we have nearly quadrupled *Primary Energy*[i] consumption to 160 trillion kWh and CO_2e emissions have more than doubled to over 50 billion tonnes per year.

i **Primary Energy** is the amount of coal, oil, gas, biomass (and nuclear and renewable equivalent) energy consumed to provide the **Final Energy** which is actually used in the economy. Primary energy consumption is greater than final energy use, mostly due to losses from electricity generation and distribution.

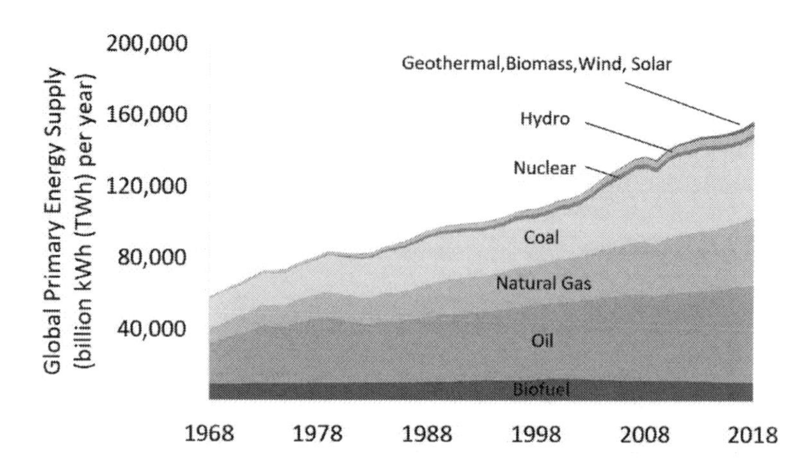

Primary energy demand per year (including losses from electricity generation) in billion kWh.[4]

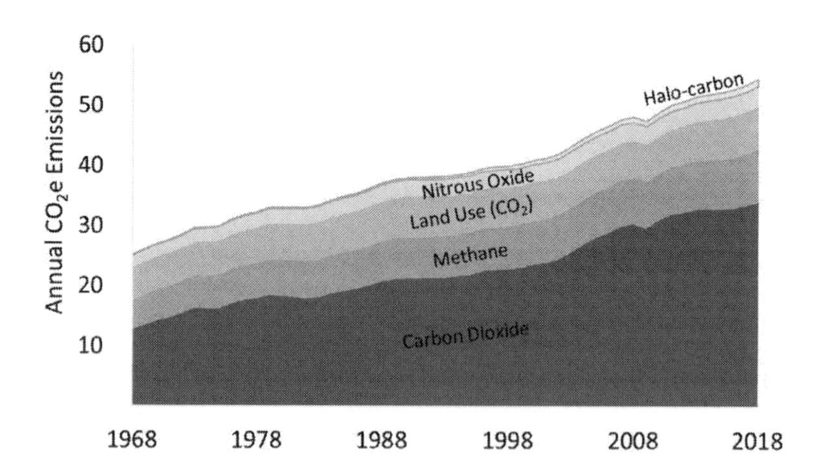

Greenhouse gas emissions (in CO_2 equivalents) from human activity.[5]

Fossil fuel electricity generation and distribution converts just one third of the available combustion energy into electricity so 60 trillion kWh of primary energy is wasted.[6] The *Final Energy* use in transport, amenities, industry, and agriculture equals 100 trillion kWh per year, excluding these losses.

Fossil fuels account for 81% of final energy, traditional biomass (wood burning) 12%, hydro power 3%, nuclear 2% and wind and solar 1% each. Energy use is evenly distributed between transport, amenities (heating, cooling, cooking, lighting etc.) and industrial manufacturing.

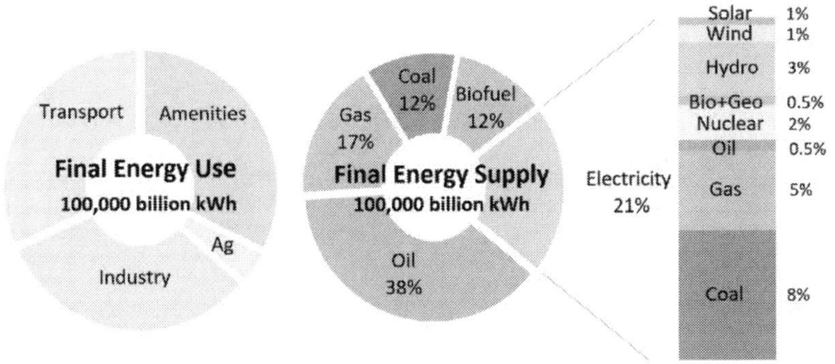

Transport Amenities
Final Energy Use
100,000 billion kWh
Ag
Industry

Coal 12%
Gas 17%
Biofuel 12%
Final Energy Supply
100,000 billion kWh
Oil 38%
Electricity 21%

Solar 1%
Wind 1%
Hydro 3%
Bio+Geo 0.5%
Nuclear 2%
Oil 0.5%
Gas 5%
Coal 8%

Left: Final Energy (excluding electrical generation losses) by end use per year.[7] *Middle: Final Energy by source per year Right: Breakout of electrical generation as a percent of total final energy.*[4]

Burning fossil fuels for energy creates 32 billion tonnes of CO_2 per year. Another 20 billion tonnes of CO_2 and CO_2e come from non-energy sources which includes the leakage of methane from natural gas and oil systems (2.5 billion tonnes CO_2e), CO_2 from agricultural land clearing and burning (5.5 billion tonnes CO_2), methane from cow and sheep burps (2.5 billion tonne CO_2e), nitrous oxides from agricultural manures and fertilisers (3 billion tonnes CO_2e), and CO_2 emissions from the reaction used in cement making[8] (2 billion tonnes CO_2).

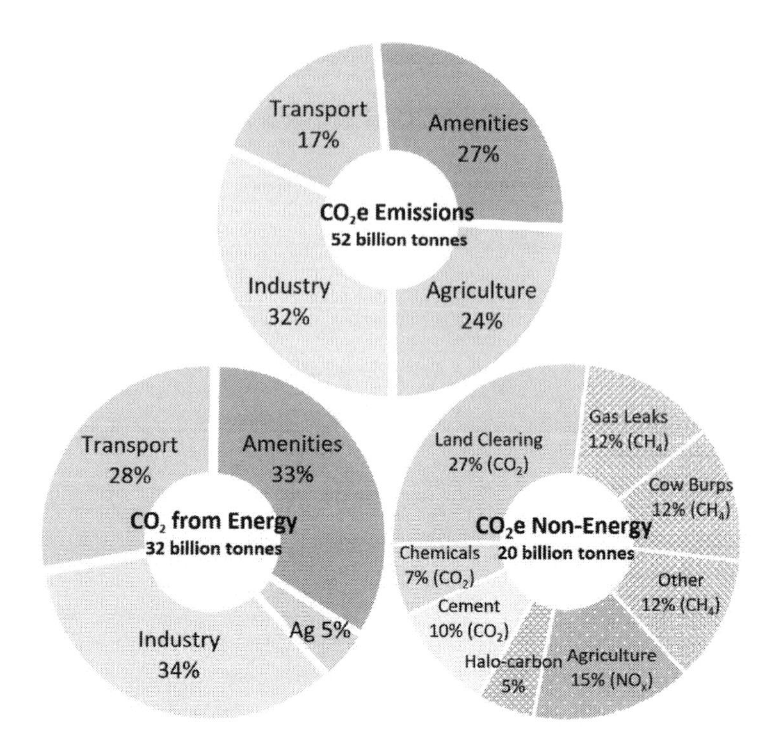

Top: Total CO_2 equivalent emissions per year including energy supply and non-energy emissions of all greenhouse gases.[4] *Bottom Left: Annual CO_2 emissions from energy generation. Bottom Right: CO_2 and other greenhouse gas emissions (in CO_2e) not from energy generation.*[3]

Emissions Per Person: The Link Between Affluence and GHGs

The world now consumes 100 trillion kWh (or 100,000 TWh[ii]) of final energy every year. If we divide by 365 days and 24 hours per day, the average power needed to run the planet is 12 billion kW (12 TW). Divide by 7.8 billion people and this equals 13,000 kWh per person per year, 36 kWh per person per day, or an average power demand of 1.5 kW per person – the equivalent of having 1.5 kettles constantly boiling water for each person on the planet.

Humans now emit 40 billion tonnes of CO_2 and another 12 billion tonnes CO_2 equivalent greenhouse gases every year. That equals 5 tonnes of CO_2 or nearly 7 tonnes CO_2 equivalent per person per year. That is

ii **Unit Prefixes:** K (kilo) = 1000 (thousand), M (mega) = 1,000,000 (million), G (giga) = 1,000,000,000 (billion), T (tera) = 1,000,000,000,000 (trillion).

one hundred times our own body weight and the volume of 30 double decker buses[iii] in gaseous emissions.[9]

However, global averages don't tell the whole story. There are large differences by region and lifestyle. Energy consumption typically increases with income or wealth. A more affluent lifestyle uses more heating, cooling, light, gadgets, and goods along with bigger houses, more food, greater global travel, and more consumption creating more emissions.

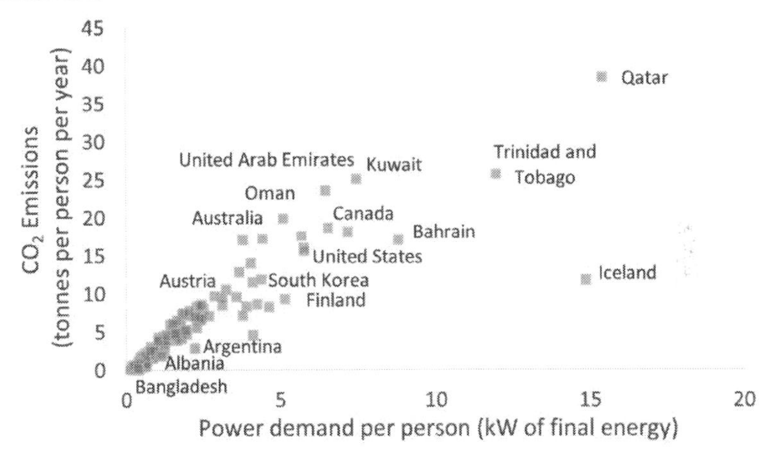

CO_2 emissions per person versus power demand per person in kW [10,11]

Towards the top end of the scale are the US and Australia, both affluent countries with a relatively low population density. Fewer people with lots of land and money leads to more car journeys, bigger houses, and more stuff. Americans have an average power demand of 6 kW each (6 kettles boiling constantly) and emit nearly 16 tonnes of CO_2 per person per year, the equivalent weight of eight Ford F150 trucks.

At the bottom end of the scale India, Bangladesh, Pakistan, and countries in areas such as sub-Saharan Africa use less than 0.5 kW power and emit less than 2 tonnes of CO_2 per person per year. That's six times less than the global average and 15 times less than the US.

iii At atmospheric pressure CO_2 occupies 0.5 cubic metres per kg.

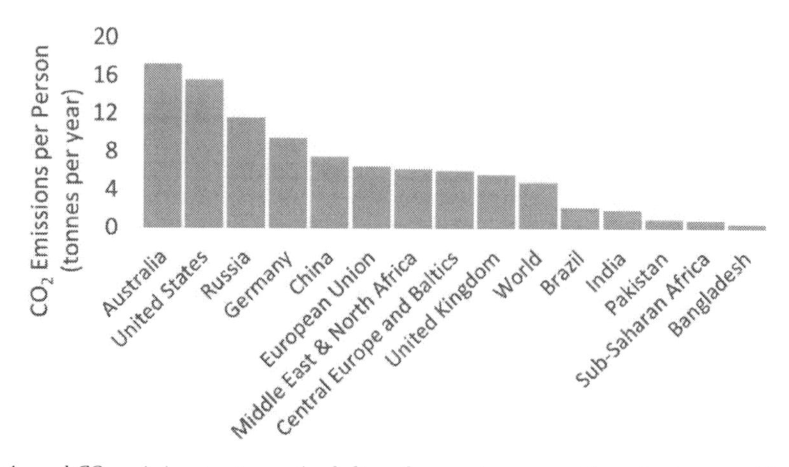

Annual CO_2 emissions per person (excluding other greenhouse gases) by region in tonnes [4,11]

Planetary Caretaker: The Three Strategies to Address Climate Change

So, should we finally accept the position of global caretaker, our first decision is to define the responsibilities and skills necessary to fulfil the role. But how do we balance the demands of society, the economy, and the environment, to create the best possible outcome? Do we listen to the dire warnings of Malthus and Ehrlich and reduce the population? Do we place our faith in the World3 model and the Club of Rome and bring an end to consumption growth? Or did Julian Simon and the cornucopians have it right? Will markets adapt and technology solve our problems?

If a global planner existed, what strategy might they adopt? Let's look at our three options: adaptation, climate engineering, and mitigation.

Adaptation: Letting the World Warm and Living with Change

An adaptation only strategy is a course of action whereby the burning of fossil fuels continues unchanged. We live with the fact that more CO_2 will accumulate in the atmosphere, thus increasing global temperatures and creating a more acidic ocean. This course of action avoids the upfront time, cost, and resources of dealing with climate change now and assumes that future generations can deal with the problem due to greater global wealth or a technological breakthrough. Instead of attempting to prevent further significant temperature increases, humanity simply adapts to the higher temperatures as they happen. This would mean relocating coastal communities, deploying more air conditioning, cooling shelters, irrigation, desalination of sea water, protecting against storms and flooding, creating a more resilient power system, creating contingency funds, changing farming practices and relocating farming areas, developing drought resistant crops, attempting to limit the spread of disease and providing robust disaster planning, education, and behavioural learning. Just the tip of a very long list of actions.

We can soften the blow of climate change and, if done well, reduce death, suffering, and economic damage. Some level of adaptation will be necessary no matter what – after all we have already locked in near 1.5ºC warming at current CO_2 concentrations. Adaptation planning for vulnerable poorer countries should be central to all climate change strategies. At its most fundamental this means accelerating development to improve incomes, health, financial, and social safety nets, providing stable government and diversifying the economy away from vulnerable activities such as farming.

However, to advocate an ***adaptation only*** strategy you must believe that the discounted future socio-economic damages will be lower than the upfront cost of preventing climate change. The economy, technology, and social structure must improve faster than the detrimental impact of climate change. Adaptation only strategies rely on manageable change – predictable rising damage and faster increases to human progress. It also relies on getting our adaptation measures right. Maladaptations are possible – well-intentioned actions gone wrong – such as sea walls built too low, resilient coastal roads creating more demand for vulnerable coastal infrastructure, insurance schemes which facilitate risk taking, or biofuels that increase food prices. Plus, there are many limits to adaptation such as the irreversible melting of ice, thermal limits of crop growth, and sea creatures' abilities to tolerate more acidic waters. Should we get the adaptation measures wrong or underestimate the physical change and stray too far from existing environmental, social, or economic systems, then dangerous tipping points could create unpredictable damage somewhere down the line.

The estimated cost of bare minimum adaptation based on work by the UNFCCC, World Bank, Stern Review, and Oxfam is in the range of tens to hundreds of billions of dollars per year over the next 30 years.[12] This doesn't include tipping points and there is general agreement that these costs rise exponentially with temperature due to the need for ever more transformational changes, which make for harder and more costly solutions. For example, we already calculated the $3 trillion per year price tag of desalinising sea water to increase agricultural irrigation as the frequency of droughts increases.

Adaptation may seem like the better option because it is initially cheaper, and the benefits are localised and visible, however pursuing an adaptation only strategy may have a low upfront cost but it risks unknowable damages towards the end of this century as change becomes more difficult to manage and tipping points become ever more likely.

Climate Engineering: Removing CO_2 from the Atmosphere or Blocking the Sun

Climate engineering refers to several proposed strategies which either directly remove CO_2 from the atmosphere (negative emissions technology) or limit the amount of solar energy absorbed by the planet (solar radiation management). These strategies could either allow us to continue to burn fossil fuels without many of the damaging climate impacts by controlling the sun's energy reaching Earth or could be used as a tool to slow change by removing CO_2 from the atmosphere.

Negative Emissions Technology: Removing CO_2 from the Atmosphere

Reducing Deforestation, Reforestation or Afforestation requires stopping the destruction of forests, restoring past forests, or growing new ones and speeding up the land based biological carbon cycle. Trees absorb carbon dioxide using photosynthesis through the day and expel just half as much CO_2 through respiration overnight when growing. The average tree draws half a tonne of CO_2 from the atmosphere as it grows over 25-75 years and locks it away in the trunk, branches, and root systems.[13] There are already 3 trillion trees on Earth covering 42 million square km or 28% of all dry land.[14]

- ***Reducing Deforestation:*** Every year the world loses around 80,000 square km of forest[iv] from unsustainable logging and clearing for agricultural land. That's a land area the size of Austria and, with 50,000 tonnes of CO_2 bound up in each square kilometre of forest, this releases four billion tonnes of CO_2 emissions from the burning and rotting wood. End deforestation and you eliminate nearly 10% of all CO_2e emissions. The cost is estimated at less than $5 per tonne of CO_2 to compensate

iv Equivalent to 8 million hectares, 8 billion trees or 0.2% of global forests.

farmers[v] for lost income on potential agricultural land and to ensure the policy is enforced.[3]

- ***Tree Planting:*** Scientists estimate there is enough space on the planet to add up to another nine million square kilometres of forest or 0.7 trillion trees without impacting agricultural output.[15] This is an area bigger than the United States. This could remove more than 400 billion tonnes of CO_2 over the next 50-100 years at a rate of up to 7-8 billion tonnes per year. It costs between $0.5 and $6 to plant and establish one tree depending on the land,[16] a total cost of about $2 trillion to plant all potential land.[vi] Reforestation could work out at less than $10 per tonne of CO_2 removed[vii] or an added cost of $0.5c per kWh[viii] to offset the emissions from fossil fuel energy.[17]

Since the 1990s, China has planted nearly one million square kilometres of new forest, covering 0.7% of all the world's dryland at a cost of over $100 billion ($3 per tree).[18] Scaling this up to every country in the world, full forest potential could be reached over the coming decades.

Tree planting is relatively cheap, effective, and low risk. It also comes with added co-benefits of soil preservation, flood control, and water retention. But it only covers a maximum of 15% of current annual emissions and eventually this strategy runs out of land unless the wood can be sustainably harvested and stored or used.[ix] Planting trees is effective, cheap, and low risk, but it can only offset a fraction of emissions.

Enhanced Weathering is a route to speed up the slow carbon cycle. Remember that rock will absorb CO_2 from the atmosphere and transport it to the bottom of the ocean. The problem is the process takes thousands of years. Enhanced weathering involves crushing silicate rock such as

---

v Global agriculture generates $2-4 trillion per year with roughly 10% profits and using 49 million square km land for crops and grazing. It would cost $3 per tonne of CO_2 sequestered to compensate farmers for lost profits over 50 years.

vi Assuming an average of $1.5 per tree, the total cost of adding 0.7 trillion trees would be $1 trillion upfront or $2 trillion with a 3% rate of return.

vii The IPCC AR5 report highlighted 3-4 billion tonnes of CO_2 mitigation potential from tree planting at a cost of less than $20 per tonne.

viii The average energy price today is $5.5-6c per kWh. If planted on agricultural land another $5 per tonne or $0.25c per kWh is needed to compensate farmers.

ix 8 billion tonnes of CO_2 per year is about 3.5 billion tonnes of wood – the equivalent of 5,000 million cubic metres or 4,000 Wembley stadiums worth of storage space. The wood could be buried or used in long lived products. 1,500 million cubic metres of wood is used in industry every year already. Find further uses for this store of carbon such as replacing a portion of the 7,000 million cubic metres of concrete we use every year and sustainable carbon sequestration from forests can continue.

basalt and spreading the rock powder onto farmers' fields, deserts, or shallow coastal areas to speed up the weathering process.

Studies conducted by Thorben Amann from the University of Hamburg show that the rate of weathering is between 5-50 tonnes of CO_2 per square km per year. Warm and wet areas are best suited for enhanced weathering, so the Continental Shelf (28 million square km), and about half of global agricultural land (19 million square km) are suitable for this purpose.[19] Spread rock over these areas and we could remove 1-2 billion tonnes of CO_2 per year, though again, unfortunately, that's just 5% of emissions.[20]

Every tonne of CO_2 sequestered in these ways, requires 3 tonnes of basalt rock. With an average price of $10 per tonne of crushed rock, this brings the cost of advanced weathering to over $30 per tonne of CO_2 or $2c per kWh to offset fossil fuel combustion.[21] Enhanced weathering is low risk but can only offset a fraction of emissions at best and is three times more expensive than tree planting.

Ocean Nutrition is a proposed strategy to accelerate the rate at which algae in the ocean sequester carbon dioxide from the atmosphere by speeding up the ocean biological pump. Algae already absorb 36 billion tonnes of CO_2 by photosynthesis every year. Some scientists believe that artificially adding fertiliser such as iron, phosphate, or nitrogen to the ocean will encourage more algae growth, draw down more CO_2, and deposit the carbon on the ocean floor to be locked away for thousands of years. There are enough fertiliser resources on the planet to sequester thousands of billions of tonnes of carbon dioxide, however, much of the CO_2 will return to the atmosphere before it can be locked away[x] in ocean

x Over 40% of the 370 million square km of global ocean surface is undernourished due to a lack of mixing with the deep ocean. Those pictures of clear blue water on holiday brochures are actually large areas of ocean devoid of algae life because the water is too warm.

To get an idea of how much fertiliser we would need to nourish the waters we can use the Redfield ratio, aptly coined by Alfred Redfield in 1934 when he realised that nearly all ocean algae have the same ratio of carbon to nutrients. Every tonne of carbon locked into the biomass of algae requires 150 kg nitrogen, 10 kg phosphorous and 100g of iron. Every tonne of algae carbon removes 3.7 tonnes CO_2 from the atmosphere.

Nitrogen costs $300 per tonne, phosphorous and iron $100 per tonne – dividing through by the required ratios gives a cost of $46 of fertiliser per tonne of carbon or $12.5 per tonne of CO_2 removed from the atmosphere by algae. Deduct the 0.5 tonnes CO_2 released from fertiliser production and the cost doubles. Assume only 25% sinks to the deep ocean and the rest re-enters the atmosphere and the costs climb towards $100 per tonne or $7c per kWh.

sediment. Providing all the required nutrients for algae growth would cost over $100 per tonne CO_2, so realistically ocean nutrition would target areas which are just lacking key nutrients. Best cost estimates by the Ocean Nourishment Corporation which is trialling this approach are around $20 per tonne of CO_2.[22]

In 2012, US self-proclaimed environmental entrepreneur Russ George convinced a local village in west Canada to support his scheme to dump iron powder into their waters to boost salmon production and to try and sell carbon credits. His start-up, Planktos, dumped 100 tonnes of iron sulphate into the ocean before returning to shore where he was arrested and accused of breaking international law by illegal dumping. George maintains that the record salmon harvest in the following year was down to his intervention and that if his data had not been confiscated by the authorities, he would have proven the carbon reduction potential of his ocean nourishment scheme.[23,24,25]

More rigorous scientific experiments using iron fertilisation have typically resulted in large algae blooms which grew so quickly that they depleted the surrounding waters of oxygen, destroying aquatic life, killing themselves, and releasing all the carbon back into the atmosphere.[xi]

Overall, ocean nutrition is more expensive than reforestation but does offer a significantly greater carbon capture potential.[26] However, the complexity of ocean systems and ocean chemistry mean this option on a large scale could also come with significant unintended consequences.

Direct Air Carbon Capture (DAC) and Storage involves deploying technology to filter CO_2 straight out of the air using large fans and chemical reactions – much like an artificial tree. The CO_2 can either be stored deep underground or turned into other products.

The biggest hurdle for direct air capture is the second law of thermodynamics which governs a process referred to as entropy that rests on the recognition that everything in the universe naturally moves towards a state of disorder or randomness and requires energy to reverse and reorder the process.[27]

Separating CO_2 from a disorderly mix of other gases in the air requires reordering and the lower the concentration the more energy is required. As Bill Gates puts it, "Pick one molecule at random out of the atmosphere and the odds that it will be carbon dioxide are just 1 in 2,500". The physics of filtration mean that at theoretical maximum efficiency, with

---

xi The Ocean Nourishment Corporation is now focussed on using nitrogen rather than iron and believes it can better control the growth rate of the algae using this method.

cheap wholesale electricity[xii], Direct Air Capture running costs would stack up to around $4 per tonne of CO_2 removed.[xiii] Unfortunately, real systems can only achieve 5-40% efficiency, so the real running cost is multiple times higher, plus the equipment, storage, and labour costs must also be included.[28,29]

A company called Carbon Engineering is one of Bill Gates's many climate change projects. This Canadian company is operating direct air capture pilot plants at the tonne scale today but believe they can scale up the facilities to capture one million tonnes of CO_2 per year.[30] Based on Carbon Engineering's scaled up costs, carbon capture and storage would end up at over $120 per tonne or $10c per kWh equivalent offset.[xiv]

Bill Gates expects with some innovation "we can realistically expect it [DAC] to get down to $100 per ton".[31] Let's say we go one step further and assume the efficiency can reach 40% and the build costs can be cut in half, this would bring the total buying and running cost down to $70 per tonne of CO_2 or $5c per kWh fossil fuel carbon offset.

A company called Global Thermostat believe they can get the process down to $50 per tonne of CO_2 once fully scaled up,[32,33] yet DAC would still prove one of the most expensive negative emissions options. Direct air capture is a very flexible solution, but the physics of the process may ultimately limit the economics and the scale at which this technology can be deployed.

Biomass with Carbon Capture and Storage (BECCS) involves growing biomass like trees or crops as fuel to provide heat or electricity, whilst capturing the CO_2 emissions from the process. The CO_2 is then stored underground. The net result makes possible a continual CO_2 drawdown from the atmosphere, whilst providing a source of energy. The technology already exists and there are a number of such plants operating around the world. Unfortunately, the extra energy and equipment to capture and store the CO_2 adds around $40 per tonne CO_2 or $2-3c per kWh to already expensive biomass electricity, creating a cost that is a significant

---

xii We use $2c per kWh as the future cost of cheap wholesale electricity generation based on wind and solar, we will explore why in the following chapters.

xiii For 410 ppm or 0.04% of CO_2 molecules in the air it requires a theoretical minimum of 0.13 kWh energy per kg to capture, and 0.07 kWh per kg to compress, the CO_2 gas.

xiv Carbon Engineering estimate the facility will use around 1.8 kWh of energy per tonne (which is 11% of theoretical best). A one million tonne per year facility will cost $700 million. Using cheap solar electricity ($2 per kWh), the energy cost would be $36 per tonne CO_2, the build cost is around $50-60 per tonne over the life of the facility and operating costs add another $25 per tonne. Plus $5-10 per tonne for transport and storage.

hurdle to widespread adoption.[34] Other issues include competition for land, CO_2 which escapes capture (>20%), and difficulties or risks around building out sufficient storage infrastructure.[35]

Assuming we could pump CO_2 back into the ground as fast as we can extract liquefied oil and gas, then 31 million cubic metres of CO_2 could be injected per day (25 Wembley stadiums) – up to a maximum of 7 billion tonnes CO_2 per year or 14% of today's annual emissions.

Solar Radiation Management: Limiting Solar Absorption

These are a range of strategies which could reduce the amount of solar energy absorbed by the planet, reduce radiative forcing, and keep temperatures cooler.[36]

Stratospheric Aerosol strategies propose pumping aerosol particles into the upper atmosphere to reflect the light before it hits the planet. Sulphur dioxide or hydrogen sulphide are the leading candidates. They are both gases that would oxidise into fine sulphate particles in the atmosphere. Sulphates have already proven effective at cooling the planet after large volcanic eruptions. The last major eruption was Mount Pinatubo in the Philippines in 1991 which ejected 20 million tonnes of sulphur dioxide and dropped global temperatures by 0.5°C for the next two years.[37] The lower stratosphere already holds a layer of sulphate aerosol, so the challenge is to increase the concentration and reflect enough sunlight to offset the increases to radiative forcing.[xv] Sulphur is cheap, and this is important as the process would require 10-20 million tonnes per year. Planes could also deliver the gas into the stratosphere at a cost of less than $30 per kg. So, with a global annual cost in the hundreds of billions of USD this strategy is extremely cheap – the equivalent of $1c per tonne of CO_2 equivalent or $0.0005 cents per kWh of fossil fuel energy offset. However, stratospheric aerosols will not prevent ocean acidification and the process must go on forever. If for any reason – such as war, terror, or technical problems – the program had to be halted, however, rapid temperature increases would occur. It is also unknown how regional effects and weather patterns would be impacted, or even what the sulphates might do to the ozone.

Solar Reflectors could be deployed into orbit above Earth and used to reflect the sun's energy before it hits the planet. This could be an array of 55,000 low orbit space mirrors (100 metre squared each), 2 billion tonnes of reflective dust in a ring formation around the planet,

---

xv Radiative forcing will be around 7 watt per square metre or about 2% of incoming solar radiation by the end of this century in our baseline of inaction.

or 3 million square km of sunshades in higher orbit. These proposals are highly speculative, however, and would take decades to deploy and again, if for any reason they stopped working, rapid warming of the planet would ensue. This method would likely also cool the tropics more than higher latitudes so again create changing weather systems. Oceans would still become more acidic. The cost of deployment would be high upfront, though if nothing went wrong, this approach would likely be cheaper than many other options over time.

Cloud Whitening involves pumping fine particles into the air just above the ocean. These so-called cloud-condensation-nuclei attract water and form reflective clouds above the ocean surface. The fine particles could be formed from sea-salt, delivered using modified boats. To reverse warming this century it is estimated that over 1,500 ships would need to continually navigate the ocean pumping out salt. The strategy could be deployed relatively quickly and adjusted as required; it could also be optimised by region and should prove just as cheap as stratospheric aerosols or space mirrors (a few cents per tonne CO_2 offset). But again, there are caveats: this route does not prevent ocean acidification and given the complexity of cloud formation and interaction, it has many unknowns.

High Albedo Surface increasing the reflection of the sun's incoming energy from land surfaces is another option. Urban areas cover less than 2% of land on Earth so painting everything white would reflect just 0.18 watts per square metre or 2.5% of the energy needed to keep the planet cool this century and would prove extremely expensive. Choosing more reflective crops and increasing the albedo of grassland and savannah may fare a little better as these areas cover over 30% of land, with estimates of up to 1 watt per square metre cooling (14% of required reflection), but this system would completely alter the natural biodiversity. Desert reflectors offer the prospect of more reflection still. Deserts cover 6% of land but receive a higher proportion of sunlight through the year: cover these areas in shiny sheeting and you could reflect enough sunlight to offset a doubling of CO_2 (half required reflection this century). The cost would be comparable to ocean nutrition or enhanced weathering at $40 per tonne CO_2 equivalent.

Negative Emissions Technologies range from low risk, low cost, but limited scale options like tree planting, to potentially large scale but high cost and uncertain options like direct air capture or ocean nutrition. Negative emissions strategies will certainly play a role in our response to climate change, but the mix of options and scale of deployment are still to be decided.

Solar Radiation Management is likely significantly cheaper than negative emissions technology but provides only a partial solution, one which may offset temperature increases but not ocean acidity. Reflection of the sun's energy from land is the lower risk option but more expensive and probably of minimal impact. Atmospheric or space-based reflection is in principle cheap, effective, and scalable but also very high risk. We don't know if these strategies can be executed or if they will work as expected. We have shown how difficult it is to model even the most basic elements of a complex system like planet Earth. Imagine trying to predict all the permutations of engineering the climate. We have no way of rigorously testing these ideas and the first real experiments would not be in a lab but would take place through large scale global deployments in the wild. No test runs, no trial and error, we would have one shot to get it right. As climate scientist Kate Marvel puts it, "if the future is to be unmitigated greenhouse gases countered by sun-blocking particles, if we mask the hangover but continue the bender, we have to be resigned to losing even more of the world we know".[38]

Whilst climate engineering appears to be a quick and cheap fix, it runs seriously high risks and neither route solves air pollution, nor does solar management solve ocean acidification. If for any reason the engineering failed – due to political squabbles, terrorism, or technical problems – the Earth would warm to the temperature it would otherwise have reached but within just 10-20 years, not nearly enough time for humans or the natural world to adapt.

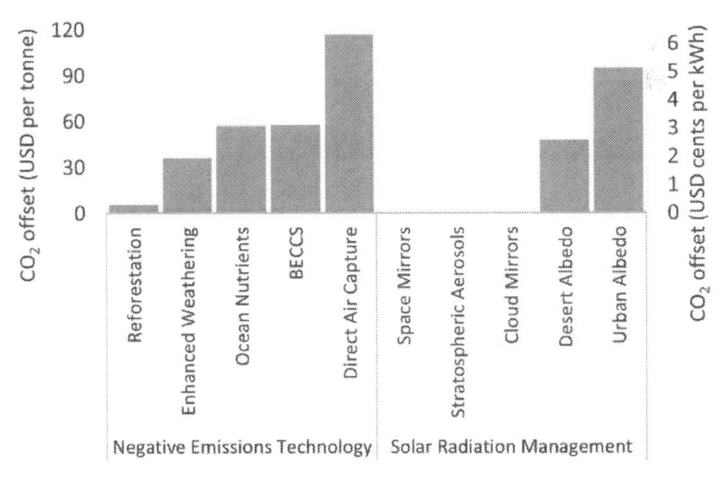

Estimated costs of Negative Emissions Technology (NET) and Solar Radiation Management (SRM) in USD$ per tonne of CO_2 reduction (left hand axis), or in USD$ cents per kWh energy offset equivalent (based on today's average of 1,850 kWh final energy provided per tonne of CO_2e) (right hand axis). Our calculations on NET and data from The Royal Society[36] *for SRM.*

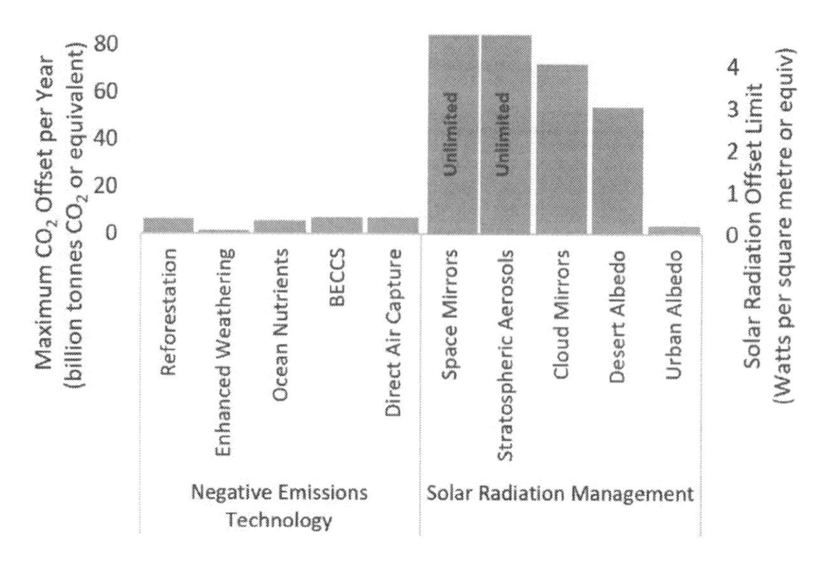

Estimated limit of NET and SRM. Left Axis: Annual CO_2 offset in billion tonnes for NET or equivalent CO_2 offset for SRM (average of next 80 years). Right Axis shows limit for solar radiation offset for SRM and equivalent for NET (80 years average). Our calculations on NET and data from The Royal Society[36] for SRM.

Mitigation and the Kaya Identity: Reducing and Eliminating Greenhouse Gas Emissions

Mitigation is our third and final strategy which, unlike adaptation or climate engineering, assumes prevention is better than cure. Mitigation is the process where we make the necessary changes to reduce and eventually eliminate ongoing greenhouse gas emissions. This slows and eventually stops the accumulation of CO_2e in the atmosphere which in turn slows global warming and ocean acidification to within safe limits.

But how do we begin to choose the best route to mitigate greenhouse gases? To better understand our options, we can use a formalised version of Paul Ehrlich's I=PAT equation called the Kaya Identity.[39] The expression simply states that emissions of greenhouse gases are the product of the population, GDP per person, energy efficiency, and emissions intensity.

Over the last 50 years annual CO_2e emissions have increased nearly 2.5 times, rising at the same speed as population. However, break the contribution down into the elements of the Kaya Identity and you quickly realise emissions could have grown significantly faster than population. Through the period consumption increased 4.5 times per person, so if it were not for large efficiency improvements and gradual declines in emissions intensity, then CO_2e release would be significantly higher.

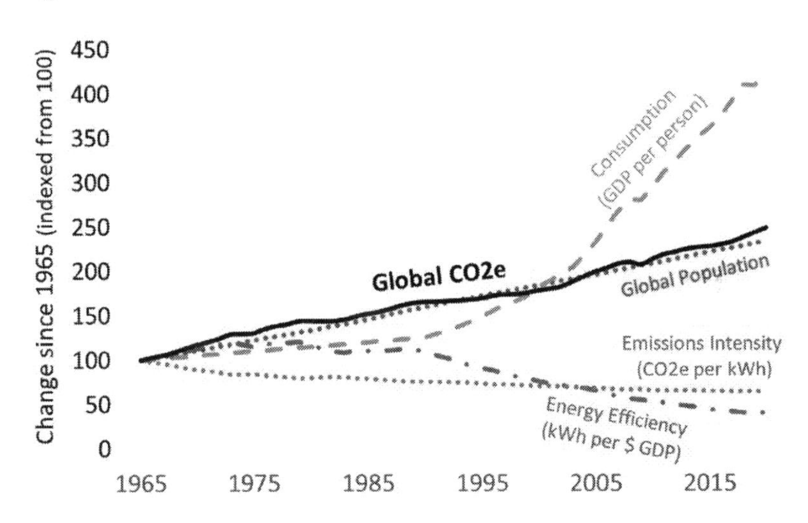

Change in population (dotted), GDP per capita (dashed), energy per $GDP (dash-dot), CO_2 emissions per unit energy (small dot) and total CO_2 emissions per year (solid) for the world since 1965. Calculated from data sources[40,41,42,43]

So, which of our levers will be the most effective route to mitigating worse case climate outcomes in the future? Population and consumption have continuously gone the wrong way. Energy efficiency has been the biggest offset but has physical limits to improvement. Declining carbon intensity has proven only a small offset so far, but it is the only solution capable of reaching zero emissions. Let's take our model from chapter three, which calculates global temperature increases based on human emissions, and pull each lever to get a better understanding.

In a world of exponential consumption with 10 billion people consuming 2% more every year, but with no improvement to technology, global warming could reach 10ºC by the year 2200. This will serve as a simple benchmark to explore the ability of each of the Kaya levers to control emissions and limit future temperature rises:

Population Controls and the Kaya Identity

First let's look at population growth. Statistics show that the number of children a woman will bear over her life is not a function of culture or religion but most closely correlated with wealth. Rising incomes create better education, health, nutrition, social care, and the potential for greater equality[xvi], thus reducing the risk of child mortality and the need for extra labour or social care in later life. So long as the money finds its way to those who need it, fertility rates will begin to decline about one generation after income begins to increase.

Wealthy countries already have stable population numbers and if wealth in poorer areas of the world continues to increase, more families will climb into the middle-income brackets where reproductive rates rapidly drop towards two children (above $5,000 income) and populations stabilise. The world is already nearing peak children with two billion below the age of 15 and the United Nations estimates a global peak population of between 9 and 13 billion people by the end of this century.

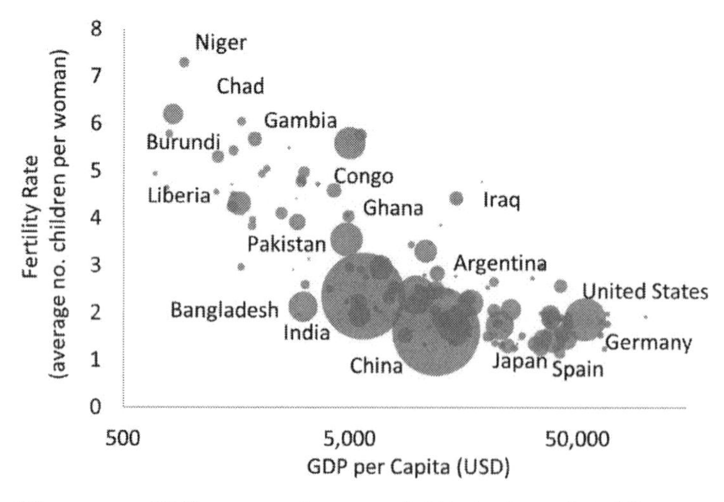

Fertility rate versus GDP per person by country (bubble size is population).[44]

xvi The relationship between equality and wealth is more complex than many other measures of social progress. Thomas Piketty notes, in his book Capital in the 21st Century, that although Malthus and Marx may have devised their own theories on inequality, the first rigorous analysis came in 1953 when Simon Kuznets pieced together the evolution of income distribution in the US since 1913. The Kuznets curve showed a bell-shaped relationship of decreasing then increasing equality through industrialisation and economic development. However, Piketty argues that deeper and broader analysis of global records shows a more nuanced story where wealth and education have the potential to drive greater equality, but warns the process is easily derailed if growth slows and no actions are taken to limit capital returns.

But what if we could speed up the process? How effective is population control for avoiding disastrous climate change outcomes? Historically, direct intervention has proven ineffective and is fraught with moral dilemmas.

India is an interesting case study on population and has also weighed in heavily as an influence on both Paul Ehrlich and Julian Simon's views of the world. Ehrlich visited Delhi in the mid-60s at the height of a food crisis, writing, "The streets seem alive with people. People eating, people watching, people sleeping. People visiting, arguing and screaming. People thrusting their hands through the taxi window, begging. People defecating and urinating. People clinging to buses. People herding animals. People, people, people, people." This became Ehrlich's vision of a world with unchecked population growth and writing in *New Scientist* magazine he warned that the US should not send food aid to countries such as India without attaching population control conditions, "where dispassionate analysis indicates that the unbalance between food and population is hopeless".[45]

Concerns over the growing size of India's population had already been resonating for many decades and India was the first developing country to adopt a state-sponsored family planning program in 1952. The policy began as a voluntary, information-led campaign but in the first two decades had had little effect on slowing down population increases. Following growing international concerns stoked by Ehrlich and others, the Indian government was under pressure to implement stricter population control in return for continued food aid.

In 1969, the National Population Council of India invited a young academic over to help champion the benefits of family planning. Julian Simon had spent the first decade of his academic career at the University of Illinois in the advertising department where he had started combining his ideas of population economics with marketing. His early convictions supported the benefits of a smaller population and were well aligned with Ehrlich. Simon threw himself into understanding the cost benefits of family planning in India and studied more and more population data. But the deeper he delved the less sure he became of his initial assumptions concerning relations between income, education, and birth rates. By the 1970s Simon had flipped the idea of population control on its head and declared that "moderate population growth produces considerably better economic performance".

However, India had ploughed ahead with stricter controls and by the early 1970s the voluntary family planning campaign had evolved into a forced sterilisation program for men with two children or more.

The policy proved hugely unpopular and after the government fell, population planning was firmly off the political agenda.

Around this time China also began targeted family planning – in the 1960s and 70s – initially targeting two children per couple by incentivising maternity leave, childcare, housing, and increasing availability of contraceptives, sterilisation, and abortion. However, by the late 1970s, newly completed surveys indicated that the population had already hit one billion, 10% more than the government had thought. The policy was quickly shifted to one child per couple in a radical attempt to actually reduce population. The campaign was enforced not only through incentives, but also with steep fines, and led to many abuses of human rights, including forced abortions and female infanticide. Opinion is mixed on the results given that China's population has grown at a remarkably similar rate to many other Asian countries.

But for a moment let's ignore the moral and practical implications and simply assume that from now on each woman in the world bears just one child. The population would decline to around four billion by the start of the next century, emissions would decline proportionally and, because of the decrease in food demand, nearly half of all agricultural land could be reforested. This would slow warming, but if each remaining person continued to increase consumption at 2% per year, the difference would be slight. Emissions would still climb to hundreds of billions of tonnes CO_2e per year and temperatures would still reach 6-7°C higher by the year 2200.

Consumption Controls and the Kaya Identity

So, if direct population control is ineffective, and population will limit itself anyway, then maybe setting a limit on consumption might be the better idea? We could reduce today's average global income by one quarter and distribute the wealth evenly across the planet. Everyone lives on $14,000 per year income.[xvii] For the top 10% this means *reducing* our demand by driving less, flying less, heating or cooling our homes less, buying fewer goods and services, and *switching* expenditure to lower emissions goods and services. Assuming no further improvements to technology, this action could hold emissions flat. The planet would still warm 4°C by the end of the next century. Better than population controls, perhaps, but we still risk dangerous tipping points and widespread damages.

---

xvii $14,000 in purchase price parity down from $18,000 today. $9,000 in absolute USD down from $12,000 today though this is less relevant because prices should equilibrate in an equal world.

Efficiency Gains and the Kaya Identity

What about technology? Can we go on exponentially consuming more goods and services but rely on markets and technology to become more efficient? Energy efficiency has been the biggest emissions offset for the last 50 years; in fact, we have improved efficiency by 1-2% per year. The fundamental basis for efficiency improvements is to maximise the conversion of energy to useful work rather than wasteful work, and to reduce waste resources as we produce goods, food and services. However, even if we project continued efficiency improvements long into the future, emissions eventually increase 2-3 times over with concomitant warming of 5ºC by the end of the next century.

Emissions Intensity Controls and the Kaya Identity

That leaves carbon intensity. What if we switched to zero emissions production of energy, goods, and services over the next 30 years whilst still letting consumption grow exponentially? We could replace fossil fuel energy with net-zero emissions energy from renewables, fossil fuel with carbon capture, or nuclear. We could transform our appliances to run on that net-zero energy. Assuming emissions from transport, industry, and basic amenities could reach net-zero in three decades, while emissions from the use of agricultural land continues until peak population and peak food consumption, then global warming would be limited to 2.5ºC. The best single option it would seem, but still one that risks dangerous tipping points.

There is no single lever which keeps global warming below the temperatures that risk unmanageable change. Population control is too slow, morally difficult, and practically challenging and as Nobel laureate Amartya Sen writes in 'Development as Freedom', "The solution of the population problem calls for more freedom, not less." Improve the lives of the poorest and population will reach a natural limit.[46]

Limiting consumption is perhaps a better option but will always leave residual emissions and it requires enormous economic, political, and cultural transformation. Even combining consumption limits and population control would still lead to 2-3ºC of warming.

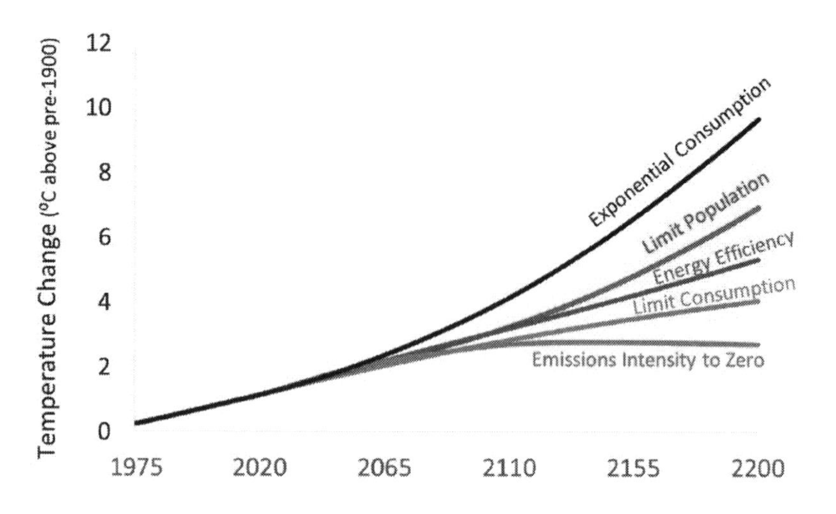

Different temperature increase forecasts based on our simplified radiative forcing model from chapter three and applying the four levers of our Kaya Identity. Exponential consumption is the benchmark with peak population of 10 billion, 2% compound consumption growth and no change to efficiency or emissions intensity. Limit Population: Fertility rates are set to 1 and population halves by next century, agricultural land is freed up and reforested, consumption growth continues per person at 2%. Energy Efficiency: Energy and agricultural efficiency are improved at a rate of 1.5% per year, consumption growth still compounds at 2%. Limit Consumption: GDP per person is set at $14,000, 10 billion people, no improvement to efficiency or emissions intensity. Emissions Intensity to Zero: emissions intensity drops to net-zero in 30 years, agricultural land use still grows until peak population, peak food consumption.

Energy efficiency has proven the biggest offset to date, but efficiency improvements decline exponentially with physical limits and will always become overwhelmed by uncapped exponential consumption increases. Cutting emissions intensity to zero through technological change is the closest we come to a low-risk solution, but 30 years is faster than any energy transition in history (that is currently 50-100 years) and without the accompanying changes to agricultural practices, land management, and food consumption, dangerous limits will still be breached.

The path of least resistance will use a combination of all the Kaya levers to create an optimal mitigation solution. Population should self-limit with increasing wealth. Energy efficiency trends must continue and accelerate to buy us more time. Consumption reduction is required in areas such as agriculture. Reducing emissions intensity provides full decarbonisation.

Mitigation using all levers of our Kaya identity provides the only complete and low risk strategy to climate change and air pollution. Any transition plan to net-zero must make best use of each.

Climate Duty and Net-Zero: The Evidence for Action

Nothing in science can ever be irrefutably proven, but the evidence for human led climate change is about as close as it comes. Scientists have provided the fundamental theory and computational models of the greenhouse effect, gathered robust evidence of the physical changes, and have reached near complete consensus that humans are the cause.[xviii] Economists have added up the socio-economic impacts and agree that warming above 1-2ºC will lead to exponentially increasing death, damage, and destruction. We have the tools to solve the problem and regardless of political targets, carbon budgets, or locked-in warming, it is never too late to improve the outcome. Acting on climate change is no longer optional but our moral duty.

We must ensure the best possible outcome in the second half of this century and beyond, for ourselves, our children, and our children's children. Just as we have a moral duty to each person alive on the planet today, so we must recognise our moral duty to those in the future. We wouldn't sit back and let a nuclear bomb obliterate a third of the world today just because it was happening somewhere else, and so we shouldn't let climate change create an inhospitable planet for future generations. And just as we can't precisely estimate the damage from an atomic bomb, so we can't exactly predict the outcome of climate change, but neither is a credible reason for letting the destruction happen. And just as Fermi's back-of-the envelope estimates could gauge nuclear blast strength, so we can gauge the best course of action on climate change even without perfect information and foresight.[47]

But who do we need to fill the role of planetary caretaker? Do we need town planners to start shifting high risk communities as we adapt? Do we need Silicon Valley to re-engineer the climate? Do we need governments to reconfigure the agricultural system? Or industrialists to improve efficiencies and build out a net-zero energy system?

We need all of the above and much more. We have already locked in nearly 1.5ºC warming unless CO_2 concentrations are brought down, so further adaptation is already necessary. Climate engineering solutions could help us buy more time but offer no solution for ocean acidification or air pollution. Adaptation and engineering may appear cheaper and easier but so is doing nothing, and all three risk unknowable damage. Mitigation is the only low risk strategy to offer a complete solution. It

xviii 97% or more of actively publishing climate scientists agree that warming trends over the past century are extremely likely due to human activity. Nasa, "Scientific Consensus: Earth's Climate is Warming", 2019.

is the safest and most effective route, but one that also bears the biggest upfront cost and equally upfront socio-economic change.

In the following chapters we'll build out a more detailed solution to climate change and air pollution. The transition to a net-zero emissions system will be led by lower emissions intensity and supported by efficiency improvements and targeted consumption changes. Negative Emissions Technology will be used as an option of last resort for hard-to-mitigate areas of the economy given the physical limitations or costs.

We will examine the technical feasibility, the relative cost, and the speed at which the global economy can transition to net-zero. We will weigh the upfront cost of mitigation against the long-term damages and risks of inaction.

Curriculum vitae polished, interview nailed, and it is day one on the job. Time to roll up our sleeves and get to work . . .

Chapter 7: Sustainable Energy
Supply Options with Net-Zero Emissions

In this chapter we will review all possible sources of energy supply; from fossil fuels, to nuclear, to renewables across both electricity and heating, to better understand how we can create a sustainable energy supply with net-zero emissions. We will compare lifecycle CO_2e, land-use, safety, reliability, remaining resources, efficiency, water-use, and costs both today and once each energy supply technology has been scaled up.

Fundamentals of Energy

Energy is the capacity for doing work. It comes in many forms such as the gravitational potential of water in reservoirs, the kinetic energy of a spinning fly wheel, chemical energy stored in fossil fuels, thermal energy of vibrating molecules, or atomic energy used in nuclear power. Energy can flow between different forms and can be put to work where needed for lighting, heating, mechanical movement, and in building the proteins which create life. The equation fundamental to understanding all energy is the most famous equation in modern science: $E=MC^2$. Einstein showed us energy equals mass times the speed of light squared.[1] Energy and mass are interchangeable: one can be converted to the other.

The sun is a great example of this special relationship. A big nuclear reactor with large quantities of gas under extreme temperature and pressure. The sun squeezes hydrogen atoms together and through nuclear fusion creates helium. The outgoing helium weighs slightly less than the ingoing hydrogen. The lost mass is converted to pure energy and released into the universe as electromagnetic radiation. This is the very same radiation which warms the Earth and is essential to life on this planet. Every one-kilogram of hydrogen mass releases 180 million kWh of energy.[2] If we were able to harness fusion reactions on Earth, it would take less than 1,000 tonnes of hydrogen to power the world each year.

The sun is also responsible for most of the energy we source here on Earth. The sun's energy drives rainfall for hydroelectric dams and creates temperature differences which drives the wind for turbines. Sunlight is also the source of energy harnessed by plants over millions of years which formed the coal, oil, and gas we extract and burn.

The total power of sunlight reaching the planet's surface is over 85 trillion kW which is 7,000 times more than the 12 billion kW used by humans to power everyday life.[3] To radiate this amount of power across the universe, the sun consumes 600 million tonnes of hydrogen per second, creating the equivalent of 4 billion tonnes of light energy, an efficiency of 0.7%.[i]

Nuclear reactors on Earth use fission rather than fusion. The reactions begin with uranium-235 which splits into smaller atoms of krypton and barium and in the process transforms 0.1% of the mass into energy.[4] Fusion is seven times less efficient than the sun's fission but still generates 22 million kWh of energy per kg of U-235.[5] It would take less than 10,000 tonnes of pure uranium-235 to power the planet each year.

Burn fossil fuels and you are also taking advantage of Einstein's $E=MC^2$. During combustion, carbon atoms in gas, petrol, or coal, combine with oxygen to form carbon dioxide and 0.00000005% of the carbon mass is converted to energy, releasing 13 kWh per kg. This is 14 million times less efficient than fusion and is the reason we burn through 12 billion tonnes of fossil fuels each year.

Despite the low conversion efficiency, combustion does have other attributes in its favour. Fossil fuels have proven readily available, energy dense, easily transportable, and with a relatively easy route to release and harness the energy (combustion). Despite the low efficiency, fossil fuels supply 80% of energy today.

Whilst early humans relied upon sun, wind, and water for energy, the industrialised world today relies mostly on coal, oil, and gas. But as knowledge has exponentially increased, so has our ability to harness other energy sources more effectively. The US army's Manhattan Project, led by Robert Oppenheimer between 1939 and 1945, developed the first transportable atomic bombs used to devastating effect in Hiroshima and Nagasaki, Japan, in 1945. But advancements in the understanding of particle physics and nuclear fission had also helped Enrico Fermi and colleagues at the University of Chicago develop the first self-sustaining

---

i Sunlight hits the Earth's surface with an average of 170 watt per square metre after reflective losses. The Earth's surface area is 510,000 billion square metres so the total power reaching the planet is over 85,000 billion kW. Every second the sun consumes 600 million tonnes of hydrogen and transforms it into 596 million tonnes of helium and 4 million tonnes of light. That is 100 billion-billion kWh energy per second. Luckily, the sun is so large it will still shine for another 5 billion years.

nuclear reactor in 1942.[6] During the same decade, Bell Laboratories demonstrated the first practical silicon solar cell designs which found their first commercial applications in outer space on the Vanguard 1 satellite in 1958.[7,8] The oil shortages of the 1970s were the driving force behind the development of modern wind turbines. The efficiencies of these technologies have steadily improved as costs have exponentially declined. Today we have an abundance of competitive energy technology options. But the challenge of tomorrow is how to transition towards a more sustainable energy system.

Can we use the knowledge and technologies afforded us by burning fossil fuels over the last 200 years to transition to a net-zero future? This chapter will explore the main possibilities for energy generation from atomic, to fossil fuel, to renewables, in order to build an understanding of how we can switch the world to a zero-carbon energy supply and at what social and economic cost. Before we look at each technology, let's take a moment to run through two key concepts for understanding energy economics: levelised cost and experience curves.

Levelised Cost of Energy (LCOE): Different Technologies on a Like for Like Basis

In chapter five we ran through the idea of time value of money and discounting future damages. These same concepts must also be applied to the cost of energy supply. The levelised cost of energy represents the total cost of an energy installation in today's money (net present value) divided by the total useful energy it provides through its life.

There are three main parts to calculating the cost of an energy installation: capital costs, operating costs, and fuel cost. Capital cost is the one-off total cost of building and financing the facility. Operating costs are the ongoing cost of maintenance, labour, insurance, and overheads paid, regardless of how often the plant is run. Fuel is the cost of the energy source and the amount required depends on the output of the plant.[9]

Levelised Cost of Energy calculations are used to compare the cost of different technologies. For example, a coal electricity plant is five times cheaper to build than a nuclear plant but will last only half as long – LCOE can tell us which is cheaper. Solar or wind farms cost twice as much as natural gas power plants upfront but have no ongoing fuel costs because the wind and sun are free. LCOE lets us compare these different technologies on a like for like basis and represents the wholesale cost of generation which utilities companies charge us in our energy bills.

Experience Curves: Estimating the Future Price of Energy based on Scale not Time

The term **experience curve** was coined by the management consultant BCG in 1966 while they were performing a cost analysis for a major semiconductor manufacturer.[10] The theory held that every time the accumulated volume of computer chips doubled, the company's unit production cost would decline by 20-30% because of the accumulated "experience" of manufacture. It turned out that not only did the experience curve work for the semiconductor industry, but the concept could also be applied nearly anywhere.

For most industries, the cost of production declines as the scale of production increases, mostly thanks to improved manufacturing technologies, the division and specialisation of labour, and the lower allocation of overhead costs per unit production. A larger industry also attracts more competitors vying for a greater share of the market which pushes companies either to improve products or lower prices and recoup the difference through better production techniques or accepting lower profit margins.

The idea of an experience curve can be applied to most manufactured products but the rate at which the price drops will vary. Generally, cost of production will fall faster for highly standardised, repeatable, modular products. It will fall slower where there is a greater degree of variability, customisation, or batch processing in the design. The price will follow the cost of production unless supply and demand imbalances or consumer choice or marketing ploys intervene.

Prices decline exponentially, with the greatest falls up front and then price reductions getting smaller and smaller as they trend towards a floor price. This represents the rock bottom cost of extracting and processing the materials with efficiencies near physical limits.

We will use the idea of experience curves to explore the price of energy technologies at greater scales of production. **The idea that prices decline with accumulated experience and not just the passing of time is key to understanding the economics of a net-zero transition.**

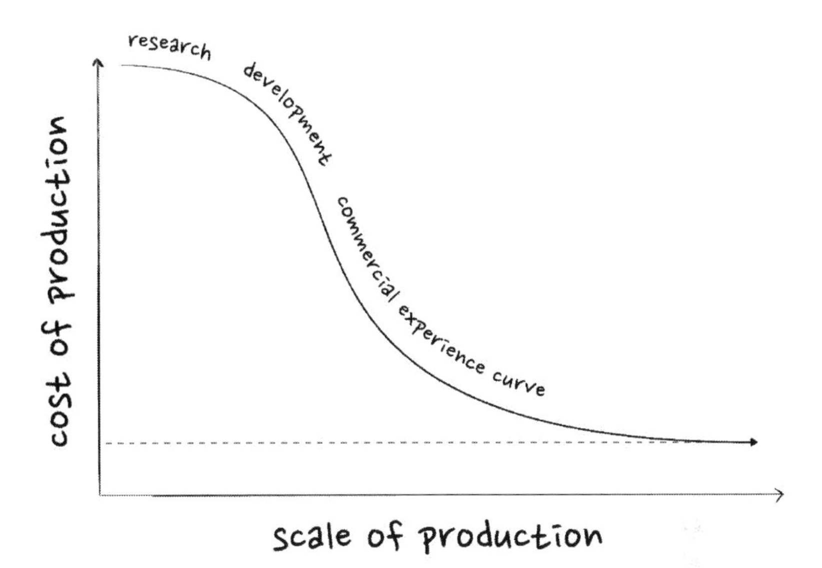

Energy Sources and Technology Options for a Net-Zero Carbon Energy Supply

Energy is fundamental to all life. Three basic elements of a successful supply network are reliability, safety and cost. In this section we will summarise the key characteristics of different energy sources to understand which can fulfil our needs whilst decoupling carbon emissions.

Here are our options:[11,12]

Coal energy originates from the sun, captured by the plants and animals that die, become buried, and transform into bonded carbon atoms. We harness the stored energy by breaking these carbon bonds and reforming carbon dioxide using the combustion reaction. The heat energy released can be used directly or can be used to heat water into steam, spin a turbine and create electricity. Over three quarters of coal is used in electricity generation and the remaining quarter is used directly in industrial heating/furnaces and steel making.

Natural Gas is a fossil fuel comprised mostly of methane (CH_4) formed from decomposing plant and animal matter under extreme heat and pressure over millions of years. Natural gas is usually found associated with coal and oil deposits or trapped in underground rock formations. Natural gas can be burnt directly for heating or can be used to generate electricity. Gas power plants inject the natural gas directly into the turbine which allows for a higher temperature of operation and higher efficiency. Half of natural gas is used in electricity generation, 25% in industrial manufacturing and 25% in space heating, water heating, and cooking.

Crude Oil is another fossil fuel containing carbon and hydrogen atoms. The carbon chains form in various lengths and can be separated into different fuels such as gasoline, diesel, and kerosene. Oil can store lots of energy in a small volume.[ii] This high energy density means oil is mostly used for powering cars, trucks, ships, and planes so they can travel as far as possible on a single refill. Oil energy is harnessed using the internal combustion engine where the reaction with air in a combustion chamber generates hot gases of CO_2 and water. The expansion of these gases drives pistons which are used to turn the wheels of the vehicle. Over two thirds of oil is used in transport.

Nuclear Fission derives its energy from the splitting of Uranium-235 into smaller atoms, releasing neutrons, which further propagate the reaction, releasing more and more heat in the process. The heat energy is used to boil water into steam and drive a turbine. The reaction is controlled by rods of carbon which can be lowered into the fuel to absorb more neutrons and slow the process. Uranium-235 makes up less than 1% of naturally occurring uranium so must be enriched or purified to create nuclear fuel. Nuclear fission is nearly 2 million times more efficient than combustion so just one kg of enriched Uranium (3% U-235) can deliver one million kWh of thermal energy. This is the equivalent of 600 barrels of oil, 100 tonnes of coal or 100,000 m^3 of natural gas. Nuclear power is predominantly used for electricity generation.

Hydroelectric Power derives its energy from the flow of water. Hydropower or water wheels have been used for hundreds of years to perform mechanical tasks like grinding and pumping. The first hydroelectric design was developed by William Armstrong in Cragside, Northumberland in 1878, and was used to power electric lights for his art collection. Hydroelectric power can be captured in a variety of ways – simply using the flow of a river to turn a water wheel or more complex dam systems which channel the rainwater into a reservoir which, when needed, flows down a tunnel to drive an electric turbine. These methods take advantage of the gravitational energy of water at height.

---

ii Oil stores over 10,000 kWh of potential combustion energy per cubic metre compared to coal at 7,000 kWh per cubic metre and liquid natural gas at 2,000 kWh per cubic metre.

Geothermal is energy extracted from heat below the Earth's surface. The heat originates from conductive transfer from the Earth's hot centre, from radioactive decay in the Earth's crust, and from solar radiation from above. The highest temperature areas get extra heat from nearby volcanic activity. High temperature ground can provide steam for electricity generation and low temperature heat from rock or soils in most locations can be concentrated to provide hot water and heating for residential properties.

Solar, like most other forms of power, derives its energy from the sun. However, unlike fossil fuels or hydroelectric, solar power cuts out the middleman. There is no intermediate process like growing trees, forming fossil fuels, driving rain or wind. Solar uses the sun's energy directly. There are three types of modern solar energy: **Solar Thermal** uses black panels to absorb the entire spectrum of the sun's energy to heat and deliver hot water. **Concentrated Solar Power** uses large arrays of reflectors to focus the sun's energy to heat a central tower that is used to generate steam and drive an electric turbine. **Solar PV** uses two layers of silicon to create a one-way junction for electrons. When the sun's visible light strikes the junction, it provides enough energy to force electrons across the layers and drive an electrical current.

Wind Power is generated when the sun's energy is distributed unevenly across land, sea, and mountains, creating cold, dense high-pressure areas and warm low-pressure areas. The molecules in the air move to balance these differences creating anything from a light breeze to gale force winds. Wind turbines harness the kinetic energy of moving air using large blades which rotate, spin a turbine, and generate electricity. The first wind turbine was built by James Blyth in Glasgow 1887.

BioEnergy is derived from biological sources. Biomass includes any organic material which has captured energy from sunlight in the recent past and stored it in chemical form. Biomass includes wood, crops, crop waste, non-edible plant, and manure. Biofuel is the biomass in a final form before consumption, such as logs of wood or ethanol from corn. The energy of the fuel is harnessed through combustion. We will class bioenergy in four camps. ***Wood Fuel*** or wood burning has been used by humans for 400,000 years and we still burn 1.3 billion tonnes of dry wood each year, 7% of final energy use. ***First Generation Biofuels*** include corn ethanol, sugarcane ethanol, and biodiesel from products like soybean oil (fat). Corn or sugar is grown on farmers' land, harvested, crushed, and put through an enzyme process to produce ethanol (similar to brewing beer). Biodiesel is produced by a transesterification process to convert fatty oils into diesel.[13] These are classed as first-generation technologies because they consume otherwise edible crop products (food vs fuel) and use valuable cropland. ***Second Generation Biofuels*** are made from non-edible biomass such as crop, food, or forest waste, or feedstock like switchgrass which can be grown on non-farm or non-forest land with minimal fertiliser and water needs. The biomass is converted to fuel by heat, chemicals, or enzymes. ***Third Generation Biofuels*** use algae, CO_2, sunlight, and nutrients to produce algae oil which can be turned to biofuel. The algae can be cultivated in either pools of water or photobioreactors in nearly any location and are even tolerant to saltwater or sewage. Third generation technology does not compete with food production and can provide significantly higher energy output than second generation.

Nuclear Fusion is an experimental form of power generation that uses similar reactions to those taking place in the sun. Most reactor designs use hydrogen isotopes contained by lasers or magnetic fields under extreme temperature and pressure to initiate and sustain a fusion reaction to form helium. Benefits of nuclear fusion include no carbon emissions, less radioactivity in operation, little nuclear waste, and essentially unlimited fuel supplies. Research has been ongoing in nuclear fusion since the 1940s, but as yet no commercial reactors have been designed that can sustainably hold the reaction steady for more than a few minutes or generate more energy out than in.[iii] A long running joke amongst scientists is that fusion energy has been just 20 years away for the last 100 years.

---

iii Nuclear fusion remains an active area of research with a 35 nation, $25 billion-dollar, government funded research program called the International Thermonuclear Experimental Reactor (ITER) project ongoing, plus private start-ups from the likes of Jeff Bezos (Amazon), Bill Gates (Microsoft) and Peter Thiel (Paypal, Palantir Technologies) passing the $1.1 billion investment mark. Others view fusion as either impossible or unnecessary. As Elon Musk (Tesla and Space X) puts it: "We've got a giant nuclear reactor in the sky. It shows up every day very reliably. If you can generate energy from solar panels and store it with batteries, you can have it 24 hours a day". From iter.org. and Jonathan Tirone, "Billionaires Chase Space X Moment for the Holy Grail of Energy", Bloomberg News, October 2018.

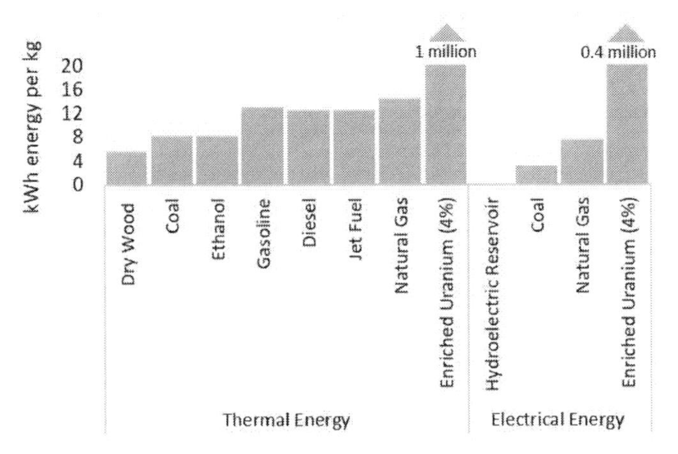

Left of chart: Thermal or heat energy released from the combustion of various carbon-based fuels or the fission of 4% enriched uranium per kg.[14,] *Right of Chart: Electrical energy from 1 kg of water in an average hydroelectric reservoir, 1 kg of coal or gas combustion, or 1 kg of uranium fission.*

Energy Supply: Greenhouse Gas Emissions and Carbon Offsetting[15]

Nuclear, hydro, wind, and solar deliver the most favourable carbon credentials with small, embedded emissions that will trend towards zero once produced from zero carbon energy. Biofuels are considered net-zero but must be approached with caution given the associated use of fertiliser, water, processing energy, and land. Fossil fuels are the primary source of carbon emissions with coal emitting the most CO_2 per kWh but once methane leaks are included, oil and natural gas are not far behind. The only way to keep burning coal, oil, or gas is to offset emissions at-point with carbon capture and storage or using negative emissions technology. None of the offsetting options provide enough scope to cover all of today's fossil fuel emissions and all come with added cost and risk.

Emissions

Burning coal for heat releases 350g of CO_2 for every kWh of thermal energy. Natural gas and oil fare better with around 200g CO_2 per kWh because the added hydrogen atoms release energy but form water not CO_2. But add the leaks[iv] from oil and gas networks and emissions climb

iv Methane or natural gas leaks are estimated at around 1-5% of production. Above 3-4% leakage and natural gas contributes more to global warming than burning coal. From IPCC AR5 WG3 p527.

towards 280g CO_2e per kWh. If the heat energy from coal or natural gas combustion is used to generate electricity then total emissions increase towards 1,000g of CO_2 per kWh electricity because half of the thermal energy is wasted in generation and another 5-10% is lost in electrical distribution.

Nuclear, solar, wind, and hydro emit no CO_2 from generation, but they do require resources and energy during processing and construction: they have embedded emissions. Nuclear requires energy to process the uranium.[16] Solar requires electricity to manufacture the panels.[17,18,19,20,21] Wind turbines use large quantities of concrete, steel, and carbon fibre in construction.[22,23,24] Hydroelectric reservoirs flood vegetation and release methane from the rotting plant matter.[25] However, even assuming the energy of production comes from fossil fuels and including all embedded emissions these technologies emit less than 30g CO_2 per kWh over their lifetime. If we source the production energy net-zero, then lifecycle emissions trend towards zero. In other words, if you power the solar factory with solar power then clean energy creates more clean energy.

Biofuels are, in theory, net-zero carbon. Biomass has grown in the recent past, sequestering CO_2 from the atmosphere which is re-released when burnt. But the use of processing energy, fertiliser, water, and land creates indirect emissions.[26,27] First generation ethanol emits 180g of CO_2 per kWh from fertiliser and energy use. Second generation fuels eliminate the fertiliser but require more energy to break down the tough plant waste, so emissions fare no better. Third generation algae oil biofuels require less processing energy than second generation, but they do use fertiliser. Third generation emissions are around 140g CO_2 per kWh, though yield optimisation could reduce this to 70g.[28]

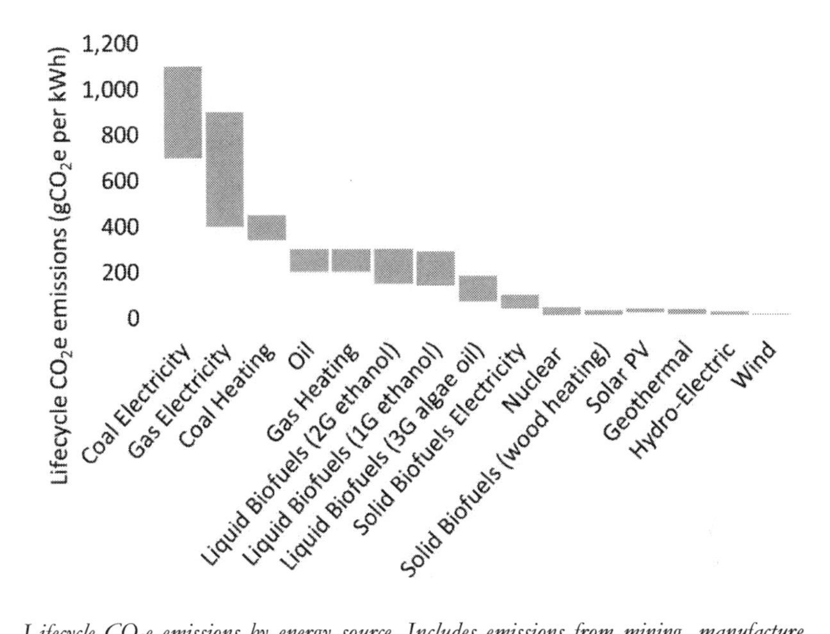

Lifecycle CO_2e emissions by energy source. Includes emissions from mining, manufacture, construction, fuel, losses, waste disposal, and decommissioning where relevant. Based on current fossil fuel centric manufacturing systems and energy sources.[v] *Author's calculations, cross referenced with NREL data.* [29]

---

v Solar PV production uses 4 kg of silicon per kW power. Refining the silicon from quartz emits 18 kg CO_2 from the reaction with coal. Purifying the silicon to thin wafers uses 1,300 kWh of electricity which, on today's global grid (0.5 kg CO_2 per kWh), emits another 650 kg of CO_2. The fixtures, fittings, and electrical equipment add another 700 kg, bringing total emissions to 1,400 kg per kW of installed solar (about 3 panels). The panels will generate 1,500 kWh electricity per year for 30 years which brings total lifecycle emissions to 30 grams of CO_2 per kWh or a payback time of 1.5 years.

A typical 2 MW wind turbine uses 18 tonnes of carbon fibre for the blades, 250 tonnes steel for the tower, 120 tonnes metal for the gearbox and 1,400 tonnes of concrete to secure it in place. Carbon fibre emits 28 tonnes of CO_2 per tonne of material: this ratio is 2 for steel and 0.25 for concrete, a total of nearly 2,000 tonnes of CO_2 emissions. The turbine will run at around 35% of capacity and last for at least 25 years – producing over 150 million kWh of electricity. Lifecycle emissions of 11 grams of CO_2 per kWh electricity. Data from Wind Turbine Models, "Gamesa G90", Data sheet, [online], wind-turbine-models.com, accessed 2020.

First generation (1G) corn ethanol can produce 600 gallons ethanol or 12,000 kWh energy per acre of corn. Growing and processing the corn uses 200 kg of fertiliser and 7,000 kWh of electricity and natural gas. A total of over 2 tonnes of CO_2 per acre or 180 g of CO_2e per kWh. Second generation (2G) cellulosic ethanol made from plant waste, instead of corn, uses 16.5 kWh of processing energy to produce one gallon of ethanol containing 22 kWh energy. So fossil fuels used to process the 2G biomass emit 190g CO_2 per kWh of bio-ethanol energy.

Carbon Offsetting

We ran through Negative Emissions Technologies such as reforestation, enhanced weathering, and air capture in chapter six. These strategies could be used to offset a limited amount of emissions but come with added cost and risk. Another option is to capture the emissions before they hit the atmosphere using at-point-of-use carbon capture and storage.[30]

At-point-of-use carbon capture technology requires between 20-40% of the primary energy to drive the reaction.[vi] Once captured, CO_2 gas must be compressed, moved to a suitable location, and stored for thousands of years.[31] Depleted oil and gas reservoirs offer the best locations. There are enough depleted reservoirs to store 900 billion tonnes of compressed CO_2, the equivalent of 25 years of current emissions. Porous rock formations called aquifers offer a more expensive and riskier[vii] storage option for a further 5,000 billion tonnes of CO_2 or 140 years of current emissions.[32,33,34]

If you store the CO_2 in liquid form at 100 bar pressure, every cubic metre of storage space will hold 0.6 tonnes of carbon dioxide. It would require 160 million m^3 of storage per day to capture all CO_2 emissions across the world. For comparison, today's oil and gas industry extracts liquid fossil fuels at just one fifth of this pump rate with 14 million m^3 of oil and 18 million m^3 of natural gas extracted each day.[viii]

Total cost of at-point-of-use carbon capture and storage works out at around $30-50 per tonne of CO_2 or $2-4c per kWh of energy offset. More expensive than most other negative emission technology options today, but next generation technology has the potential to reduce the efficiency loss to 10%. This could bring total costs of at-point-of-use

vi At-point carbon capture technology falls into three camps: pre-combustion which strips away the carbon from natural gas using the pure hydrogen as fuel; post-combustion which uses a chemical scrubber to remove the CO_2 from the flue gas (3-30% CO_2 concentration) after combustion; and oxy-combustion which involves burning the fossil fuel in recycled CO_2 and pure oxygen, creating an exhaust stream of CO_2 and water which is easily separated. Pre-combustion has better efficiency but higher capital costs and cannot be retrofitted onto existing plants. Post-combustion is the cheaper option and can be retrofitted but has the lowest efficiency. Oxy-combustion can be retrofitted but is more difficult and costly.

vii Care must be taken around the rate at which the CO_2 is pumped: done too quickly and the porous rock can fracture and release the gas.

viii CO_2 rigs are used today to force out the dregs of oil and gas from depleted reservoirs and can inject about 1 million tonnes per year, so a total of 35,000 sites would be required to cover all global CO_2 emissions. This is twenty times the number of oil and gas rigs in existence today.

carbon capture and storage towards \$20-30 per tonne[ix] or an extra \$1.5c per kWh.

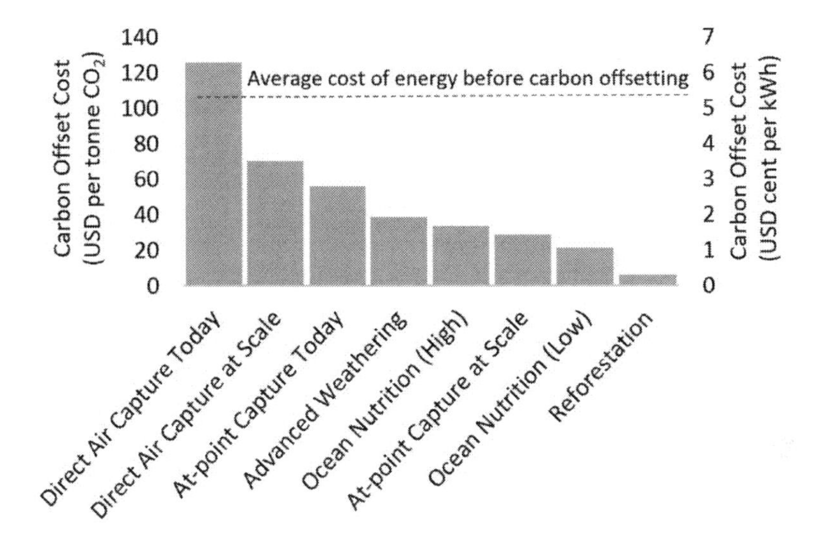

Cost to offset $CO_{2}e$ emissions in USD per tonne $CO_{2}e$ and in equivalent USD cent per kWh final energy supply (based on today's average of 1,850 kWh final energy per tonne of $CO_{2}e$). Average energy prices are \$5.5-6c per kWh today (weighted average of all energy types). Author's calculations based on references in Chapter 6.

Continue to burn fossil fuels as part of the energy supply in a net-zero system and the carbon emissions must be offset. Reforestation is the cheapest option but is limited to seven billion tonnes of CO_2 per year. After accounting for other claims on carbon offsetting from agriculture and industry and it's clear there isn't enough tree planting land available. Fossil fuel energy supply would require at-point-of-use carbon capture on all facilities and would need to fund reforestation and rock weathering to offset the 20%+ emissions that escape capture. Carbon capture and storage could probably handle a maximum of seven billion tonnes of CO_2 per year, limited by the speed at which the gas can be pumped underground.

ix Carbon capture technology can extract 65% of the flue emissions but reduces energy output by more than 20% for capture and another 7% for compression. The equipment costs are \$400-800 per kW and so the total costs are \$30-50 per tonne of CO_2 for capture, plus another \$1-10 per tonne to transport and store. A cost of \$2-4c per kWh to offset carbon emissions at-point.

Energy Supply: Power Density and Potential Output

Power density is a measure of the average power output per unit area used in generation. Once we know the power density of each energy source, we can understand which technologies are limited by space. Centralised power generation using fuel extracted from deep underground takes a negligible land area to provide today's energy supply. Wind and solar are next best, despite having an order of magnitude smaller power density they could power the planet with less than 2% of land. Try powering the planet with biomass and you run into space constraints. It would require half of the world's forests managed for fuel wood or half the world's land for growing biofuel.

Power Density

A typical centralised power plant will occupy around 100,000 square metres of land with a power output of 500 MW[x]. This gives centralised coal, gas, and nuclear[35] energy supply a power density of thousands of watts per square metre.[36] Include the land area used to source the fuel and store the waste,[37,38] and this brings nuclear and gas power into the hundreds of watts per square metre.[39] Coal operations can drop to below 100 watts per square metre if sourcing coal from open cast mines.[40] Still, powering the planet using centralised thermal power plants requires less than 0.1% of land which is why you rarely see a power station today.

Solar power density is a factor of one hundred times lower than thermal power. Sunlight reaches the Earth's surface with an average of 170 watts per square metre and solar PV panels can turn 18-20% into electricity (30 watts per square metre). Once you include the spacing between panels (to avoid shading) the power density is less than 20 watts per square metre.[41] Solar has a significantly lower power density than centralised thermal power, yet solar as the sole source of generation could power the world's energy needs with less than 0.5% of land on Earth. There is more than enough suitable space to power the planet on solar PV and the land can still be used for fruit and vegetable growing, grazing, sustaining pollinators (bees), and biodiversity.

The power density of wind is a factor of ten times lower again. The density of wind turbines is limited by the spacing between units. A gap of five times the rotor blade diameter is required to avoid destroying the wind flow. The Alta wind centre in California is one of the largest

x A 1 kW kettle can boil 1 litre of water in 5 minutes. A 500 MW power plant could boil the 2.5 million litres of water in an Olympic swimming pool in 30 minutes.

onshore wind farms in the world and has a power density of 4 watts per square metre.[42,43] Offshore fares a little better due to stronger winds at sea: Walney in the UK is one of the world's largest such installations and has an average power density of 6 watts per square metre.[44,45] Wind turbines would require a few percent of land or coastal areas to provide enough energy for the global population[xi]. There are more than enough wind resources to power the world and the land or waters can still be used for farming and fishing.

High temperature geothermal provides a similar power density to wind and solar. However, the potential to scale up geothermal for electricity generation is limited by suitable locations which are situated near areas of volcanic activity. Iceland, West Coast US, West Coast Latin America, and East Coast Asia have good siting however the global technical potential is around 200 GW which is less than 2% of final power. Low temperature geothermal can be sourced anywhere by digging 6 metres underground but can only be used for low temperature heating and hot water.[46]

Bioenergy has even lower power density than solar due to the low efficiency of sunlight conversion by plants. Whilst a solar PV panel can convert 20% of the sun's energy into electricity, vegetation typically converts less than 0.5% into biomass providing less than 0.8 watts per square metre. Convert the biomass into liquid fuels and you waste at least half again. Switching the world's energy supply to wood fuel would require 13% of land cover or about half of all the forests on Earth managed for fuel.[47] Bioethanol would need half of all land on Earth for growing corn or switch grass.

Third generation algae oil offers some hope of improving bioenergy power density with strains being developed that capture up to 7% of the sun's energy. If optimised, algae oil could provide the world's energy needs with 3% of land.

Hydro comes bottom of the pile on power density. The Three Gorges Hydroelectric Dam in China is the largest power plant in the world with 22,500 MW peak capacity (equivalent to 50 coal facilities).[48] The

---

xi The Alta wind centre in California covers 80 million square metres, has peak capacity of 1,500 MW but runs on average at 23.5% of peak – a power density of 4 watt per square metre. Walney offshore wind farm covers 73 million square metres, with peak capacity of 1,000 MW, averages 43% of peak and provides 6 watt per square metre power density.

The Earth's surface is 517 million square km with over two thirds ocean and less than one third land. The 150 million square km of land is composed of one third forest, one third agriculture, 20% sparse land, 10% snow or water and 2% urban development.

reservoir covers 1,000 square km, which is an area larger than the UK and it holds 40 billion tonnes of water, hundreds of metres high. The Three Gorges Dam is so large it creates a bulge on the planet which slows the spin of Earth and extends our day by six hundredths of a second.[49] If you include the water catchment area the whole network is nearly 1 million square km.[50] Hydro yields a power density of less than 0.1 watts per square metre.

The low power density and environmental damage caused by building dams will limit hydroelectric potential. Although many existing dams could be converted to produce low carbon power.[xii]

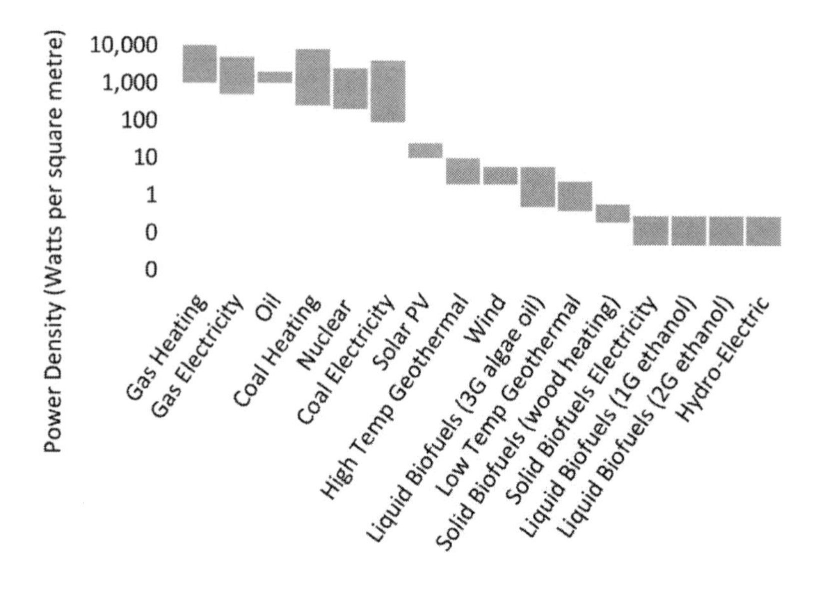

Power density measured in watt power per square metre by energy source. Fossil fuels include wells and mined area. Electric sources refer to utility scale power plants, wind farms and solar farms. Gas and coal heating is based on industrial operations, oil based on motor vehicles, biofuels based on the area required to grow the biomass and hydroelectric the catchment area and reservoir. Author's calculations, cross referenced with John van Zalk et al. [51]

xii The US has 84,000 dams and just 3% have generators.

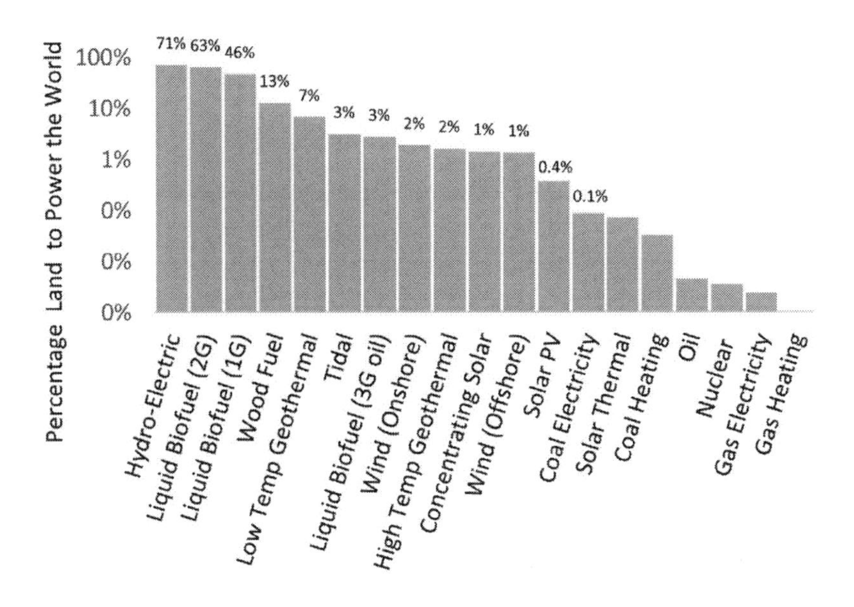

Percentage of global dry land required to provide 12 TW of power by technology. Author's calculations based on average power densities. Land area from FAO.[52]

Energy Supply: Remaining Resources and Sustainability

The cornucopian idea of interchangeable resources has led humanity from burning trees, to burning coal, to burning oil and gas. The switch from wood to fossil fuels preserved what was left of Earth's forests but has come at the expense of Earth's atmosphere and environment. Half a century ago M. King Hubbert pronounced, "It is hard to know which is the more remarkable – that it took 600 million years for the Earth to make its oil, or that it took 300 years to use it up".[53] However, Hubbert's peak oil predictions have yet come to pass, the world has continued to unlock ever greater resources of fossil fuels through technical innovation and has so far ignored the destructive impacts of carbon dioxide.

A particularly inventive, or perhaps reckless, case of innovation was thought up in the late 1960s when American concerns about running low on natural gas were building and the atomic energy commission was desperate to find a more peaceful, practical use for nuclear energy. Enter project Rulison. The Austral Oil Company devised a plan to access the hard-to-reach shale gas deposits deep in the Rulison fields of Colorado by detonating nuclear bombs deep within the mountains. With the greenlight from President Nixon, who was particularly excited at the prospect of nuclear drilling, and after the evacuation of 36 families within five miles of the blast radius, the Atomic Energy Commission dropped a 40-kilioton nuclear weapon – nearly three times as powerful as the Hiroshima bomb – down the well head and detonated. The blast registered 5.5 on the Richter scale. Austral Oil waited patiently while the site cooled before they returned to inspect their not-so-hard work. Initial analysis confirmed the flow rate was nearly ten times greater than conventional natural gas stimulation. The only problem, the resulting radioactivity, meant the product was unusable and the eye watering cost of the nuclear bombs rendered the process uneconomic.[54]

It would take another two to three decades before substantial progress on extracting hard-to-reach shale gas was made. Texas oil man George Mitchell of Mitchell Energy was convinced of the potential bound within the depths of these porous rock formations and would succeed where all the major oil and gas companies had failed. Mitchell's engineer, Nick Steinsberger, perfected a recipe for fracking fluid consisting of water, sand, biocide, acid, and skin cream. Steel casing was cemented down a drill hole and the secret recipe was pumped through the steel casing under high pressure to fracture the rock, whilst the sand propped open the fractured network to let the natural gas flow back to the surface. The process was a success. Gas flowed freely, the pores remained open, and Mitchell Energy production numbers swelled as they acquired land and opened up more shale gas deposits across Texas. Soon the big operators started to take interest and in 2002 Devon Energy acquired Mitchell's business for $3.5 billion. They combined the recipe for fracking fluid with their new technology for horizontal drilling and kick started the shale revolution in the US, unlocking vast resources, driving natural gas prices four times cheaper than the rest of the world, and playing a major role in the economic boom of the US economy over the last 20 years.[55]

But the environmental issues associated with fracking mean it remains controversial: horizontal drilling requires an increased use of water resources,[56] the widespread use of hydraulic fracturing has resulted in a significant increase to induced earthquakes,[57] the process can lead to significant methane leakage, and the fracking fluids can contaminate surrounding groundwater. The industry is pushing to reuse water, avoid earthquake prone areas, and create benign fracking formulations to appease environmental concerns. In 2011, Dave Lesar, the CEO of Halliburton Energy Services, took this to an extreme when, during a keynote speech, he called upon one of his executive team to drink a glass of their newly developed fracking fluid to demonstrate its safety.[58] Aside from a rather stale taste in his mouth the executive in question went unharmed, but opponents remain unconvinced.

Hubbert's peak oil forecasts may have proved premature due to non-conventional oil and gas deposits like shale, but we know fossil fuel resources are ultimately finite. There are an estimated 1,000 billion tonnes of coal, 200 trillion cubic metres of gas, and 1,700 billion barrels of oil *reserves* which have been located and can be extracted at today's cost.[59] Based on current consumption, coal would last 130 years, oil 60 years, and gas 50 years.

Adding all the remaining fossil fuel *resources* – which include deposits not yet located or more difficult to reach – and coal supply increases eight times, oil four times, and gas five times. A total of 10,000 billion tonnes of fossil fuel carbon.[60] Coal alone could power the planet for another 400 years. Add oil and gas and this produces a total of more than 500 years of energy at today's consumption levels.

Uranium is another finite resource with just four million tonnes of identified reserves, but with only 70,000 tonnes consumed each year there is enough to last another 70 years.[61] Find a route to economically extract uranium from sea water and you multiply resources by a thousand-fold – enough to power the entire planet for thousands of years.[62,63]

Hydro, wind, solar, and biofuels are sustainable or renewable energy sources. They will go on providing energy so long as the sun shines – another 5 billion years. And whilst a finite supply of materials is available to create the dams, turbines, and panels in order to harness the energy there is more than enough. Silicon is the second most common element in the Earth's crust and the rare Earth metals, required to produce permanent magnets for generators and motors, are not so rare. The available neodymium on Earth could support sufficient wind turbines to power the planet three times over.

As Thomas Edison, the father of electricity, said to Henry Ford in 1931, "We are like tenant farmers chopping down the fence around our house for fuel when we should be using nature's inexhaustible sources of energy – sun, wind and tide… I'd put my money on the sun and solar energy. What a source of power! I hope we don't have to wait until oil and coal run out before we tackle that."[64]

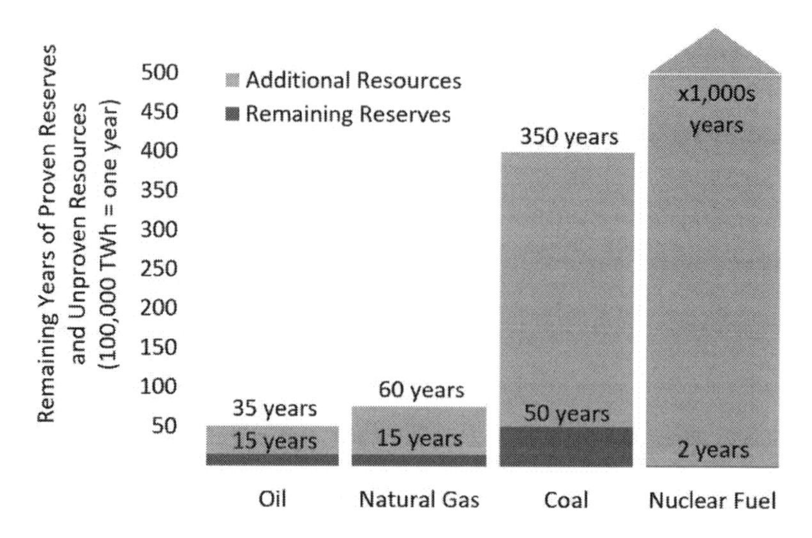

Reserves are oil, gas, coal, and uranium already discovered and which can be extracted using today's technology at reasonable cost. Resources show estimated coal, oil, gas and uranium on Earth which have yet to be discovered or require better technology or more cost to extract. Units are 100,000 TWh or 1 years' worth of final energy, so bars show how many years each resource could power the whole planet. Author's calculations based on data referenced in the preceding text.

Energy Supply: Safety and Well-Being

We can compare the safety and impact on health of each energy supply using deaths or serious illness per billion kWh (TWh) of energy supplied. Surprisingly, nuclear ranks as the statistically safest form of power generation and wood burning as the most harmful due to the indirect deaths from air pollution. Even more surprisingly, nuclear disasters don't even rank as having the highest direct death toll, because coal mining and hydroelectric facilities claim that undesirable top spot.

The failure of the Banqiao Dam in China is the worst energy disaster on record. The dam began construction in 1951, designed to prevent flooding and to generate electric power for the surrounding region. However, early on concerns were raised by prominent engineers on the build quality and the design of the dam. Soon after completion, cracks appeared in the construction and the concrete had to be reinforced. Later, concerns were raised about the low number of sluice gates in case of intense flooding. In August 1975 Typhoon Nina hit the region and dumped two metres of rainfall in two days: the equivalent of more than one year's worth of rainfall. The water level quickly rose above the 117-metre wave protection wall, the sluice gates struggled, and the system failed. Seven hundred million metres cubed of water were released over a six-hour period creating 10-metre-high waves travelling at 30 miles per hour. The torrent of water wiped out 750 square km of land below and killed up to 230,000 people, a devastating loss of life.[65]

From the deadliest catastrophe to the costliest. The BP Deepwater Horizon disaster in 2010 was the largest spill in history, releasing five million barrels of oil into the Gulf of Mexico after the concrete cap, used to seal the well for later use, ruptured and the blow out protector failed. Investigations found the oil company BP, rig operator Transocean, and contractor Haliburton, guilty of gross negligence and reckless conduct. Cost-cutting, poor safety systems, and systemic issues across the whole industry were identified as the root cause.[66] It turns out that although oil companies may be proficient at pumping oil out of the ground, they are largely lacking in remedial plans should something go wrong. In the follow up safety hearing in US congress, the then CEO of US Oil giant Exxon, Rex Tillerson, admitted their Gulf of Mexico oil spill response plan included the emergency contact of a long deceased scientific advisor, and information on dealing with walrus populations, which were the same errors found in the BP plan. As Chairman Markey bluntly put it: "there aren't any walruses in the Gulf of Mexico; and there have not been for 3 million years". It turned out that all the major oil companies had the same deficient disaster planning drawn from a shared source. "The only technology you seem to be relying on is the Xerox machine", concluded Markey.[67] After fines and claims, BP has paid $65 billion in compensation. However, despite the huge impact of this spill on the environment and the tremendous clean up and economic costs, the accident itself resulted in 11 deaths and 17 injuries.

Whilst any loss of life is terrible, the deadliest energy disaster and the costliest energy disaster on record killed just 3% of the number of people who die from air pollution every year. Although we associate energy related deaths with headline grabbing catastrophic events, most deaths are actually a result of long-term inhalation of particulate matter and toxins released from the combustion of wood, dung, and fossil fuels.

Wood burning has the biggest health impact of all energy sources with 250 premature deaths per billion kWh of energy. Developing countries – where open wood-burning stoves or fires are used indoors – account for most of the 3-4 million lost lives. Lifestyle trends are driving higher installations of wood burners in developed countries such as the UK, where particulate matter from wood burning now outnumbers that from cars and trucks by two or three times. One wood burner is the equivalent of having seven idling lorries continuously outside your home.[68]

Coal ranks as the most dangerous fossil fuel, with 50-100 premature deaths and up to 250 cases of serious illness per billion kWh. With coal supplying over 20% of final energy, the detrimental health impact of particulate matter and sulphur dioxide emissions leads to 1.5-2 million deaths per year.

Oil or liquid biofuels are nearly as bad, with 50 deaths and over 150 cases of serious illness per billion kWh. Gasoline, diesel, and jet fuel accounts for over 35% of final energy, emitting harmful particulate matter and nitrous oxides which kill nearly two million people per year from diseases such as asthma, chronic bronchitis, heart disease, and lung cancer.

Natural gas burns far cleaner than wood, coal, or oil with 5-10 deaths per billion kWh. Wind, solar, and hydroelectric emit no harmful products during energy generation and have recorded less than two deaths per billion kWh.[xiii]

Nuclear is statistically the safest energy source with no dangerous atmospheric emissions, despite four major disasters. The SL1 explosion in Idaho, US in 1961 killed three operators. This was followed by the worst nuclear accident in US history, Three Mile Island in 1979, where the radiation leak dosed two million people in the surrounding area with 0.0014 mSv of radiation – the equivalent of one tenth of an x-ray each. No adverse health consequences have been recorded or are expected.[69]

The most notorious nuclear disaster was Chernobyl in 1986, Russia, which killed two in the explosion, 30 from radiation poisoning, and 600,000 people in the surrounding area were exposed to 10-20 mSv of radiation. This is the equivalent of a CT scan each and increases the risk of cancer by 3-4% over their lives.[70,71]

The melt-down of Fukishima in Japan in 2011 killed two workers immediately, but the evacuation of 170,000 people led to another 2,000 indirect deaths of elderly people due to medication interruption and poor living conditions.[72]

Any loss of life and suffering is tragic but since 1954, nuclear energy has generated 90 trillion kWh of electricity for the world with direct deaths

---

xiii Wind turbines are often accused of killing large numbers of birds and bats due to collisions with the blades and the data suggests between 0.3 and 4 deaths per GWh, compared to 5 deaths per GWh for fossil fuel electricity generation (from habitat loss, pollution, and collisions). The total number of bird deaths from all forms of energy generation are relatively small when compared to deaths from cats, which total 25 times greater. Scientists have recently discovered painting one turbine blade black can significantly reduce bird deaths. From Mark Z. Jacobson '100% clean renewable energy and storage for everything'.

totalling only seven and radiation poisoning killing another 40. The impact of having to evacuate local populations adds another 2,000 to those who lost their lives, and elevated cancer risk adds a further 4,000 premature deaths over 80 years. This is a death rate of 0.07 per billion kWh of energy supply: the total number of deaths caused by nuclear energy in 70 years makes up less than 0.1% of the deaths caused by air pollution in just one year.

Whilst the operation of nuclear energy facilities is statistically safer than most people assume, the weaponisation threat of nuclear material cannot be ignored. The world has so far managed to navigate 40 years of Cold War, to avoid mutually assured destruction scenarios, and has mostly agreed on the non-proliferation of nuclear weapons.[xiv] However, an increased prevalence of nuclear energy could make enriched fissile materials and weapons technology more accessible.

Iran has been under economic sanctions since 2006 after failing to comply with UN resolutions that demand the country halt its uranium enrichment program. Iran maintains that the nuclear programme is for electricity generation, but allegations of covert weapons development have long persisted. The Iranian nuclear program took an unexpected turn in 2010 when Belarusian cyber security software company VirusBlokAda discovered a malicious computer worm called Stuxnet. The worm was designed to infect Windows operating systems and spread over a local network until it located a specific software found in the logic controllers of the Siemens' uranium centrifuges used in Iran. The centrifuge model was likely identified from an Iranian promotional video on the internet. A USB stick containing the Stuxnet worm was plugged into the isolated network and it re-wrote the programme code, disrupted the speed of centrifuge operation, and ultimately destroyed 1,000 machines, thus setting the Iranian nuclear enrichment programme back years. Expert analysis of Stuxnet has shown it was probably the costliest and most sophisticated malicious code ever created, with strong allegations of US and Israeli government origin being made.[73] Despite the Iranian nuclear complex being off-line, Stuxnet somehow made its way onto the internet and infected hundreds of thousands of computers. Utilising so-called zero-day software which attacks vulnerabilities then completely unknown to Microsoft, Stuxnet has opened the door to an increasing threat from cyber-physical attacks around the world.

---

xiv The US, Russia, UK, France, and China are designated nuclear states under the Non-Proliferation of Nuclear Weapons Treaty. India, Pakistan, North Korea are not part of the treaty and have openly developed and tested nuclear weapons. Israel likely has nuclear weapons but has not openly affirmed its position. The US is the only country to have used nuclear weapons as an act of war.

Whilst this version of the worm slowed the Iranian nuclear programme, potential adaptations to this type of software, which is now out in the wild, create an enhanced risk of cyber-physical attacks which could target nuclear energy or weapons facilities with more deadly consequences. It is the enormous, concentrated power of nuclear fission or fusion which makes the technology an attractive next step in the evolution of energy, but this same characteristic also brings significant threats. Until the world can develop stronger international laws and robust global cooperation, there will always remain a low probability, but high damage threat, from the weaponisation of nuclear power.

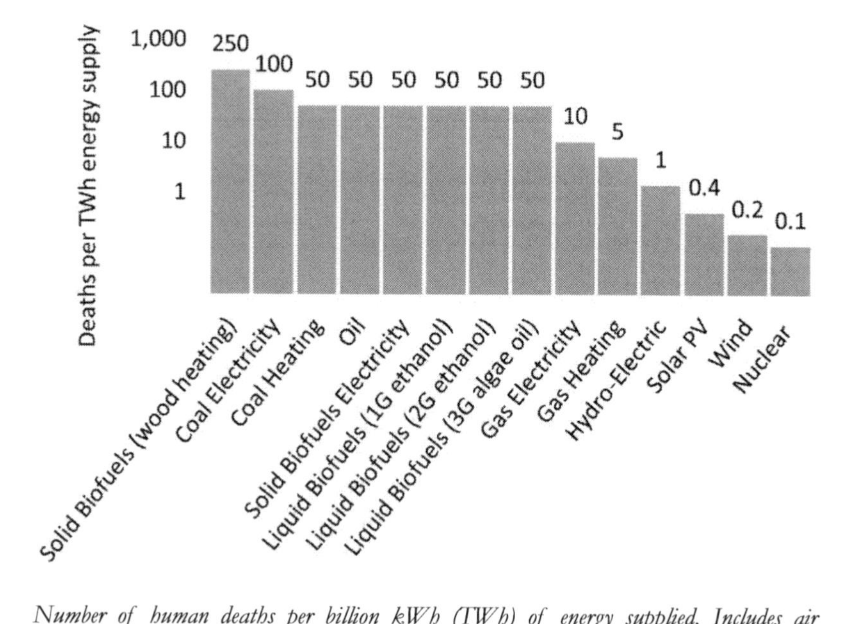

Number of human deaths per billion kWh (TWh) of energy supplied. Includes air pollution, disasters, and accidents. Logarithmic scale. Author's calculations, cross referenced with sources.[74,75,76,77]

Energy Supply: Reliability, Predictability, and Variability

Renewables such as wind or solar are exceptionally reliable but inherently variable and unpredictable. We know exactly how much energy they will provide through the year, but that energy will vary through the seasons, and is intermittent in the short term. Overcoming variability and unpredictability requires seasonal and short-term energy storage solutions. Fossil fuels, on the other hand, are predictable but inherently unreliable. Fossil fuel energy is easily recovered, transported, and stored, and readily switched on and off when needed. But regional supply inequality means that a small number of companies or nation states control access to reserves. Fossil fuels have fundamental political and market risk, creating volatile prices and shortages.[78] The world endured multiple energy crises in the 1970s, when western nations temporarily hit peak oil production, and in 1990, with the oil price shock following the Gulf War, and, more recently, the natural gas shortages in China. Political risk, energy shortages, and price volatility are a recurring theme. Power struggles and energy security have always been a dominant feature of fossil fuel markets.

When Edwin Drake struck his first oil well in Titusville, Pennsylvania, in 1859, a young bookkeeper named John D. Rockefeller was working in neighbouring Ohio at a small trading house. Rockefeller was an ambitious young man who delighted in hard work and reportedly listed his greatest ambitions as earning $100,000 and living 100 years. He formed his first partnership with Maurice B. Clark and during the Civil War their business went from strength to strength, organising food and supplies for the army. Rockefeller paid close attention to the newly developing oil market and in 1863, at the age of 23, took his first steps into the industry.

Rockefeller was not seduced by the risky fortunes of oil prospecting but recognised the compelling business case for refining crude oil. He knew the large mark-up on kerosene for lamps could earn even a mediocre refinery more than 50% profits and if he found markets for the other components of crude oil (rather than dumping them in a nearby river), he could boost profits even higher. Rockefeller's business acumen, penny-pinching, and fiercely competitive nature, meant that by 1868 the company owned the largest refinery operations in the world. He went on to buy out his partners, reform the company as Standard Oil, and either buy up or destroy most of the competition. Standard Oil made Rockefeller the world's first billionaire and at its peak he was worth more than $300 billion in today's money, nearly twice as rich as Amazon's Jeff Bezos. Standard Oil controlled most of the world's resources up until 1911 when the monopoly was forcibly broken apart by the US government.[79] Many of Standard Oil's splinter companies ended up forming the Seven Sisters cartel which continued to control most of the world's oil until the 1970s, eventually broken up into BP, Shell, Chevron, and Exxon. As Javier Blas, Bloomberg energy correspondent, writes, "It would increasingly be the traders, and not the large oil companies, who would determine who could buy and sell oil, empowering the new petrostates."[80] And as the Seven Sisters cartel was broken apart, a new one formed, with major oil sovereign states in the Middle East, North Africa, and South America creating OPEC which to this day controls most of the global oil supply.

Power has changed hands and nations continue to try and adapt. The US has managed to transition from energy importer to exporter with the discovery of shale gas. Russia is attempting to diversify oil and gas supply from reliance on Europe with pipelines into China. And the Middle East has set out a plan to diversify into petrochemical products. Crude oil is the world's premium fossil fuel and one of the most important resources in our global energy system and yet an international cartel still controls nearly half of production and four fifths of remaining reserves. OPEC members collude to control supply and influence market prices – actions which would be illegal in any nation state but because of weak international laws, go unchallenged.[81]

Political turmoil, fighting, and market manipulation regularly create mismatches between the supply and demand for fossil fuels and they have driven volatile pricing[xv] over the years.[82,83,84] This can make life very difficult for governments forecasting their economies, companies planning budgets, and individuals estimating their bills.

Renewables, on the other hand, can be viewed as highly reliable in many ways but also inherently variable and unpredictable. Where fossil fuel energy relies on the dependable supply of coal, oil and gas, renewables source their fuel from sun, wind, or water. There is no political risk, no supply chain shortages, or possible embargos.

The average solar, wind, and rainfall through the year is very dependable, though it does vary with the seasons and is unpredictable over the short term.[85] Cloudy or rainy days may produce less than one fifth of the solar energy compared to bright sunny days. For the most populated areas on Earth, the sun is three times stronger in summer compared to

---

xv Oil has traded anywhere between $10 and $150 per barrel ($0.6c per kWh to $9c per kWh), coal has traded between $20 and $120 per tonne ($0.2c per kWh and $1.3c per kWh), and natural gas between $1.5 and $20 per MMBtu ($0.5c per kWh to $7c per kWh).

winter.[86] Wind can stop completely with little forewarning and summer provides only half the wind energy of winter.[87] Hydroelectric has its own inbuilt storage so can handle short term unpredictability, but variability in rainfall through the seasons limits the average load.

Fossil fuel plants or nuclear facilities can be run with a high load factor or near fully rated power capacity so long as fuel is available. Renewables depend on the weather and will require energy storage solutions if they are to deliver a majority proportion of final energy.

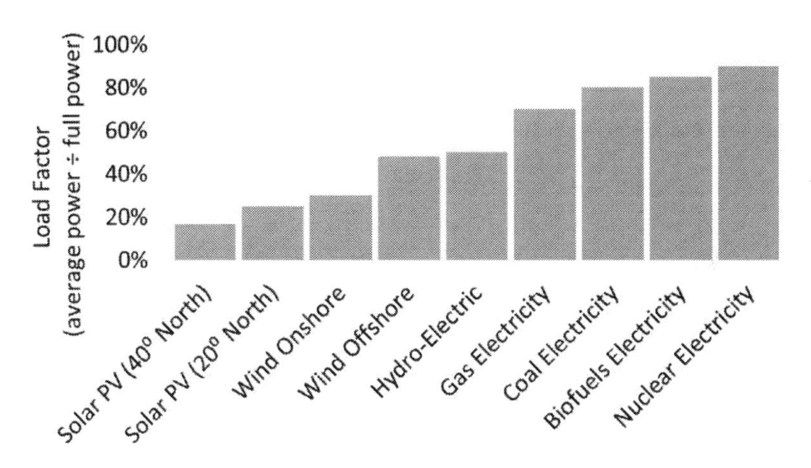

Typical load factors for electricity generating installations. Nuclear and coal plants are designed to be run at a high percentage of their maximum capacity due to high capital costs and cheap fuel. Hydro is limited by seasonal rainfall. Wind is limited by how often and how strong the wind blows. Solar is limited by the hours and intensity of sunlight, a solar farm in Germany will generate a 10-12% load factor, a similar farm in Chile may deliver 25% and add a tracking system to follow the sun through the day and the load factors can increase to 35% but with extra investment. About 90% of the global population lives between 20 and 40 degrees north of the equator. Based on references in preceding text.

Energy Supply: Technology Efficiency

The first two laws of thermodynamics provide two key concepts governing the efficiency of energy generation: the best you can do is break even, but you won't break even. Real world efficiency is a balance between theoretical limits governed by the laws of physics, and the complexity or cost of getting there. But is there room for improvement on today's technology? Unsurprisingly, fossil fuel technology is already deep on the experience curve and potential improvements are minimal. Wind can boost power output with taller turbine designs accessing stronger wind speeds, but the real game-changing improvements rest with solar PV and third generation biofuels. Both technologies are early on the experience curve, with plenty of upside to theoretical limits.

Burn coal, oil, or gas for thermal energy and up to 90% of combustion energy is directed into useful heating. Transform that heat into other forms of energy and efficiencies quickly decline. Combustion engines turn just one quarter of oil energy into the motion of a vehicle.[88,89]

The efficiency of thermal power plants and electricity generation are governed by the Carnot Cycle, which states that the hotter the temperature of operation compared to the exhaust, the higher the efficiency.[90] Coal-powered steam turbines can heat pressurised water up to 600ºC with maximum 50% useful energy output. Combined Cycle Gas Turbines burn gas at 1,500ºC and use the exhaust gases for a secondary steam cycle boosting overall efficiency to 65%.[91] Unsurprisingly, after one hundred years of deploying fossil fuel technology at scale, most equipment is near its real-world limit.

Nuclear thermal power achieves similar efficiencies to coal power plants. Next generation nuclear designs are looking to increase operating temperatures to boost efficiencies, but the big improvements may come from redesigning the fuel cycle.[92,93] An example is fast breeder reactors which are designed to enrich more fuel during operation.[94] Bill Gates's company TerraPower, founded in 2006, is attempting to redesign a nuclear reactor with a type of fast breeder core called the travelling wave.[95] The reactor technology will use a small core of enriched uranium-235 to breed plutonium-239 from surrounding layers of depleted uranium waste. If successful, fuel cycles would last decades and the design has the potential to produce 50-60 times more energy for the same amount of enriched uranium, with a corresponding reduction in nuclear enrichment and waste.[96] Given that most nuclear power plants are still based on 50-year-old designs, bringing this technology into the twenty-first century would most likely create large efficiency improvements.

Commercial single junction solar PV panels convert 18-20% of the sun's energy into electricity. The theoretical maximum for silicon is 34%, limited by the range of the sun's spectrum which silicon can absorb; 20% of the sun's spectrum is too high energy, 30% is too low energy and 16% represents unavoidable losses after the light is absorbed. Improving on the theoretical limit requires adding further layers of different materials.[97] So, for example, a three-junction cell could reach a theoretical 49% efficiency. Given that solar PV is still early on the experience curve and far from theoretical limits, there are likely sizeable gains to be made.

Wind turbines already convert 50% of wind energy into electricity.

The theoretical maximum is 60%, beyond which the wind loses so much energy that the air accumulates behind the turbine and slows generation.[98] Increasing the power output of wind turbines won't come from efficiency improvements, but through accessing more wind energy. The kinetic energy of the wind is proportional to the wind speed cubed. Double the height of the turbine and you add 10% to average wind speed and one third to power output. Onshore wind turbines (land based) have grown from 100m to nearly 200m high and have increased power from 1 to 5 MW per turbine over the last 20 years.[99] Offshore wind turbines (at sea in shallow waters) can go even bigger, up to 220m and nearly 10 MW power.[xvi]

First- and second-generation biofuels capture just 0.1% of the sun's energy. Third Generation algae oil is more promising, with commercial efforts able to produce up to 2% efficiencies. If third generation algae biofuel could double growth rates and extracted oil yield, this would boost energy output to a more respectable 6.5% of the sun's energy.[xvii]

However, as technologies improve efficiencies, climate change may hamper progress. Every 1°C increase in temperature reduces the efficiency of thermal power plants by 1-2% and creates enhanced risk of disruption to cooling water supplies.[100] Changing weather patterns create changes to wind speeds, clouds, rain, and water run-off, all of which will influence the future performance of wind, solar, hydroelectric, and biomass.

---

xvi The DEA is funding a massive turbine project in the US looking at the possibility of a 500m high installation, taller than the empire state building, with double the power output of your average turbine.

xvii If third generation algae biofuel could double growth rates from 25 to 50g dry algae per square metre per day and extracted oil yield from 25% to 50% of algae mass this would boost energy output to a more respectable 6.5% of the sun's energy.

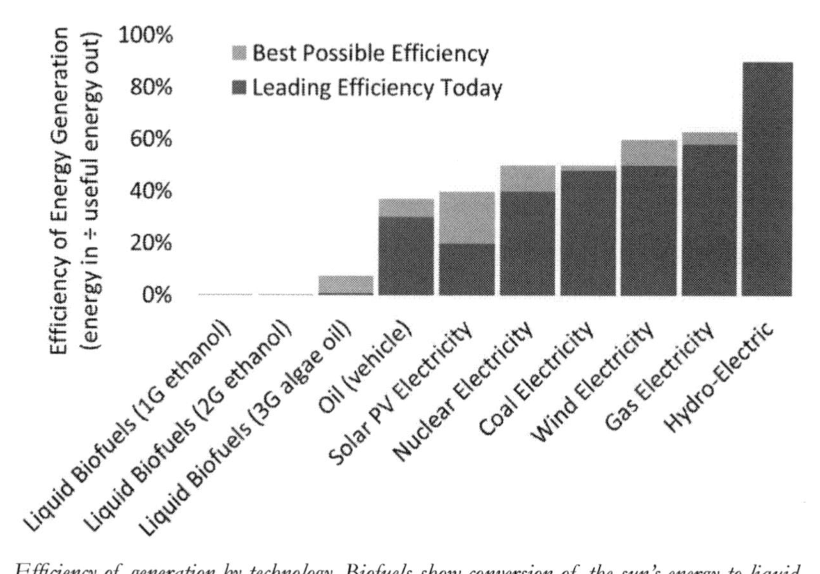

Efficiency of generation by technology. Biofuels show conversion of the sun's energy to liquid fuel. Oil shows thermal combustion to vehicle motion. Solar PV, sun to electricity. Wind, power of the wind to electricity. Nuclear, coal, and natural gas show thermal fission/ combustion to electricity. Hydro, gravitational potential of water to electricity. Based on references in preceding text.

Another way to think about efficiency is to quantify the amount of final energy delivered compared to the energy required to extract, process, transform, and distribute that energy. This is called Energy Return on Invested Energy (EROIE) – the ratio of useful energy out per unit of energy in. The world uses around 100 trillion kWh of final energy per year and 10% of the total is used to generate that energy. This is the energy expended in drilling, mining, refining, purifying, trucking, and distributing of fossil fuels. EROIE ratio can range from as high as 80:1 for coal, to as low as just 3:1 for tar sands oil or shale gas. As fossil fuels are depleted, the energy required for extraction from hard-to-reach deposits becomes ever greater and the EROIE declines. Oil has dropped from an average of 30:1 to 10:1 over the last 30 years.

Renewable electricity requires no fuel extraction or transport, but the manufacturing process does use energy. Wind and solar deliver a similar EROIE to the average of fossil fuels today. Switching from coal, oil, and gas to wind and solar will still require 10% of global energy expended in the energy industry itself.

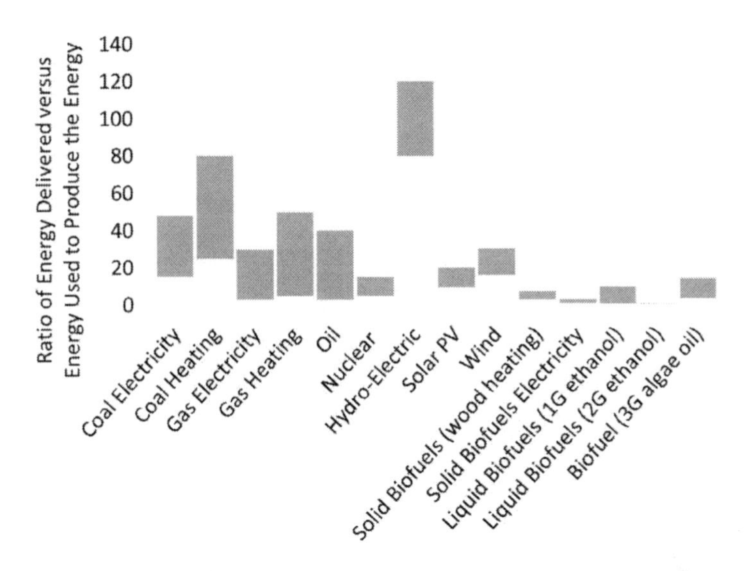

Energy return on invested energy (ÆROIE) – the ratio of energy output for every unit of energy input for each type of energy technology adapted from Hall et al. 'EROIE of different fuels and the implications for society'.[101]

Finally, we can think about the quantity of materials extracted from the earth in order to produce the fuel or equipment used to generate the energy out. Wind, nuclear, and solar deliver 5 to 60 times more energy (over the useful life) per tonne or cubic metre of mining compared to fossil fuels.

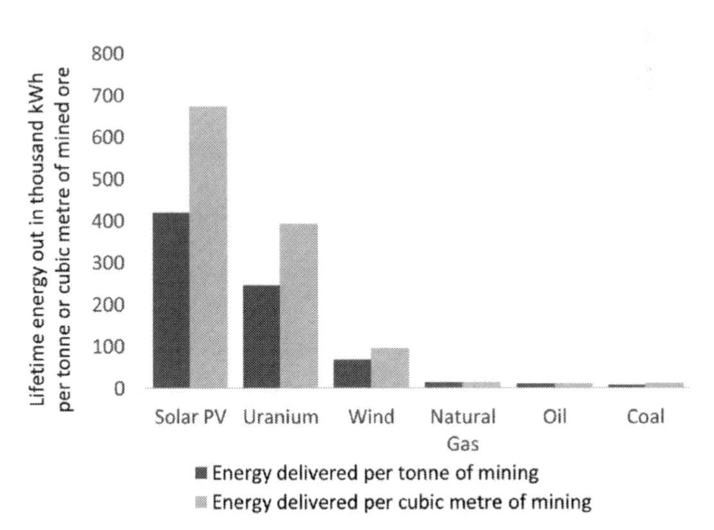

Energy delivered over the life of the fuel or equipment (in thousand kWh / MWh) per tonne or cubic metre of material, fuel, and ore mined (accounts for % ore grades and conversion losses in processing).

Energy Supply: Water Consumption Comparison

The world currently uses 4,000 cubic km of fresh water each year which is roughly one third of our renewable supply. A growing global population and warming planet will put further pressure on already vulnerable fresh water supplies and growing energy production could add to the problem.

First generation biofuels consume the largest quantities of freshwater for irrigation of the fuel crop. Solid biofuel electricity may consume similar quantities of water depending on the irrigation needs of the biological fuel. Second generation biofuels require only steam for processing and third generation algae should be able to use wastewater or salt water.

Hydroelectric reservoirs lose a significant amount of water due to evaporation during operation. Thermal power plants withdraw around ten gallons of water for cooling per kWh of electricity generation and another gallon is consumed during operation. The withdrawn water is routed back into the watercourse but there are limits on the return temperature.

The remaining energy sources consume fresh water during resource extraction, processing, and equipment build, but this is minimal if averaged over the lifetime of operation.

With our current fossil-fuel-based electricity system it takes about 11 gallons of water to produce one kWh of electricity. So for every cup of water you boil in the kettle, it takes another two cups to generate that electricity.

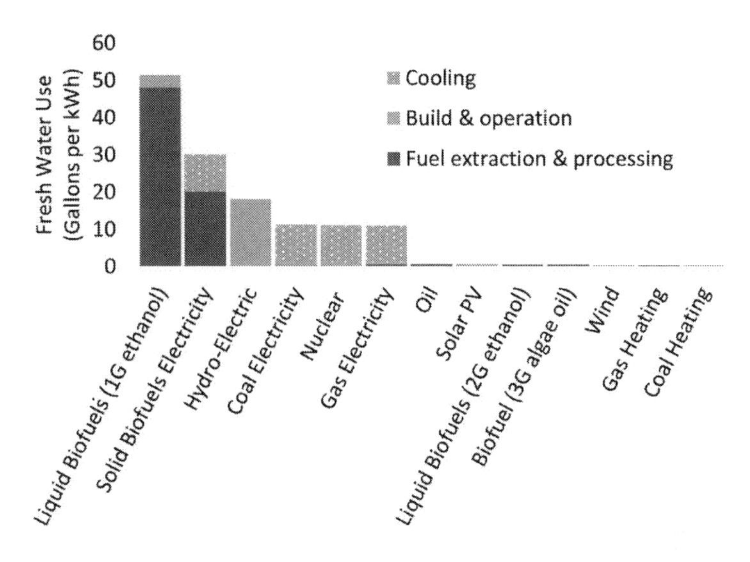

Fresh water used to generate different forms of energy. Solid Dark: Fresh water consumed in growing crops or extracting and processing fuel. Solid Light: Fresh water consumed to build energy equipment or during the operation of the equipment. Dotted: Fresh water withdrawn to cool thermal power plants and returned to the watercourse. Collated from sources. [102,103]

Energy Supply: Levelised Cost of Energy (LCOE)

Burning coal, gas, or wood for heat costs less than $2 cents per kWh and will likely remain the cheapest source of low-grade energy given the high efficiency and low equipment costs. Delivering higher quality energy in the form of electricity or liquid fuel comes at a higher price.

The incumbent technologies for electricity generation are centralised thermal power plants run on coal, gas, or nuclear fuel and deliver wholesale electricity between $2.5c and $8c per kWh with little room for improvement. But over the last 20 years, the cost of wind and solar electricity generation has dropped by over 90% and now competes with the cheapest fossil fuel electricity in most regions. Wind and solar account for just 2% of final energy, so if we scaled up generation towards two thirds of global supply, installations would be required to double a further 4-5 times. With solar costs falling by 20%, and wind by 10%, every time installations double this implies renewable electricity could be delivered at less than $2-3c per kWh in all regions. Wind and solar wholesale electricity would become cheaper than fossil fuel electricity and close to competing with coal or gas heating. Some solar contracts in sunny areas are already being struck at these prices.

Oil is the premium energy source for transport, packing more energy into a smaller volume to power cars, ships, and planes over greater distances on a single refill. At around $5c per kWh energy, you pay for the privilege. 3G algae oil costs double that of crude oil today, but once optimised could reach a similar price point.[xviii]

LCOE At Scale

I have been careful not to use the term "future costs" but rather "at scale" cost because it's not the passing of time that brings down the price for commercialised technologies, but the amount of installed capacity and the experience curve.

Fossil fuels technologies are already fully scaled, near efficiency limits,

---

xviii **Levelised Cost of Energy Today**

A 500 MW coal fired power plant can cost anywhere from $1 to $4 billion to build and is designed to run at an average load factor of 80% for 30 years. This brings levelised capital costs (including financing at 3%) to between $1c and $4c per kWh. Add another $1c for running costs and $1-3c for coal and the total levelised cost comes to $3-8c per kWh of electricity. A similar 500 MW Natural Gas facility is far cheaper to build at less than $1 billion dollars, but gas is slightly more expensive. The total levelised cost is around $2.5-6c per kWh.

Nuclear facilities tend to be at least double the capacity of coal or gas, at 1,000 MW, and will cost anywhere from $5 to $20 billion to build and eventually decommission. The facilities are designed to run at 90% and to last for over 60 years. The levelised capital cost including financing is between $2.5-7.5c per kWh plus another $1c to run and $0.5c for fuel. Total LCOE of $4-8c per kWh electricity which includes the cost of decommissioning. Arguably the economic costs of major disasters such as Fukishima and Chernobyl could be added to the estimated future price of nuclear energy, these disasters cost around $500 billion which over the 90,000 TWh of nuclear electricity generated throughout history adds an extra $0.5c per kWh.

Solar PV has reached a price point where it is competing with fossil fuel electricity. A 1 MW installation costs around $1 million and will last 30 years or more. The average output depends on the location, closer to the equator and the panels receive more sun all year round. With over 90% of people on Earth living between 20 and 40 degrees north, the average load factor will be between 15-25% of rated power. This brings capital costs to just $2-2.5c per kWh plus another $1-2c for maintenance. A total LCOE of $3-4.5c per kWh.

Wind power has also achieved significant price declines. A one MW wind installation costs $1.5-3 million to build but will run at 30-45% of rated power on average (onshore and offshore respectively): total capital costs of $2-4c per kWh plus maintenance brings wind electricity to $3-6c per kWh.

Oil and derived products such as gasoline, diesel, and jet fuel, cost anywhere from $2c to $7c per kWh depending on the price of oil ($25-120 per barrel); first generation corn ethanol is a similar price. But run these fuels through a combustion engine and only one quarter of the energy is turned into motion of the vehicle.

Bio algae oil is the most promising biofuel. A 20 million square metre facility could produce 0.5 million MWh of algae oil per year (57 MW power) at a cost of $400 million to build. That's $3-4c per kWh capital costs, another $3-4c per kWh for fertiliser and $3-4c per kWh running costs brings 3G oil to double the cost of gasoline today.

and deep on the experience curve. Coal and oil should expect minimal further cost reductions, but prices will remain volatile depending on market supply and demand. Natural gas may trend towards the lower end of its cost range with increased global trade of liquefied gas and adoption of higher efficiency combined cycle designs.

Future designs of nuclear are looking to improve safety, waste, and sustainability of uranium resources; however, capital costs will likely remain high. Smaller Modular Reactors (SMR) built in factories and unloaded on-site are looking to reduce costs but remain unproven[xix].

Significant cost improvements can still be made in solar, wind, and 3G biofuels as manufacturing costs decline, power output scales, and efficiencies improve.

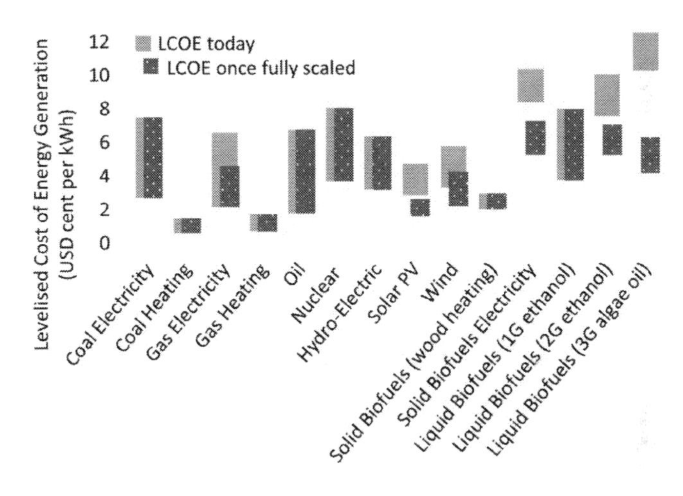

Levelised Cost of Energy supply today and at full scale excluding inflation. Combined capital costs, operating costs, and fuel costs per kWh of useable energy product – electrical or thermal. Represents the wholesale cost of generating energy before distribution, overheads, and taxes. We use a standard 3% discount rate / post tax weighted average cost of capital excluding inflation (equivalent to ~5% nominal post tax, and 6.5% nominal pre-tax) for all technologies both today and at full scale – this is the same as our damage discounting and also represents the mid-point of global renewables financing today[xx] (7% nominal post tax) and best in class financing today (2.5% nominal post tax). Author's standardised LCOE calculations, using data sources. [104,105,106,107,108,109,110,111]

---

xix Small Modular Reactors could generate around 300 MW of power output or about half of an average coal plant.

xx We calculate real LCOE which is fully comparable between different sources and with today's prices. One of the most widely quoted LCOE calculations are published by Lazard Investment Bank but give the average nominal price including inflation through the life of the project: long lived nuclear projects will include more inflation than shorter life solar projects and cannot be fairly compared using this method. Lazard use ~9% pre-tax nominal cost of capital which equates to 4.5% post-tax real when adjusted for tax and inflation (compared to our 3% post-tax real).

Since the year 2000, installed solar capacity has increased from 500 MW (500,000 kW or equivalent to one coal power plant) to over 700,000 MW today (1,400 coal plants). Every time the installed capacity of solar has doubled, the installation cost has declined by 20%. Solar PV installation costs have dropped from $12 million to less than $1 million per MW over the last 20 years. The price of solar energy per kWh has fallen even faster with 30-40% reductions every time installations have doubled, because not only has the build cost come down but the financing has grown steadily cheaper.

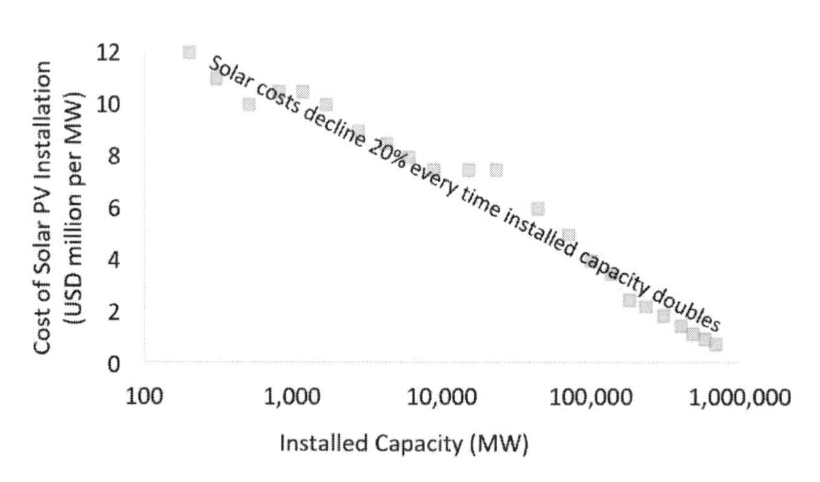

Solar PV experience curve shows the decline in the average cost of installing 1 MW of solar farm capacity, as installed capacity has increased. Each time capacity has doubled, costs have dropped 20%. Plotted from data source.[104]

The average cost of capital[xxi] for a solar installation in Germany was 8% in the year 2000 – banks wanted a 6% return per year on their loans, partners wanted over 10% on their equity stake.[112] Today, cost of capital has dropped to less than 2.5% (5% equity, 1-2% debt) because solar projects have proven low risk and reliable and interest rates have fallen across the world. This represents a best-in-class financing system with a competitive and well-regulated banking industry and certainty on revenue streams. The average cost of capital across the world is

xxi Cost of capital is the return which investors require to make an investment, and the rate depends on the level of risk involved. Equity and debt are two ways a company or project is funded. Interest and repayments on debt are paid from project cash flows first and anything left over is reinvested or returned to the equity holders. Equity is paid after debt so carries higher risk which is why equity yields a higher expected return. The average company will use one third equity and two thirds debt funding, but low risk wind or solar projects with predictable cash flows can often use less equity and more debt.

closer to 7% (~5% excluding inflation)[xxii] but continues to trend down as developers become more confident with the performance of the technology and financing structures evolve.[113,114] Every 1% reduction in the cost of capital or discount rate reduces the cost of renewable power by 10-15%.

Solar PV still represents less than 1% of final energy consumption leaving plenty of room for further cost declines as it scales up. If Solar PV were to grow to one third of final energy, installations need to double a further five times. Following the same (20%) experience curve implies costs drop towards $0.25 million per MW or less than $2c per kWh without violating the floor price of production.[115] Projects are already being bid at less than $2.5c per kWh for large solar farms in the Middle East where peak sun hours are very high.

Wind makes up just over 1% of final energy supply so the production and installation of turbines has not reached full scale economics. The experience curve has delivered a 10% cost reduction for every doubling of capacity. Slower than solar, because the installations are less modular, but still rapid. If wind were to scale to one third of total final energy needs, cumulative capacity will double another four times over. Build costs could decline from $1-3 million to $0.7-2 million per MW and bring the total cost of wind electricity to between $2-4c per kWh.

Third generation algae oil is more than double the price of oil products today, but the process is yet to be optimised. Take growth rates and yields to best-in-class levels and costs have the potential to more than halve.[116]

---

xxii Analysis of one third of the solar farms in the US showed that the actual generation from the installations were within 6% of that forecast by the developers. As the data collection and expertise improves, renewable assets should perform to within a few percentage points of forecasts and this increased predictability makes the investment safer and reduces the cost of capital.

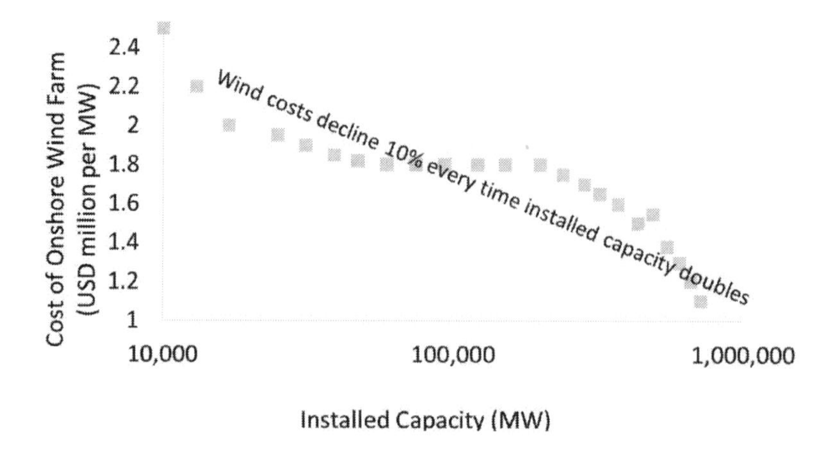

Wind experience curve shows the decline in the average cost of installing 1 MW of wind farm power capacity, as installed capacity has increased. Each time cumulative capacity has doubled, costs have dropped 10%. Plotted from data source[104]

Energy Supply Comparison

Having reviewed all possible energy supply options, the case for renewables is compelling. Solar PV and wind power are cost competitive, sustainable, safe, reliable, and have high enough power density to provide the majority of future low carbon energy supply. The only drawback of wind and solar is the unpredictability and variability of generation which will require energy storage solutions to balance the supply and demand.

The next chapter will explore the possible options for energy storage to help us understand whether a high penetration of wind and solar is possible, affordable, and optimal.

Chapter 8: Energy Storage and Distribution

Building a Safe, Reliable, and Cost Competitive Network

In this chapter we will explore the inherent variability of wind and solar pv to understand the storage requirements of a high penetration renewables energy supply. We will run through possible short term and seasonal energy storage options and compare the physical attributes, engineering requirements, and costs both today and once the technologies are scaled up. We will finish the chapter by running through what is required of a safe and reliable energy network and assemble a fully loaded cost of energy which compares generation, storage, transmission, distribution, and carbon offsetting for all possible sources of energy.

Short-Term and Seasonal Energy Storage Requirements

Electrical power grids are designed to cater for the fluctuation in demand through the day by increasing and decreasing supply when necessary. Peak electricity consumption tends to be in the afternoon and early evening when people are returning from work or school, whereas consumption declines by one third overnight before ramping back up in the morning.[1]

Today, grid generation is made up of base load, load following, peaking, and grid reserve power sources. Base load is typically large nuclear or coal fired power stations which are cheap but only cost effective when running near full capacity and typically cover the minimum power overnight. Load following are mid-sized coal or gas plants designed to track the extra demand through the day and adjust the power output accordingly. Peaking power generation may only switch on for a few hours in the late afternoon and is typically comprised of hydroelectric, small cheap gas turbines, or diesel generators which can turn on and off very quickly and act as back-up. Grid reserve is the excess capacity to ensure a consistent supply.[i] Perfectly orchestrating the output of these different power sources into the grid means power operators can match generation with use. In today's electricity system, the energy is stored in the fuel, ready to be turned to electricity at the flick of a switch.

Now, let's imagine a future energy supply made up of half Solar PV and half wind, with enough capacity to cover all the world's energy needs.[ii] solar PV and wind are highly complementary technologies because the sun shines in the day and the wind tends to blow turbines harder at night when undisturbed by turbulence from the sun's heat. High-pressure weather systems tend to be sunny with little wind and low-pressure systems windy, with more cloud.[2] Summer gets more sun and winter more wind. When one technology is generating less, the other is usually working harder, but the results are not perfect.[85,86,87]

Remember that solar power follows a bell-shaped curve which peaks midday and drops to zero output at night. So, the majority of solar energy is delivered through the middle of the day. This means that even on a perfectly average sunny and windy day there is still a mismatch of energy supply and energy demand. If you overlay the two profiles, the mismatches add up to around 25% of generation which must be stored in the middle of the day and "time-shifted" for use overnight.

Now imagine a day when the wind doesn't blow, and the sky is cloudy. Not only does the demand pattern not match the supply pattern, but the total generation is much lower than required, creating a 75% shortfall.

In other words, to cover the day-night mismatch or to cover a dark, still day, we need 75% of daily demand ready and waiting. Now this is an overly simplistic way to think about storage because we are assuming all the world's generation is in one location. Solar and wind can be interconnected between different regions and countries which lowers the chance of completely still and dark periods. On the other hand, still and dark periods may last longer than one day in any region. Overall, as much as half of renewable generation will require daily storage, which requires up to 140 billion kWh (TWh) of storage capacity or 0.14% of annual energy use that can be cycled 365 times per year to store a total of 50 trillion kWh of energy (half of all final energy use).

---

i Grid reserve ranges from the very short-term response (seconds/minutes) of frequency reserve which regulates the quality of grid electricity, up to spinning, supplemental, and replacement reserves which can cover minutes or hours of extra supply.

ii Global demand is 12 TW (billion kW) final power or 100,000 TWh (billion kWh) final energy demand per year. We will need 30 TW of rated solar capacity with 20% average load factor, generating an average of 6 TW. Plus 15 TW of wind capacity, with average 40% load, generating another 6 TW.

Researchers use decades of historic weather data in specific locations to get a more precise storage requirement by calculating how much electricity a fully renewable grid would have produced at any given time. They use computer models to optimise the grid by stitching together different generating areas and interconnecting supply and storage resources until supply and demand is suitably matched over time. The typical standard is no more than 1 days' worth of outages over a ten-year period. There is still no consensus on whether 100% renewables are optimal, but most studies conclude that a high penetration of renewables is certainly possible and a good solution. A review of studies by Zerrahn et al in 2018 showed that most models predict short term storage capacity requirements between 0.01% and 0.14% of annual energy[3] demand once renewables make up more than two thirds of electricity. Our back-of-the-envelope estimate (0.14%) is at the high end of this range and serves as a conservative assumption.

Short term or daily storage solutions solve intermittency of generation, but seasonal variability is another issue. Our wind farms will generate double the output in winter compared to summer. Our Solar PV farms will deliver triple the output in summer compared to winter.[4] Wind and solar can counter-balance one another through the seasons, but maybe not perfectly. In many regions there will likely be a mismatch between supply and demand through the year. In our 50:50 wind:solar example, the total seasonal mismatch adds up to about 4% of total annual energy demand or 4,000 billion kWh (TWh). This seasonal energy imbalance must be stored in summer and reused in winter.

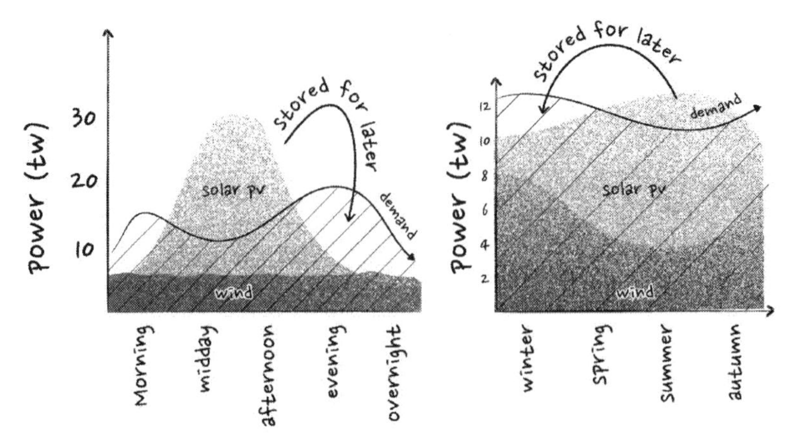

Left: Daily mismatch of electricity supply and demand under a 50:50 wind:solar system. The peak of generation must be stored for use in the evening, overnight, and early morning. Right: Seasonal mismatch of electricity supply and demand under a 50:50 wind:solar system – a portion of the energy must be stored in summer and reused in winter. Author's calculations based on assumptions given in preceding text.

There is currently 10 billion kWh of short-term electrical storage capacity globally. More than 90% is pumped hydro. This equals 17% of global daily electricity demand or 4% of daily energy demand. This capacity needs to increase up to ten-fold if renewables become the dominant supply of energy for the planet.

So, let's review our storage options.

Energy Storage Technology Options[5,6,7,8,9]

Flywheels are heavy spinning wheels that store kinetic energy. Incoming electricity powers a motor, which spins a heavy steel disc, when the energy is required again the spinning wheel is used to turn a generator.

Capacitors use two metal plates with insulating material sandwiched between. They store energy as positive and negative charge on each side of the insulator. Super capacitors use liquid electrolytes to increase the storage capacity.

Lead Acid Batteries are a mature technology which uses the reversible reaction between sulphuric acid and two lead plates (electrodes) to charge and discharge the battery. Lead Acid is one of the most widespread types of battery technology, commonly used to power the starter motor in cars.

Lithium-Ion Batteries work by using electricity to shuttle positively charged lithium ions through a liquid electrolyte from a metal-based cathode to carbon-based anode. When the stored energy is required again, a circuit is connected and the lithium ions flow back, producing electricity.

Vanadium Flow Batteries use two tanks of dissolved vanadium metal separated by a proton exchange membrane. The liquid vanadium solution is pumped around the tanks and past the membrane. When charging, the voltage forces positive charges to move from one tank to another. When the energy is required again the chemical reaction is reversed and drives an electrical current.

Sodium Sulphur Batteries operates at over 300ºC and use the reaction between molten sodium and sulphur to charge and discharge the battery. Sodium sulphur batteries require a constant supply of energy to maintain their temperature.

Demand Side Response is not strictly a storage technology but a demand management function. It is already widely employed today. Large industrial users of electricity have contracts to reduce power consumption at peak electrical demand on the grid. They can turn off non-essential refrigeration, mechanical, or heating processes for a certain duration and get paid to help balance the grid. With the roll out of the Internet of Things[iii], integration of both industrial and electrical appliances will become more widespread. Demand side response can become fully automated, run by your smart meter, router, and household appliances such as heaters, aircon, fridges, and electric cars to help balance the grid acting like a virtual battery.

Gravity Energy Storage involves moving weight from a low level to a higher-level using electricity and letting the weight fall back down to recoup the energy through a generator. One example comes from a company called Energy Vault who are developing a 33-storey tower with six crane arms that stacks and unstacks 35 tonnes of reclaimed concrete blocks to charge and discharge the "gravity battery".

Compressed Air Energy Storage uses electricity to power a compressor which pressurises air into a closed storage space such as a salt cavern or depleted gas reservoir. When the energy is required again, the air is allowed to flow back through a gas turbine to produce electricity.

Pumped Hydro works by pumping water from lower levels to an upper reservoir using electricity. The energy is stored as gravitational potential. When power is needed again the water flows back through a turbine to generate electrical power.

---

iii The internet of things describes a network of physical objects that are embedded with sensors, software or other technology for the purposes of connecting and exchanging data over the internet.

Thermal Energy Storage uses excess energy to generate heat and store it in underground tanks, pits, aquifers, or boreholes until required. This can be done at lower temperatures using water or higher temperatures using phase change materials such as molten salt. Possible thermochemical routes such as the reversible reaction of potassium oxide will store and release heat at 99% efficiency but are more expensive.

Electrolysis of Water to Hydrogen[10,11,12] or so-called green hydrogen is a process whereby water (H_2O) is split into hydrogen (H_2) and oxygen gas (O_2) using electricity. There are three main technologies for electrolysis: Alkaline, Polymer Electrolyte Membrane (PEM), and Solid Oxide, each with its pros and cons. The hydrogen gas can be compressed and stored in salt caverns or depleted gas reservoirs for use later. Hydrogen can be reused in a variety of ways:

- **Hydrogen could be fed into the natural gas network** and piped straight into residential and commercial boilers, fireplaces, cookers, and industrial heating equipment. According to NREL studies so far show that no major issues arise with hydrogen concentrations of up to 5-15% in natural gas. At higher concentrations all-natural gas pipes must be converted to plastic and all equipment must be retrofitted to burn hydrogen.
- **Hydrogen could be converted to Synthetic Natural Gas** which can be used in the existing gas networks and appliances. This is done by combining the hydrogen with carbon dioxide to produce methane (natural gas) and water. The chemical route involves the Sabatier reaction which uses heat, pressure, and a nickel catalyst or a biological methanation route using microorganisms called Archaea.
- **Hydrogen could be converted to electricity using fuels cells** bringing the hydrogen cycle full circle by combining hydrogen with oxygen to generate electricity and water. Alternatively, hydrogen can be fed through modified gas turbines to generate electricity.

So, we have many energy storage options at our disposal, but to understand which will work for different applications and at what cost, we must understand the fundamentals of each technology including efficiency, energy leakage, life, resource limitations, and economics.

Solutions for Short Term Energy Storage: Physical Properties and Economic Costs

Short term storage requires technologies suited to a daily charge and discharge cycle with low energy leakage, reasonably high roundtrip efficiency, durability, sufficient resources, low carbon, and low cost per kWh storage capacity.

Based on our selection criteria we can rule out several technologies straight away. Flywheels leak 15% of energy every hour so are not suitable for use over hours or days. Electricity-to-hydrogen-to-electricity has low roundtrip efficiency, so you only get back 25% of what you put in, which makes a poor candidate for daily storage.[iv] Lead acid batteries have low durability and can only survive 1,000 charge and discharge cycles and will not last long enough.[13]

That leaves us with pumped hydro, lithium-ion, sodium sulphur, vanadium flow, and gravity. Let's compare the economics. For comparison we will assume each of our storage options is charged up with cheap (fully scaled) solar electricity at $2c per kWh and cycle the system once per day.

Pumped hydro is the mainstay of electrical storage technology today and with a levelised cost of around $5c per kWh (including cheap solar charging) is still the cheapest way to time shift power.[v] Current global capacity of pumped hydro is around 9 billion kWh but a recent report, by Andrew Blakers from the Australian National University's R100 group, estimates there are another half a million sites globally which could store an estimated 22,000 billion kWh.[14] That's 150 times our required future short term storage. Pumped hydro is a good bet.

Lithium-ion batteries are becoming one of the most promising technologies for short term energy storage. The onset of electric vehicles

---

iv Electricity to hydrogen back to electricity has the lowest roundtrip efficiency. Using electrolysis to split water to hydrogen is 60% efficient. Compressing hydrogen for storage is 93% efficient. And turning hydrogen back to electricity through a fuel cell is just 45% efficient. A total round-trip efficiency of just 25% (or up to 40% once optimised). The amount of wasted energy will likely limit the use of power-hydrogen-power to either mopping up extremely cheap surplus electricity, or for use in applications which benefit from the high energy density of hydrogen.

v The largest pumped hydro facility in the world is in Bath county, Virginia, USA. The facility is considered the largest "battery" in the world, holds 100 million cubic metres of water and can deliver 3,000 MW for 11 hours – that's 24,000 MWh of energy storage. The construction was about $6 billion in today's money and equates to around $2,000 per kW of generating power, a levelised cost of around $5c per kWh including charging. From Dominion Energy, "Bath County Pumped Storage Station", [online], dominionenergy.com, accessed 2020.

has driven down the cost of lithium-ion by over 90% in the last 20 years. The experience curve is running even faster than solar with a 35% cost reduction every time installed capacity doubles. Lithium-ion technology has high efficiency, a robust life expectancy, plentiful supply[vi] and competitive costs. If we assume lithium-ion batteries serve one third of our short-term storage on the grid and the majority of passenger car needs, then the produced capacity needs to double another 6-7 times over and the experience curve predicts costs reaching below $50 per kWh without violating the floor price. Costs of cheap solar generation plus lithium-ion battery storage could fall from $10c per kWh today to below $4c per kWh at scale – cheaper than gas or coal electricity.

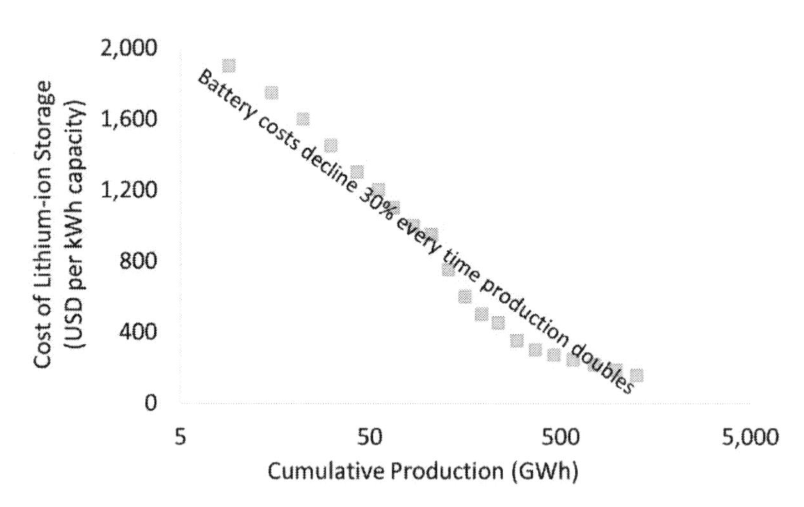

Lithium-ion battery experience curve over the last 20 years. Showing the declining cost of one kWh storage capacity as cumulative production of batteries has increased. Each time production has doubled, costs have dropped 35%. Plotted from data source[15]

One of the largest lithium-ion storage installations completed to date is South Australia's $60 million facility supplied by Elon Musk's Tesla in 2018.[vii,16] The installation started off as a Twitter bet after the Australian Prime Minister Malcolm Turnbull blamed storm related power outages

vi There are an estimated 17 million tonnes of quantified lithium reserves on Earth and another 63 million tonnes of resources yet assessed for cost of extraction. Lithium-ion batteries consumes 100 g of lithium per kWh. So, building out 140 TWh of short-term storage requires a total 14 million tonnes of lithium. If all 1 billion cars on Earth were swapped for long range electric vehicles this would consume another 10 million tonnes (1 billion x 100 kWh x 100g lithium per kWh). So roughly one third of known resources provide enough lithium for both short term grid storage and electric cars. Resource data from USGS, "National Minerals Information Centre, Lithium Statistics & Information", [online], usgs.gov, accessed 2020.

vii The battery facility in Australia has 129 MWh electricity storage capacity and can deliver up to 100 MW power. An installed cost of $460 per kWh including shipping, duties, and installation.

on renewable energy intermittency. The head of Tesla's battery division Lyndon Rive, got wind of his complaints, and claimed Tesla could build 100-300 MW of battery storage in 100 days and solve the power grid problems. Elon Musk raised the stakes over Twitter by promising to finish it in less than 100 days or Australia could have it for free. All parties agreed and Tesla delivered the installation with 40 days to spare. This installation has proved very profitable and has triggered a wave of large-scale lithium-ion battery installations across the world. A trend likely to continue.[17]

Where lithium-ion is very flexible in size, compressed air is an interesting technology option for larger storage projects. There are only a handful of facilities in the world today and the technology is still being developed.[18,19,20] Storage costs are cheap but due to technical challenges which require the use of natural gas to re-heat the cold expanding air, the roundtrip efficiency is low. Next generation designs should eliminate the need for natural gas heating and improve the overall efficiency. If the capital costs can be kept down then compressed air storage could prove an interesting option for larger scale, shorter term, energy storage facilities.

Vanadium flow batteries have a similar cost profile to lithium-ion today, are durable and highly scalable.[21,22] However, potential cost reductions are limited by the price of vanadium and global resources limit the number of batteries which could be made.[23] Vanadium flow will probably play a more minor role in grid storage.[viii]

Gravity batteries are still in the design phase but have the potential to provide an alternative to pumped hydro where the local geography cannot accommodate a reservoir. If initial cost estimates prove correct the technology can provide comparable economics but caution must be used surrounding the CO_2 emissions from large scale use of concrete.[24]

Sodium sulphur is relatively expensive today, but given the cheap materials used for manufacture, there is no reason it cannot be competitive once scaled up. The technology roadmap is not as clear as lithium-ion but if properly developed sodium sulphur could provide an interesting technology route especially for warmer locations.[ix]

---

viii To produce Vanadium flow batteries, every kWh of storage capacity requires more than 3.5 kg of vanadium. The remaining 63 million tonnes of vanadium left on the planet will only support 10% of the 140 TWh of short-term storage we might need.

ix The United Arab Emirates recently (2018) contracted Japanese company NGK to install 648 MWh of sodium sulphur battery in Abu Dhabi across 10 locations. The system is described as the "world's largest virtual battery plant". From: Andy Colthorpe, "UAE integrates 648 MWh of sodium sulfur batteries in one swoop", [online], energy-storage.news, 28 Jan 2019, accessed 2020.

Demand side response is our final option. Around 30% of total energy demand could be eligible for remotely turning on and off [x] and let's assume 50% of this energy use is accessible for 1 hour per day. That adds up to 0.6% of daily energy use or the equivalent of two billion kWh per day. A long way from the 140 billion kWh of short-term storage capacity we are looking for but a very helpful resource to call upon in times of exceptionally high demand or low supply. Electric vehicle batteries are another storage facility we could tap. Assuming there are 1 billion cars globally, each with a 100-kWh battery and their owners allow 15% cycling per day, this could provide another 15 billion kWh of energy storage, more than 10% of our required total and especially useful in countries with high car penetration.[25] Variable electricity tariffs are another form of demand response which can be used to encourage, for example, overnight electric car charging to help balance demand to best suit the grid.

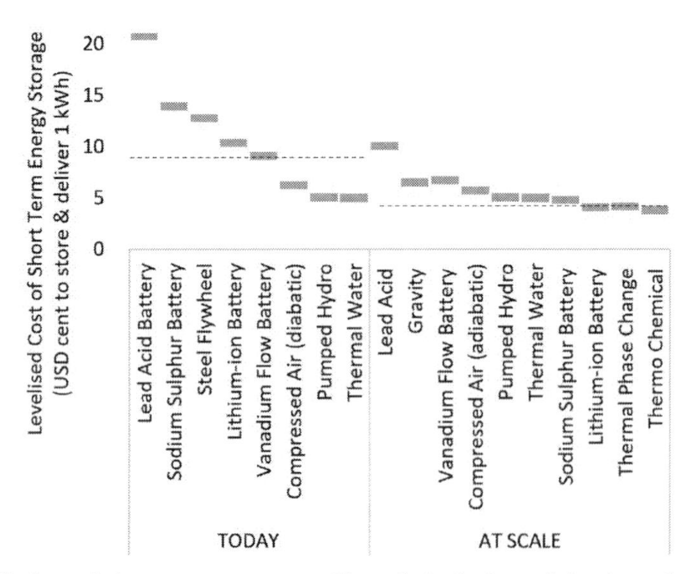

Levelised cost of short-term energy storage. Shows the levelised cost of charging and re-using 1 kWh of energy assuming one full roundtrip charge/discharge per day. Based on charging with \$2c per kWh cheap solar electricity, capital costs and operating costs including roundtrip losses, installation life, cycle life, and leakage. Same 3% post tax, real, cost of capital as all other calculations (5% nominal, 6.5% nominal pre-tax). Author's standardised LCOE calculations based on data sources[5,6,7,8,9,10,13]

x If we add up the areas of energy use which could be temporarily shut down for demand response it would include the metals, chemicals, and paper industry, alongside heating and cooling applications both commercial and residential. This adds up to 30% of total energy demand.

Overall, pumped hydro and lithium-ion battery technologies should be able to provide much of the short-term electrical storage solution. Thermal storage solutions can be used for low grade heat applications and compressed air, vanadium, sodium sulphur, gravity batteries, and demand side response can play a supporting role.

This blend of technology brings the average levelised capital cost of short-term storage today to $4-5c per kWh with a 75% roundtrip efficiency and total cost including charging with cheap solar at just under $10c per kWh. Once the technology is scaled, costs will halve and efficiencies will improve by 5-10%, bringing the total cost to generate and store renewable electricity to less than $5c per kWh – cheaper than fossil fuels ($5c per kWh) and remember less than half of renewable electricity will require storage.

Solutions for Seasonal Energy Storage: Physical Properties and Economic Costs

Longer term storage solutions require technologies suited to monthly or annual charge and discharge cycles which places a significantly different set of constraints on our technology choices. Low energy leakage becomes critically important given the longer duration; plus seasonal storage requires de-coupled power and storage solutions with extremely cheap storage.[xi]

De-coupled power and storage provides the ability to scale up storage whilst holding the cost of the power equipment steady – this is especially important for the economics of seasonal storage which requires a high storage to power ratio.

Once again, we can rule out several technologies straight away. Flywheels leak too much energy. Batteries have combined power and storage so become way too expensive when only used once per year and there are not enough storage caverns in the world to use compressed air.

That leaves thermal, pumped hydro and hydrogen.

Heating water using cheap scaled up solar PV in summer ($2c per kWh) and re-using the energy for low grade heating in winter can be

---

xi Power costs cover the equipment required to convert the incoming electrical energy into stored energy and back to useful energy. Including the grid connection equipment for batteries, the motors, pumps, and generators used in flywheels, pumped hydro, gravity, and compressed air and the electrolysers, reactors, and fuel cells, for hydrogen.

Storage costs cover the materials or infrastructure required to store the energy. This includes the chemicals used in battery cells, the reservoirs of pumped hydro, concrete blocks of gravity storage, and caverns used for compressed air or hydrogen.

accomplished with efficiencies of 50% if stored in large enough volumes and well insulated.[26] Residential and commercial space and water heating accounts for 15% of final energy use (15 trillion kWh per year) which is significantly more than the 4 trillion kWh seasonal storage imbalance. The infrastructure required to deliver the energy to buildings will add to the cost and may ultimately limit uptake. But thermal energy storage is a low-cost, low-risk, seasonal storage solution for heating which could definitely play a large role in correcting our seasonal imbalance.[27,28]

Pumped hydro can also be used for seasonal storage. Nearly all the costs involved in building a pumped hydro facility are in the power generation end, including the dam, tunnel system, and generator. The cost of preparing the reservoir is comparatively small given the right location. J. Hunt et al. estimate there are up to 17 trillion kWh worth of potential locations which could cost less than $5c per kWh for seasonal stored energy – four times our seasonal storage needs.[29]

Hydrogen ecosystems are the final part of our seasonal storage solution. Hydrogen has an extremely high specific energy and a good energy density; you can store lots of energy per kg or cubic metre of compressed gas.[30] This makes seasonal storage cheap and it means hydrogen is an extremely useful form of energy for transportation or high temperature industrial heating. Using cheap solar electricity, the cost of green hydrogen would be $7c per kWh today and $5c per kWh once the technology is scaled[xii] and optimised.[31,32,33,34]

Finding the best route to using hydrogen is more difficult. If you turn it back to electricity you add the costs and inefficiencies of a fuel cell or modified gas turbine which doubles the cost to more than $10 per kWh.[35,36]

Using the hydrogen directly in the natural gas network for heating (where heat cannot be electrified) is probably the most cost-effective route but is logistically difficult. Hydrogen is more corrosive than natural gas and burns with a hotter, longer flame. A 100% hydrogen network would require upgrades to the pipeline and all gas burning appliances. It is a

---

xii Hydrogen electrolysers use electricity to convert water into hydrogen. The maximum theoretical efficiency is 83% but real world is closer to 50-70%. The first green hydrogen was produced as far back as the nineteenth century, when Danish inventor Poul La Cour designed a windmill to generate electricity and power the electrolysis of water into hydrogen, which he used to fuel his gas lamps.

Electrolysers today cost around $500 per kW power and the efficiency is around 60%, that means producing green hydrogen (from water) using cheap renewable power would cost around $7c per kWh today and around $5c per kWh once the technology is scaled and optimised.

large undertaking but has been done before in the UK which switched from town gas (50% hydrogen) to natural gas in the 1970s and this change should add just $0.5c per kWh.[xiii]

Finally, installing an overcapacity of cheap solar power is an alternative to using seasonal storage, so that even during the seasonal low in winter they provide enough energy. Based on our 50:50 solar:wind scenario we would need to add two thirds more solar panels to cover the winter shortfall. This eliminates the need for the annual 4 trillion kWh of seasonal storage and also eliminates the need for 10% of short-term storage through the year (another 7 trillion kWh) and works out at the equivalent of $7c per kWh.[xiv] The drawback of overcapacity is it cannot support short term storage in times of extended low renewables output.

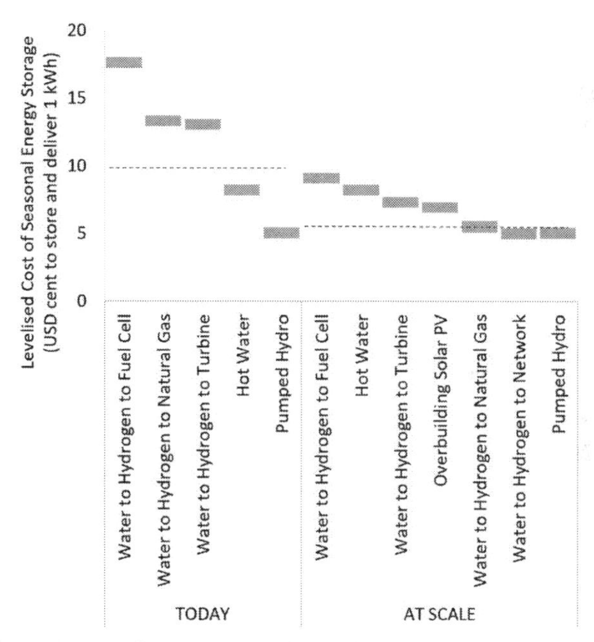

Levelised cost of seasonal energy storage. Shows the levelised cost of charging and re-using 1 kWh of energy assuming one full roundtrip charge/discharge per year. Based on charging with $2c per kWh cheap solar electricity, capital costs, and operating costs including roundtrip losses, installation life, cycle life, and leakage. Same 3% post tax, real, cost of capital as all other calculations (5% nominal, 6.5% nominal pre-tax). Author's standardised LCOE calculations, based on data sources.[10,11,12,18,19,20,26,27,28,29,30,31,32,33,35]

xiii Based on cost estimates from the H21 NoE plans to switch part of the UK to hydrogen, the levelised cost would be around $0.5c per kWh to upgrade the network and appliances to hydrogen, plus the $5-7c per kWh to produce and store the hydrogen..

xiv The annual cost of installing an extra 4 TW of full-scale solar power is $700 billion per year or $7c per kWh.

Seasonal storage will likely be a mix of thermal, pumped hydro, and hydrogen serving heat, electricity, and transport or industrial needs, all helping to balance seasonal variation in renewables and supporting short term storage during extended poor weather conditions. The levelised cost range is large today but should trend towards $5-10c per kWh once the technologies are fully scaled. Whilst more expensive than short term storage, energy products like hydrogen add significant value in replacing oil or gas-based products in transport or industrial use and could act as last resort electricity generation using cheap open-cycle gas turbines for back-up power in the short term.

Transmission and Distribution of Energy Products

In addition to the costs of generation, storage, and carbon offsetting, we must also include the cost of transmission and distribution (T&D) of energy to arrive at a comparable fully loaded cost.

Solid, liquid, and gas fuels cost around $2c per kWh for transportation and distribution to customers. This includes the capital cost of transport infrastructure including lorries, tankers, and pipelines, energy to move the fuels plus the selling and administration costs from all the companies in the supply chain. Electricity transmission and distribution is closer to $4-5c per kWh due to the higher cost of building and maintaining the electricity network.

Electricity networks have been around in the US and Europe for over a century. It is believed the Gard du Nord in Paris was the first public building to be lit by electric light and the first public electricity supply was completed in late 1881 when Siemens installed a generator in the town of Godalming, UK, to which anyone could connect for a fee.[37] A year later, on the other side of the Atlantic, as Edison was turning on his generators at Pearl Street station in lower Manhattan, America's first electric grid was born. It also marked the beginning of a bitter war with competitor George Westinghouse. Thomas Edison was the flashy inventor funded by J.P Morgan providing direct current[xv] electricity to the nation. George Westinghouse was an army corporal, entrepreneur, and engineer who had already made his fortune pioneering the railway air brake. After being snubbed by Edison, Westinghouse teamed up with William Stanley to develop the alternating current technology which was emerging in Europe. He realised that by using alternating current he could increase the voltage and transmit electricity with

xv Direct current flows in one direction through the grid. Alternating current pushes electricity back and forward many times per second over the grid. AC current allows the voltage to be increased or decreased using transformers which use the magnetic field of an alternating flow of electricity.

smaller losses, before reducing the voltage again to make his power safe for household lighting. AC would require fewer generators, less copper, and would benefit from significant economies of scale. Edison may have had the fame and financial backing, but Westinghouse had the cheaper technology. The race was on.

The two became bitter rivals competing to electrify the US. Edison maintained AC current was dangerous, stating "Just as certain as death, Westinghouse will kill a customer within six months after he puts in a system of any size" and would electrify animals at press conferences to prove his point. Edison even played a role in the development of the electric chair recommending, of course, a Westinghouse generator. But the dangers never materialised, and The Westinghouse Electric Company gained more and more of the US market. The final blow for Edison came when Serbian-American engineer Nikola Tesla joined forces with Westinghouse and developed a motor which would work with AC current. Westinghouse could now provide not just electrical light but also mechanical movement at a fraction of the price of Edison. Westinghouse won the much-coveted contract to illuminate the Chicago World Fair in 1893 which was attended by one quarter of the US population and marked the end of the road for Edison.[38]

The technologies used to build out electrical grids across the UK, Europe, and America since the end of the nineteenth century have not changed much. The grid still relies on centralised electricity generation, alternating current, voltage transformers, and a central operator to keep all the plates spinning. Electricity remains the only major commodity in the world which must balance supply and demand in real time with no means of storage. Building out a greater share of renewable electricity will require not only storage solutions but will also create the opportunity to build a more robust grid with distributed generation, two-way flow, and digital smart control. As Jeffery Rifkin, American socio-economic theorist and activist, puts it: "The dumb centralised electricity grid will have to be reconfigured into a smart distributed digital Renewable Energy Internet".[39] This will require investment in high voltage transmission lines to create larger more resilient networks, sensors, and smart control across power lines and substations. Plus, the final leg of distribution to homes and industry will require smart meters, intelligent transformers and database control.

The extent to which transmission or distribution is built out depends upon the mix of generation. More utility scale wind and solar farms will require more transmission to deliver the electricity from more remote locations into cities or to interconnect regions. Ironically, the long-distance transmission lines in the future could use DC current, with the most efficient systems now converting the high voltage AC to DC, transmitting the power over vast distances and converting it back to AC again to step down.[xvi]

By including more distributed residential and commercial renewables on the grid (such as rooftop solar), you save on transmission, requiring only upgrades to the local grid to better handle two-way power supply. Estimates suggest 25-40% of today's power requirement could be supplied by local users "behind the meter".[40,41] Savings on transmission and distribution (T&D) costs are offset by the higher installation cost of smaller projects.

The Electronic Power Research Institute estimates a total investment of $450 billion is required to transform the US to a smart electricity grid. For each billion kWh of annual electricity consumption, it costs $70 million to upgrade the existing network and will cost $250 million per billion kWh per year to add new smart capacity. This adds a levelised cost of $0.3c to $1.5c per kWh of renewable electricity.[42]

In 2020, California found out the hard way about the importance of addressing centralised power plant vulnerabilities and the importance of sufficient storage and diversification of electrical supply[xvii]. In the midst of the global Covid-19 pandemic, with stay-at-home orders in place, the state experienced one of the strongest heatwaves on record with Death Valley reaching the highest temperature ever recorded on Earth. Wildfires were raging across the state turning the sky an apocalyptic red, with choking smoke blotting out the sun across much of the west coast. As temperatures rose, so did the use of air conditioners; electrical demand surged. California called on neighbouring states for more power, but the same extreme heat left no spare generation and blistering temperatures were reducing transmission line capacity. On 14 August as the sun started to set, air conditioners stayed on, but solar generation declined. Yet the grid operator hadn't updated their forecasting models to account for the growing share of solar power and had forgotten to include the planned shutdown of two natural gas facilities. Combined with a drop in wind speed, a lack of storage, the lower efficiency operation of thermal

xvi A transformer consists of two separate coils of wiring wrapped around an iron ring. The primary coil creates a fluctuating magnetic field as the electricity alternates back and forward which induces an alternating current in the secondary coil. By changing the ratio of loops in the two coils you can either increase or decrease the voltage. Electricity only flows in one direction in direct current so transformers will not work, and the voltage is fixed. Higher voltage transmission creates less resistive losses. Today's global electrical grids average about 10% losses which range from 2% in Singapore up to 70% in Togo.

xvii Electricity Generation in California: Natural Gas 47%, Solar PV 14%, Wind 8%, Hydro 11%, Nuclear 9%, Other 11%.

power plants due to the heat, and failures in the power bidding market which instructed some generators to *actually* export power, California was running out of juice.[43] If electrical demand outstrips supply and the grid fails, it can destroy equipment and take weeks to bring the grid fully back online.

Fearful that reserve limits of 6% were about to be breached, the operator forced rolling blackouts across the state for two consecutive days, leaving hundreds of thousands of people with no power and no way to stay cool. Those in the poorest neighbourhoods were hit hardest with sweltering temperatures magnified thanks to the mostly concrete surroundings and the highest concentration of blackouts.[44] California is the perfect example of climate change in action, the growing threat of physical impact from extreme heat and fire, the inequality of damages, the need for adaptation, and the importance of building more resilient systems. California is also the wealthiest state in the wealthiest country in the world. Imagine the destruction which goes unreported across much of the planet.

Fully Loaded Energy: Aggregated Supply, Storage, Offsetting, Transmission and Distribution Costs

Finally, we can calculate the fully loaded energy cost to include generating, storing, offsetting, transporting, and distributing energy for a net-zero system.

Coal and natural gas heating have an extremely low cost of extraction and generation at just $1-2c per kWh heat. To offset the carbon emissions adds more than 50% to energy costs and the distribution network another $1-2c per kWh. But with fully loaded costs of less than $4c per kWh fossil fuel heating will likely remain the cheapest form of energy, though use limited by CO_2 storage capabilities which will probably only cover just 10% of annual emissions at best.

Oil is the premium fossil fuel with double the cost of natural gas or coal. Corn ethanol is a proven renewable alternative to oil but cannot scale up due to the land requirements for growing corn. Third generation algae oil is significantly more expensive today but once scaled may offer a more sustainable, scalable, and zero carbon alternative at a similar price to oil.

Solar panels and wind turbines are already generating renewable electricity at a similar cost to coal or natural gas. Once scaled, and pushed down the experience curve, generation costs will become significantly cheaper. Assuming half of all renewable generation will require short term storage, this adds $2-3c per kWh cost today, but just $1c per kWh once storage is scaled.

Electricity from any source comes with another $4-5c per kWh costs for transmission and distribution to build and maintain the network.

Nuclear electricity and gas electricity with carbon capture will end up more expensive than wind-and-solar-plus-storage but may need to form a baseload to optimise the grid. The cost of coal electricity is uncompetitive.

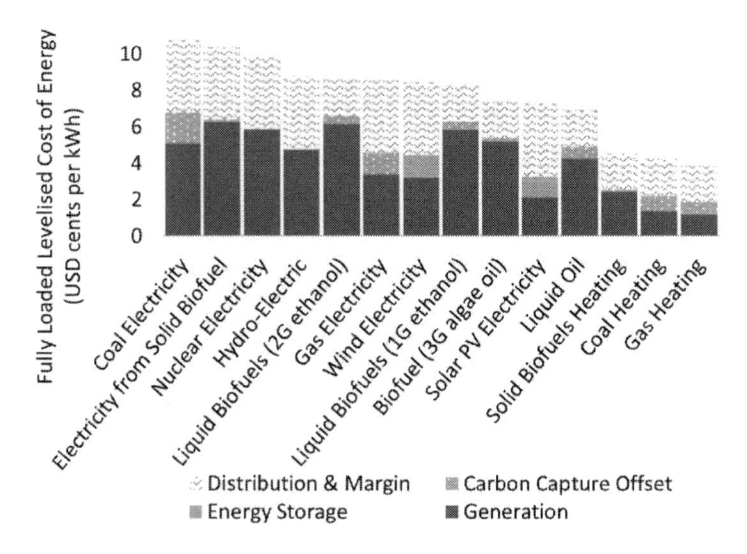

Fully Loaded Levelised Cost of Energy shows the combined cost of generation, energy storage, carbon capture, and energy distribution to provide a net-zero energy supply. Wind and Solar have 0.4 kWh of storage for every 1 kWh of generation using mostly lithium-ion and pumped hydro at $2.3c per kWh capital and operating cost. Each technology has lifecycle carbon emissions offset based on 70% Carbon Capture and Storage, 20% reforestation and 10% enhanced rock weathering costs of $24 per tonne. Electricity T&D is $4c per kWh, distribution costs for everything else is $2c per kWh. Author's standardised LCOE calculations based on extrapolated experience curves and expert analysis of floor costs for each technology, data sources referenced through the preceding text in this chapter.

Future Energy: Building a Safe, Reliable, Cost Competitive Net-Zero Energy Supply

Having run through our energy supply options we can start to form an idea of how to build a safe, reliable, cost competitive, zero carbon energy supply. However, without considering the demand side we can't yet form a full picture – that will come in the next chapters. But what we do know so far is:

- Wind and Solar PV are the two stand out technologies which can provide safe, sustainable, zero carbon energy at scale with competitive fully loaded costs. They should form the backbone of a net-zero energy supply. Electricity will have to increase from 21% of supply to become the primary source of most energy – electrify everything where possible.
- Wind and Solar PV are intermittent generators and require storage. Pumped hydro, lithium-ion, and a handful of other options provide scalable solutions. Expert opinion is still divided on specifics, but studies have shown that a small percentage of non-intermittent baseload might significantly reduce storage costs.[45] Investing in the research and development (R&D) for next generation nuclear or carbon capture and storage should provide a fall-back option in case baseload is needed down the line.
- Despite the potential cost declines of wind and solar electricity, fossil fuel energy used directly for heating will remain cheap and difficult to displace. However, the physical sequestration limits and risks of carbon capture and storage or negative emissions technology mean we must find alternative solutions where possible.
- Use of indoor dung or wood burning should be eliminated given the extremely detrimental health impacts from air pollution.
- Oil can play little role in a net-zero economy. Carbon capture for transport is not possible and negative emissions technology can cover just a fraction of oil emissions. First- and second-generation liquid biofuels production are limited by low power density and land availability. Oil must be replaced by either electricity, hydrogen, or third generation algae oil.

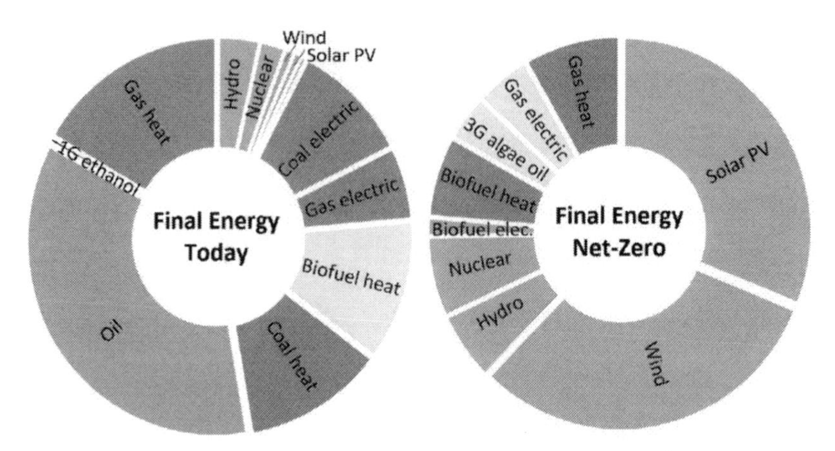

Left: Mix of final energy today, data based on IEA final consumption.[46] Right: Possible mix of final energy in our net-zero scenario.

Today's economy uses a relatively balanced mix of final energy consisting of solid biomass and coal (21%), liquid oil (42%), natural gas (15%), and electricity (21%). Final energy use, or demand, in our future economy must adapt to run on a zero-carbon energy supply. Renewable electricity with short term storage must form the backbone of our zero-carbon energy supply, supported where necessary by fossil fuels with carbon capture, nuclear, hydrogen gas, and solid or liquid biofuels.

The following chapters will run through energy use across transport, industry, amenities, and agriculture. We will examine the form in which energy is used today and where it is possible to switch to a zero-carbon alternative. We will also examine the cost of switching and any opportunities to utilise our remaining mitigation tools to improve efficiency or reduce consumption whilst improving the global quality of life.

Chapter 9: Transport in a Net-Zero Future

In this chapter we will explore how the transport sector can be re-engineered to run on a net-zero energy supply.

Transport Today: Energy Use and CO_2 Emissions Breakdown

The transport sector accounts for 33 trillion kWh of final energy consumption per year which represents 32% of global final energy use. The associated CO_2e emissions are 9 billion tonnes per year which represents 17% of total. Liquid oil products account for over 90% of the energy used (refined gasoline, diesel, and jet fuel).[1,2,3]

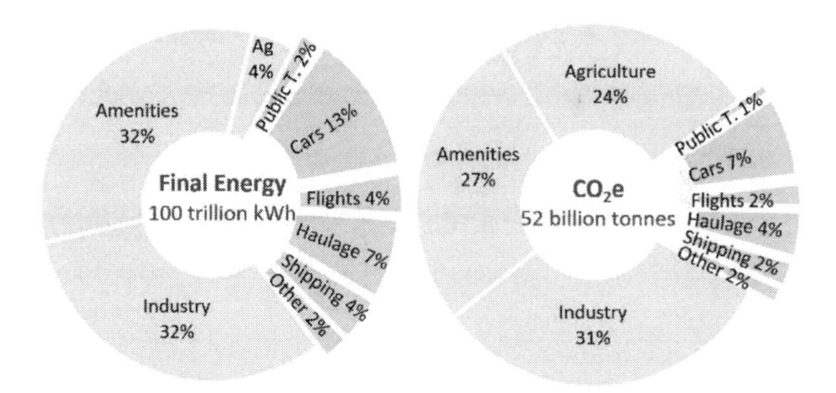

Passenger Transport is 60% of total transport energy use and CO_2e emissions with 75% from cars, 15% from air travel and the remaining 10% from rail, buses, and motorbikes. The average person on the planet travels around six thousand miles per year.[i]

i Passengers in developed countries travel up to 15 thousand miles per year and those from developing countries only a few thousand. Around 10% of the global population accounts for 80% of motorised miles travelled with much of the world's population hardly travelling at all. IPCC, "AR5 WG III", 606.

Consumer Goods Freight represents 40% of transport energy use and CO_2 emissions, with one third from shipping, one third from smaller trucks and vans, and just less than one third from heavy trucking (a small amount is from rail and pipeline). The average person on the planet consumes food, goods, and fuel every year which has travelled a total of ten thousand tonne-miles.[ii]

Transport Tomorrow – Planning for a 'Peak Travel' Future

In our peak energy world, the global population will increase from 8 to 10 billion people. Should everyone on the planet enjoy a European standard of transport, passenger-miles on land double, passenger-miles in air triple, and freight miles increase by half. This will increase transportation energy consumption to 76 trillion kWh and increase carbon dioxide emissions to 22 billion tonnes CO_2e, assuming no change to our transport system.

So how do we decarbonise peak transport? Clearly, cars, vans, buses, and trucks should be the first target areas as they represent 70% of energy use and CO_2 emissions, and are by far the most inefficient mode of travel and freight. The remaining 30% of energy and emissions are from passenger planes and freight ships which need to decarbonise longer term.

We need to take today's transport system, which is mostly run on oil products using internal combustion engines, and switch it to run on electricity, hydrogen, or biofuel. Let's review the technology options:

Cars, Vans, and Small Trucks: Net-Zero Technology Options

We can calculate the energy consumption of an average car using some basic physics.[4] The two main energy uses in driving are accelerating from standstill and overcoming air resistance when moving. The energy of driving can be calculated from the formula for kinetic energy:

$$\textbf{Kinetic Energy} = \tfrac{1}{2} \times \text{mass} \times \text{speed}^2$$

---

ii Shipping, rail, and pipelines consume less than 0.1 kWh per tonne-mile (energy to move 1 tonne one mile). Long distance road haulage uses 0.3 kWh per tonne-mile and smaller trucks and vans up to 1 kWh per tonne-mile. Again, road represents the most inefficient mode of freight travel, and although shipping represents one third of energy consumption, it moves more than two thirds of global goods tonne-miles.

A 1,500 kg car should consume about 300,000 joules or 0.21 kWh of energy per mile during urban driving.[iii]

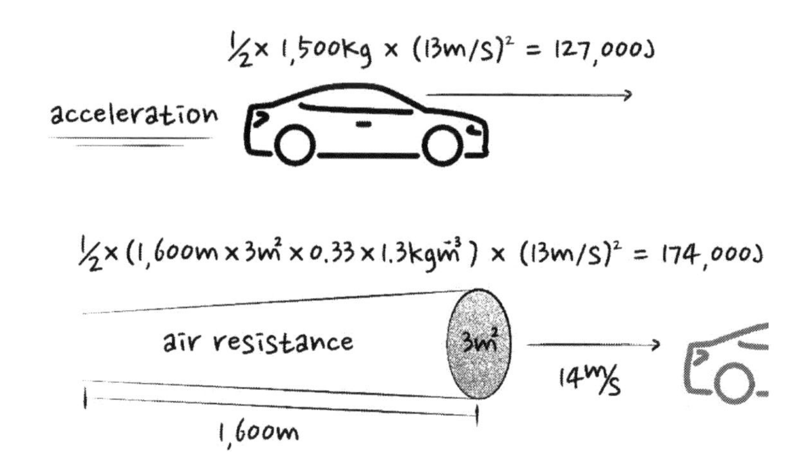

Internal Combustion Engines: Most cars on the road today are gasoline fed Internal Combustion Engines (ICE). These vehicles use the combustion of gasoline and air inside the engine cylinders to move pistons using the expansion of hot gases. These pistons then move the crankshaft and turn the wheels of the car. Modern gasoline cars have a fuel economy of 40 miles per gallon, the equivalent of 0.81 kWh per mile – four times greater than our theoretical car. Why? Well gasoline engines are only 25% efficient. Just 0.21 kWh of energy is used to move the car one mile, the other 0.6 kWh is wasted as heat and noise from the engine.

A basic gasoline car costs $20,000 of which $2,000 is the engine. The average car will last 15 years and drive 12,000 miles per year with an average of two passengers. Based on the costs, fuel efficiency, gasoline

---

iii Accelerating 1,500 kg of car from standstill to 14 metres per second (30 mph) uses 127,000 joules of energy or 0.04 kWh. Assuming 4 stop-starts per mile during city driving and acceleration will use 0.16 kWh per mile.

Once the car is moving at a constant speed, it must push a tunnel of air out of the way. The volume of air moved per mile is the front area of the car (3 square metres) times the distance travelled (1,600 metres or 1 mile) times the drag coefficient (0.33). Multiply this tunnel of air by the density of air (1.3 kg per cubic metre) and you have the mass which must be moved – equal to about 2,000 kg of air per mile. So, if driving at a speed of 14 metres per second, it requires 174,000 joules or 0.05 kWh per mile to overcome air resistance.

A total energy consumption of 0.21 kWh per mile in urban driving.

price, and our 3% discount rate, the levelised cost of buying and running a gasoline car is about $10c per passenger-mile ($7c capital cost, $3c fuel costs).

Internal combustion engines are very inefficient (25-30%) but because gasoline and diesel liquids can pack so much energy into a small space (10,000 kWh per m³) just one tank of fuel can propel a car over 500 miles. This is the reason combustion engines have gained dominance over the last 100 years. The challenge for an alternative technology is providing the required range for the vehicle without adding too much weight, volume, and cost to the car.

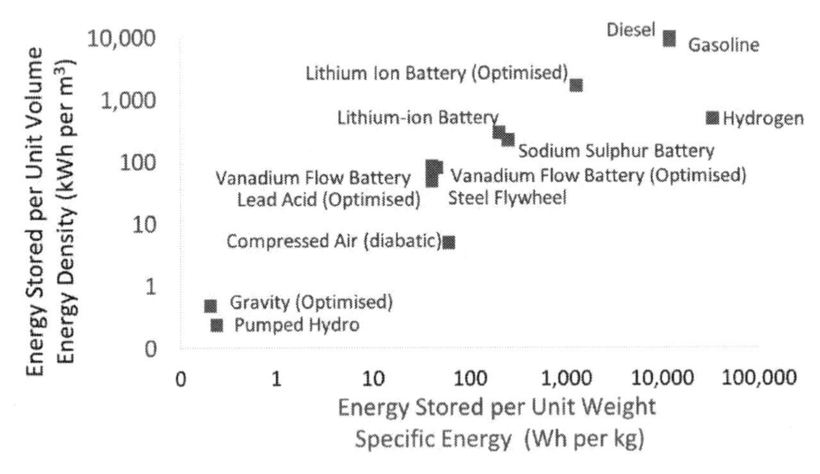

Specific energy gives the energy storage capacity per unit weight. Energy density gives energy storage capacity per unit volume. Both are important because weight and space design constraints will limit the range between every re-charge and the efficiency of the vehicle.[5,6]

Electric Cars: Roll back 120 years and the dominance of the combustion engine vehicle was less than certain. Electric cars had come to market in the 1890s and, by the turn of the century, were outselling their gasoline counterparts. Electric cars were the vehicle of choice for many city drivers with no vibration, smell, difficult gears, or noise; and car charging networks were being built up in big cities like New York by the Edison Company. But Ford's pioneering mass manufacturing techniques for the Model T dramatically reduced the cost and refined the performance of gasoline powered vehicles, so that electric innovation could not keep pace and it would take another century before competitive electric vehicles would resurface.[7]

The modern electric car replaces the internal combustion engine with a lithium-ion battery used to drive electric motors which are over 90%

efficient, with very little waste heat or noise. Moreover, some of the energy usually lost during braking can be redirected to re-charge the battery. A Tesla model 3 uses around 0.3 kWh per mile (120 mpg equivalent) which equates to around 75% efficient compared to our perfect car and three times better than a modern gasoline equivalent.[8]

Despite the better efficiency, the biggest economic hurdle for electric vehicles is the upfront cost of the battery. The levelised cost of buying and running an electric vehicle today is $12c per passenger-mile or 20% more than our gasoline car. But once electric cars are produced at scale and run on renewable electricity, the cost should drop to $8.5c per passenger-mile or 15% cheaper than a gasoline car. Fuel costs for electric cars are just $1c per passenger-mile because despite electricity being more expensive than gasoline, the efficiency of the car is three times better. For drivers that do a lot of miles, electric will already be cheaper.[iv]

Tailpipe emissions of electric cars are zero, but of course electricity mostly comes from fossil fuels today and the battery adds ~3 tonnes of embedded CO_2 emissions in production. Our average electric car, driving 12,000 miles per year, will emit 1.5 tonnes of CO_2e if run on today's grid electricity.[v] Our average gasoline car will emit 2.5 tonnes of CO_2e per year. The electric car will have made up for higher production emissions with lower running emissions in three years and over the life of the vehicle will save more than 13 tonnes of CO_2. Electric vehicles already generate less CO_2 in nearly every region of the world today, but importantly, battery electric vehicles offer a mode of transport for cars, vans, and small trucks which can efficiently use renewable electricity rather than oil, facilitating a shift to our zero-carbon economy.[9]

---

iv Lithium-ion battery packs cost around $150 per kWh storage today. So, if we want a car which can drive 300 miles on a single charge then we need 90 kWh of battery storage (300 x 0.3) which costs $13,500, weighs 500 kg, and takes up over 300 litres of space. The weight and space requirements are certainly more troublesome than incorporating a gasoline tank but ultimately do not limit the car design. The main limiting factor is the upfront cost of buying the battery. However, we also ran through the rapid rate at which battery costs are dropping as production scales up. Once the price of storage reaches $50 per kWh the battery pack costs just $4,000 per car.

v Grid electricity is roughly 500g CO_2 emissions per kWh.

Running out of battery (range anxiety) and lack of charging infrastructure are near term hurdles to wider adoption, but the increasing range of vehicles and increasing numbers of charging points (which add just $0.2c per p-mile) are addressing these concerns. I would expect battery electric vehicles to become the dominant technology for cars, vans, and small trucks over the coming years.[vi]

Hydrogen Fuel Cell Cars: Fuel cell vehicles use stored hydrogen mixed with atmospheric oxygen to produce electricity and water. The electricity is used to charge a small battery and drive an electric motor. Hydrogen fuel cell vehicles offer a route to zero carbon by using either hydrogen made from water splitting or hydrogen from fossil fuels with carbon capture. But the added step of turning electricity to hydrogen fuel and back into electricity again reduces the efficiency of the whole process towards 35% - better than a gasoline car but significantly lower than a pure electric vehicle.

The cost of fuel cell vehicle driving today is around $14c per passenger-mile (20% more than an electric car), and the cost at scale would be around $10c per passenger-mile (15% more than electric).[vii]

Ultimately the technology will always be less efficient than electric because of hydrogen production, compression, and then conversion to electricity. The use-case for hydrogen vehicles is for longer range cars and faster refuelling times. However, most users will end up charging their electric vehicles like they charge their phones – overnight. On the occasional longer trip they can use fast chargers while taking a rest break, we already have batteries that can add hundreds of miles range in 15-30 minutes. I suspect that fuel cell cars may become stuck in a chicken-egg scenario where the hydrogen refuelling stations are not built for lack of demand, but no one is buying hydrogen cars because there are very few filling stations. In the meantime, we are becoming accustomed to electric, which will prove cheaper in any case.

---

vi Other types of electric vehicles are on the market including hybrids which have regenerative braking and a small 1-kWh battery plus a gasoline engine. Plug-in hybrids have a 15-kWh battery that will drive 50 miles and a combustion engine. Both technologies are more efficient than a basic gasoline car (60 and 80 mpg respectively) but still use gasoline or diesel so cannot form part of a zero-carbon future unless the emissions are offset with negative emission technology.

vii Hydrogen fuel cells cost $50 per kW power today, and hydrogen storage tanks costs $15 per kWh. A standard 150 horsepower, or 110 kW, car with enough storage to drive 300 miles requires $10,000 of power and storage equipment in the place of a $2,000 combustion engine, or $13,500 battery. Once scaled, fuel cell costs could potentially halve.

Cars, Vans, and Small Trucks: A Transition Underway

Cars, Vans, and Small Trucks are already going through a transition to electric with nearly all major car makers announcing the expansion of electric car models and cutting back investment in developing new gas and diesel models. The move to electric vehicles has been slow coming, with many false starts, but now seems inevitable. The tipping point for the transition came about through a combination of innovation, legislation, and scandal.

First came the launch of the Toyota Prius in the year 2000 – this was the first mass market hybrid vehicle and set the standard for low CO_2 emissions and high fuel efficiency. However, it wasn't until 2011, when a combination of rising oil prices and celebrity owners who championed the eco credentials (Leonardo DiCaprio, Harrison Ford, Jennifer Aniston, and Tom Hanks) that global sales accelerated. Next came the all-electric Nissan leaf in 2010 which at $30,000 was priced for the mass market and was one of the best-selling electric vehicles in the world. Meanwhile, Elon Musk was building out Tesla Motors. Musk had made his first fortune co-founding PayPal and subsequently selling the company to eBay in 2002. In 2004, Musk bought a stake in the start-up Tesla Motors, taking over as CEO in 2008. Tesla's first car, the Tesla Roadster, was released in 2008, a high-performance sports car costing more than $100,000. Musk's vision was to build out a car company like a technology company by entering the luxury high value and high margin part of the market first (to fund and prove the technology), and then raise money through public markets (initial public offering in 2010) to fund the mass market commercialisation of cheaper models. In 2019, came the launch of the $35,000 fully electric Model 3, which has now taken the top spot in electric car sales. Tesla has become the world's most valuable car company, despite producing less than 1% of new cars: a clear sign that investors believe the future is electric.

Over the last decade, and in response to concerns growing around air pollution and CO_2 emissions, Europe, the US, and China have tightened emissions and fuel efficiency regulation into 2020 and beyond.[viii] European car companies such as BMW, Daimler (Mercedes), and VW must produce a certain number of hybrid or electric cars or face significant fines. The US must rein back gas guzzlers and the Chinese government continues to push their national champion car makers

viii Obama toughened the miles per gallon standards in the US, Europe has reduced gCO_2 per km standards for new cars by over 30%, and China now sets quotas for low emissions vehicle production.

such as SAIC, Dong Feng, and FAW into electric vehicles, attempting to build a battery car brand to export globally – something China was unable to accomplish with combustion engines.[10]

Electric vehicle launches and regulation were already pushing the market away from combustion vehicles when scandal proffered a further helping hand. In 2014, the California air board commissioned a study on the emissions discrepancies between US and European vehicles. A group of scientists at West Virginia University, who were helping with the study, detected significant differences between the emissions of VW diesel cars when tested on the road rather than in the lab. These diesel vehicles use a catalyst system which sprays ammonia (ad blue) into the exhaust to get rid of harmful nitrous oxide emissions. Effective at reducing NOx, this system also, however, reduces the fuel efficiency of the car.

VW were not satisfied by this trade-off: they wanted to market a clean and fuel-efficient diesel car. So, someone at VW decided to cheat. This involved fitting a defeat device into the vehicle design thereby enabling NOx removal when the car sensed it was being tested in a lab and was being driven at a constant rate with no wheel turns, but a device that also disabled the system when involved in normal driving. VW diesel cars could therefore beat the emissions tests whilst claiming superior fuel efficiency over competitors. But this also meant dangerous levels of nitrous oxides being released into cities all over the world. VW has recalled all affected cars and lawsuits continue to pile in from around the world. The corporate damages are estimated at up to $35 billion, one of the largest in history. VW lost 60% of its value ($75 billion of market cap) in the 6 months following the scandal. Now under control of new CEO Herbert Diess, the car maker is trying to shake off the demons of the past by pushing hard into electric cars with plans to launch 50 new models by 2025. Sensing a change in consumer attitude, many other global carmakers have followed suit. The scandal also acted as a catalyst to accelerate tightening of emissions legislation out to 2030 and the bringing in of real road testing, rather than simply lab testing vehicles.[11]

Overall, a combination of innovation, legislation, and scandal have acted to push the car industry towards low emissions vehicles. Battery technology has shown the best combination of cost and performance and the price of lithium-ion will continue to decline at a rapid rate as the experience curve drives a lower cost of manufacture. Electric car production is still in its infancy with only ~5% of new cars electric, but production is growing exponentially.

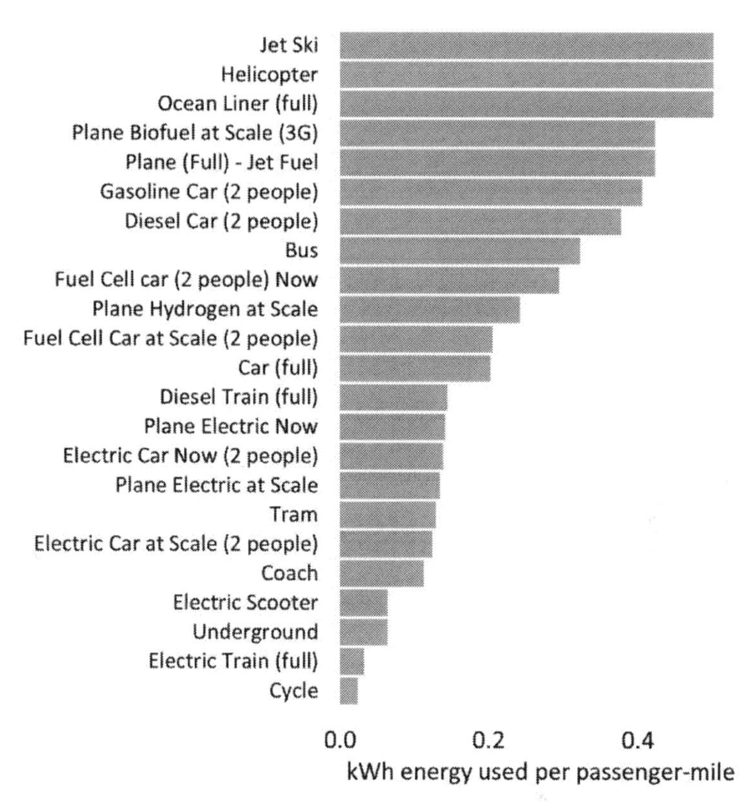

Energy Intensity of different modes of travel. kWh energy used to transport one passenger one mile.[4]

Heavy Duty Freight Trucking: Net-Zero Technology Options

Large semi-trucks used for transporting products along highways tend to spend most of their life travelling at a constant speed. To figure out the energy consumption of an idealised truck we just need to calculate air resistance. Using the same formula for kinetic energy, this equals 3.6 million joules or 1 kWh per mile which is the equivalent of 34 miles per gallon.[ix]

Internal Combustion Engines: Nearly all large trucks on the road are diesel combustion engines which achieve around 10 miles per gallon fuel

ix Based on a truck driving 30 metres per second (70 mph) pushing a mass of air weighing 8,000 kg every mile (11 square metres area x 1600 metres x 0.36 drag coefficient) this equals 3.6 million joules.

efficiency or 3.8 kWh per mile, 25-30% efficient. The maximum truck weight allowed on many roads is about 36 tonnes and each trailer can hold around 80 m^3 of goods.

A basic diesel truck will cost about $120,000 of which $5,000 is the engine. Large freight trucks last ten years and average 100,000 miles per year so fuel economy is of upmost importance. The unloaded weight of the truck is about seven tonnes, leaving space for 29 tonnes of freight. Adding this up means a fully loaded diesel truck can move freight for about $1.4c per tonne-mile ($1c fuel and $0.4c capital cost).

Electric Trucks: An electric truck is 80% efficient but a battery big enough to drive 500 miles adds $100k to the upfront cost of the vehicle today. The total buying and running cost works out at $1.4c per tonne-mile, on a par with diesel, but will decline to $1c per tonne-mile at full scale battery production, 40% cheaper than diesel.[x]

However, increase the range and the economics become less competitive because capital costs increase, and the haulage weight declines.[12] Electric trucks are best suited to operators who require less than 600 miles range (up to 8 hours highway driving) between rest stops which is a legal maximum in many countries.

Hydrogen Trucks: Hydrogen fuel cell trucks have a lower efficiency at 35-40% or 2.5 kWh per mile (14 mpg equivalent) but the capital cost of the de-coupled fuel cell and hydrogen storage may start to offer better economics for ranges longer than ~600 miles before recharging/refuelling. Around 10% of trucking haulage travels further than 600 miles range.[13] The cost of a 1,000-mile hydrogen fuel cell truck works out around $1.8c per tonne-mile today and $1.1c per tonne-mile at full scale[xi] – cheaper than diesel once optimised.[14]

As development of electric cars and progress with fuel cells continues, it should be entirely feasible to replace diesel haulage freight trucks with electric or hydrogen fuel cell equivalents at the same cost per tonne-mile, if not cheaper. Electric will be better suited to shorter range vehicles (below 600 miles) and hydrogen better suited to longer range (~10% trucks). The problem of hydrogen refuelling, or electric truck charging, should be less challenging given the set routes for most long-distance

---

x An electric truck is 80% efficient, uses just 1.25 kWh electricity per mile fully loaded (27 mpg equivalent), and has a fuel cost of just 0.3c per tonne-mile [efficiency based on Tesla, "Semi", [online], tesla.com, accessed 2020]. The cost of the battery for a 500-mile truck is around $100,000 today and $30,000 at full scale. The battery weighs up to 4 tonnes and slightly reduces the total freight capability.

xi Building a hydrogen truck with a 1,000-mile range (comparable to long distance diesel) would require 450 kW power (600 horsepower) which would cost $60,000 in fuel cells and hydrogen storage today, and $24,000 at scale.

haulage. Many of the world's largest truck makers are already developing electric/hybrid/fuel cell technology and Tesla are poised to be first to market with their all-electric Semi in 2021.

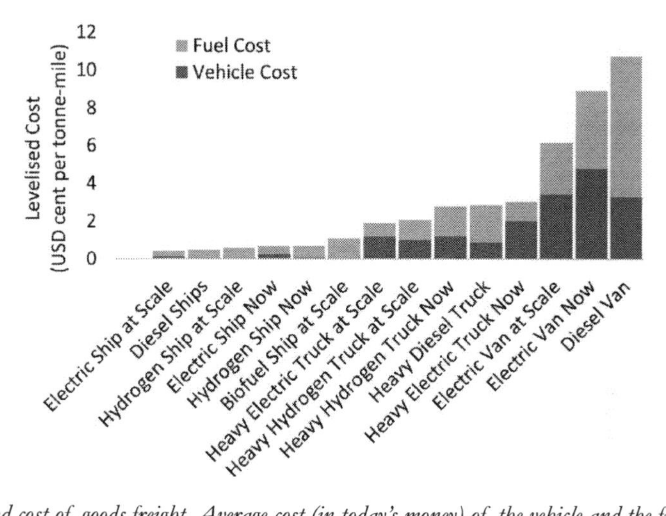

Levelised cost of goods freight. Average cost (in today's money) of the vehicle and the fuel per tonne-mile. Based on energy prices pre-tax of $7c per kWh diesel ($2 per gallon, $0.5 per litre), $8-10c net-zero electricity, $9-13c per kWh ($3-5 per kg) green hydrogen and average vehicle prices adjusted for battery or fuel cell capital costs. Author's standardised LCOE calculations based on the cost and performance estimates given in preceding text and in chapters 7 and 8.

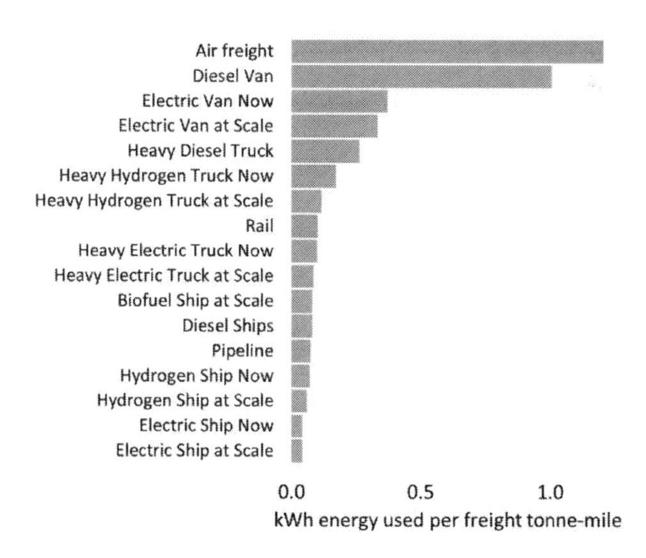

Energy Intensity of different modes of goods freight. kWh energy used to transport one tonne one mile. Based on author's estimates and from the US Department of Energy report.[15]

Air Travel: Net-Zero Technology Options

Our fundamental physics gets a little trickier with air travel. Planes push a tunnel of air out of the way in the direction of travel and simultaneously push a tunnel of air downwards to overcome gravity. There is a handy number called the drag-to-lift coefficient (about 0.035) which factors these effects, and when multiplied by the mass, the force of gravity (9.8m/s) and the distance travelled (1600 m or 1 mile), gives the required energy to fly a plane.[4] For an idealised plane this adds up to 0.5 million joules per tonne per mile or 0.2 kWh per tonne-mile.

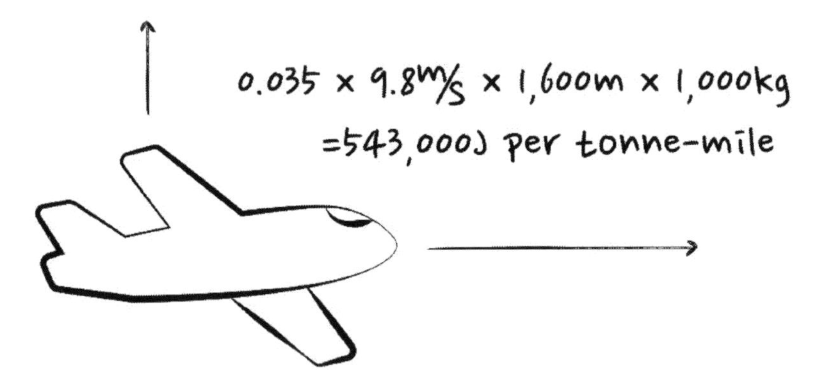

Combustion Jet Engines: Planes use jet fuel (from oil) to generate enough thrust to overcome gravity (lift) and air resistance (drag) and propel the plane forward at around 500 miles per hour. Real planes consume about 0.5 kWh per tonne-mile because once again combustion-based engines are only 30% efficient.

The total cost of flying a full plane, long haul, works out at $5c per passenger-mile ($3c fuel and $2c capital costs) or $350 per passenger from London to Singapore. But before you start asking for a discount on your next flight, that doesn't include operating and admin costs.[xii]

Long haul air travel will need to continue to burn hydrocarbons because batteries weigh too much, and hydrogen takes up too much space. Decarbonising long haul air travel requires negative emissions technology offset (planting trees or direct air capture) or the use of biofuels in place of jet fuel. Sustainable third generation biofuel algae oil

---

xii An empty Boeing 747 weighs 180 tonnes and has a maximum take-off weight of 400 tonnes. Jet fuel contains 12,500 kWh of energy per tonne. A 7,000-mile long-haul flight from London to Singapore requires 120 tonnes of fuel. This leaves 100 tonnes of spare weight for 500 passengers, crew, and luggage. A Boeing 747 costs $240 million and will last 30 years flying on average 1.7 million miles per year. [Spec from Top Speed, "1989-2010 Boeing 747-400", [online], topspeed.com, accessed 2020.]

could be produced at $6-7c per kWh if optimised – a comparable price to jet fuel today. Short haul and mid-distance flights offer other options.

Electric Planes: It should be possible to replace a plane's fuel and propellers or engine with batteries and electric motors or a modified electric jet engine.[16] Commercial planes fly 1.7 million miles per year, so fuel is the biggest cost item. Highly efficient electric planes could offer a cheap mode of air travel which runs off renewable electricity. The hurdle for electric air travel is range. To design an electric plane to fly long haul you need a battery weighing thousands of tonnes more than the maximum take-off weight.[xiii] The only solution is to reduce the range of the plane. A Boeing 747 with a battery that weighs 120 tonnes (leaving 100 tonnes for passengers and luggage) would fly 350 miles today and 700 miles once the technology is optimised. The costs work out at nearly half that of fossil fuel jet engines per passenger-mile.[xiv]

Electric planes are limited by weight to short haul flights only, but they offer a potential 30-50% cheaper option to current jet engine technology. There is already growing interest in electric planes for short haul travel with a company called Eviation having launched a prototype 9-seater electric plane that can travel 650 miles and is expected to enter service in 2022.[17] easyJet is also investing in the technology with their partner Wright Electric and expects to be flying commercial sized electric planes for short haul city flights within 10 years.[18,19]

Hydrogen Planes: Hydrogen planes could, once again, offer a middle ground between fossil fuels and electric. Renewable hydrogen could be used to power a fuel cell plane with a small battery and electric motors or modified jet engine. Hydrogen offers better fuel efficiency than jet engines, but lower than electric. Hydrogen can store more energy per tonne than fossil fuels but only one fifth of the energy per cubic metre and so space constraints limit hydrogen planes rather than weight. Adding any more than a 2,000-mile range would impact the number of

---

xiii An electric plane would consume around 0.18 kWh per tonne-mile (>80% efficient). However, batteries can store only 200-400 kWh per tonne (versus 12,500 kWh for one tonne jet fuel).

xiv The upfront capital cost of an equivalent 500-mile range Boeing 747 with a battery and electric propulsion would add just $10 million for the battery. The fuel costs would decline from $3c per p-mile (jet fuel) to £1c per p-mile (electric). So, the overall cost to buy and run an electric plane would work out at $2.5-3c per p-mile versus $5c per p-mile for a jet engine.

passengers or cargo the plane could carry.[xv]

Hydrogen offers a potential 20-30% cheaper option[xvi] to current jet engine technology and can be produced with no carbon emissions by water electrolysis or from natural gas with carbon capture. Zero Avia start-up in California is testing a retrofit small plane with fuel cell technology and they hope to start producing small aircraft with a range of up to 500 miles.

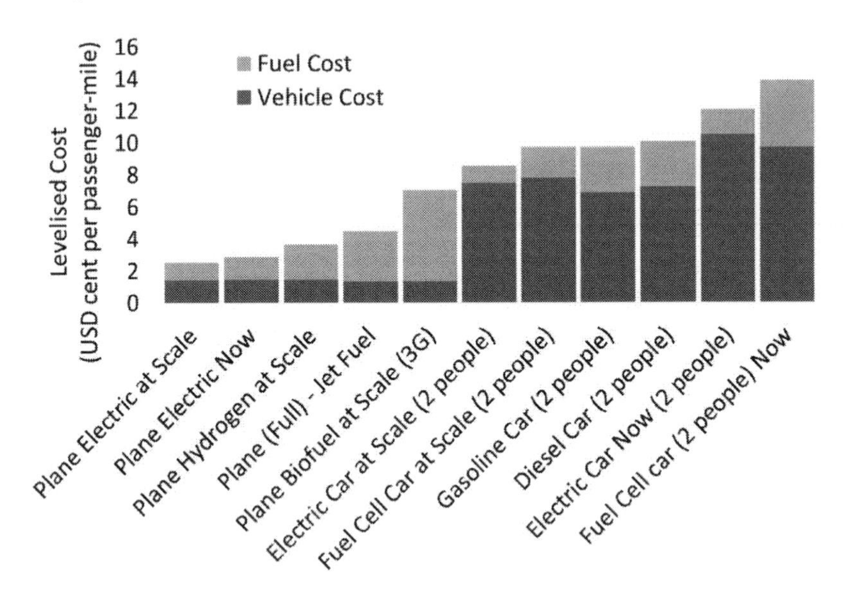

Levelised cost of transport. Average cost (in today's money) of the vehicle and the fuel per passenger-mile over the life of the vehicle. Based on energy prices including T&D (excluding government taxes) of $6c per kWh gasoline ($2 per gallon, $0.5 per litre), $7c per kWh diesel, $8-11c net-zero electricity, $9-14c per kWh ($3-5 per kg) green hydrogen and average vehicle prices adjusted for battery or fuel cell capital costs. Author's calculations for standardised LCOE based on cost data referenced in the preceding text.

xv A mid-range flight of 2,000 miles would require just 7 tonnes of hydrogen (hydrogen stores 33,000 kWh per tonne energy vs 12,500 for jet fuel) but would require over 150 cubic metres of space (hydrogen stores 1,400 kWh per cubic metre versus jet fuel 10,000 kWh per cubic metre). We assume hydrogen planes use hydrogen compressed at 700 bar. Alternatively, cryogenic liquid hydrogen stored at minus 253ºC could double the energy density and possible range but comes with added technical challenges.

xvi The upfront capital cost of a 2,000-mile range Boeing 747 with fuel cells and hydrogen storage would be $10 million more than a standard jet engine. The fuel costs would decline from $3c per p-mile (jet fuel) towards $2c per p-mile (hydrogen). So, the overall cost to buy and run a hydrogen plane would work out at $4c per p-mile versus $5c per p-mile for a jet engine.

Re-engineering Transport for a Peak Travel Net-Zero World

In a peak travel world, there could be 10 billion people on Earth enjoying a European standard of transport, travel, and goods freight. Global average passenger-miles travelled by bike, car, bus, and train will more than double and passenger air miles will triple. The average distance travelled per person will increase from 6.5 to 14 thousand miles per year. Global freight miles will increase by 50% with the average consumer needing 16 thousand tonne-miles of freight per year (from 11 now). Using today's transport systems that would increase energy demand from 31 to 75 trillion kWh, and this takes annual transport CO_2 emissions from 12 to 19 billion tonnes.

Re-engineering our peak travel world down to net-zero carbon requires rapid deployment of electric, hydrogen, and biofuel technologies for short-, mid-, and long-distance travel.

Road travel and freight accounts for 70% of total emissions and can be tackled fastest. Rolling out of better infrastructure for walking, cycling, and electrified public transport should be an immediate priority. Rapid bus transport, greater train services, and city trams are all established, proven ways to improve the efficiency and ease of public transport. Where private travel is required the easiest and most cost-effective option is to electrify cars, vans, and small trucks.

Electric vehicles use three times less energy per mile compared to combustion engines so despite the slightly higher upfront cost of the vehicle, electric cars, vans, and trucks will soon be cheaper to buy and run than their gasoline equivalents. A more extensive charging infrastructure, longer vehicle range, and a greater proliferation of designs will continue to convert consumers to electric and the industry is on the right track towards full electrification thanks to a triangulation of innovation, scandal, and regulation.

The physics and economics of batteries tell us that larger electric trucks for heavy duty freight should already prove cost effective up to 600 miles range. Once battery production is fully scaled up, electric freight trucks should work out 40% cheaper than their diesel counterparts. For trucking requiring more than 600 miles between stops, hydrogen fuel cells could prove the best technology. Once scaled, hydrogen trucks could deliver a 30% lower cost per mile compared to the best diesel trucks today. Electric and fuel cell designs are already close to hitting the road.

Passenger air travel remains one of the more challenging areas to decarbonise. Weight constraints for batteries mean electric planes may be limited to short haul only. Space constraints for fuel cell planes limit designs to 2,000-mile trips. Decarbonising long haul flights will need to rely on breakthroughs in third generation algae biofuel. Despite the difficulties, the basic maths tells us electric and hydrogen planes could work out 20-50% cheaper than today's jet engines. There are an estimated 100 companies around the world already working on developing the technology.

Finally, sea freight is the last area of transport which requires moving to net-zero. It is already the most efficient mode of moving goods but still accounts for 1 billion tonnes of CO_2. Again, short distance electrification and longer-range hydrogen power are the most feasible technology options.[20,21]

Shifting the transportation sector to net-zero carbon using ***today's*** technology costs, efficiency, and price of zero carbon energy would create 50% higher capital costs, 90% higher energy prices, but 55% less energy consumption. The total cost of buying and running a net-zero carbon transport sector would cost 20% more than sticking with the existing oil-based system.

However, experience curves mean that as the technology scales up, the costs decline. Repeating the calculation using ***full scale*** technology costs, a zero-carbon system will reduce energy consumption by 55%, increase capital costs by just 10%, and run on energy that is just 25% more expensive per unit. We can buy and run the future global transport system using zero carbon energy at 20% lower total cost than sticking with existing fossil fuel technology.

If we assume increasing penetration of ridesharing and more localised, efficient, industrial production as the Internet of Things (IoT) gains traction, this could lower total transportation system costs by another 20%.

Re-engineering the transport system for peak travel and zero carbon will cost the world up to 40% less than sticking with our current oil-based system. It allows developing economies to bring transport systems up to developed market standards and requires no compromise on quality of life in already affluent regions. More than 80% of the transport system will be cheaper using electricity rather than oil and we already have commercialised technology for two thirds of the transition. The biggest hurdle to decarbonising transport will be developing the remaining one third of technology which includes electric planes, biofuels, and hydrogen vehicles.

			Peak Transport (beyond the year 2060)			
Global Transport System (per year)		**Today**	**No System Change**	**Net Zero Today's Tech**	**Net Zero At Scale Tech**	**Net Zero At Scale + Efficiency**
Energy Consumption	k billion kWh	32	76	36	34	29
Blended Energy Price	$c per kWh	6	6	12	8	8
Levelised Energy Costs	$ trillion	2	5	4	3	2
Passenger & Freight	k miles/capita	17	29	29	29	26
Levelised Vehicle Costs	$ trillion	3	5	8	6	4
Total Costs	**$ trillion**	**5**	**10**	**12**	**8**	**6**
CO2e emissions	billion tonnes	9	22	2	0	0

Table showing the energy use, output, emissions, and total costs of our transport system now and in different futures. All estimates are based on the detailed estimates and solutions given through this section and using the energy supply estimates given in chapters 7 and 8.

Chapter 10: Industry in a Net-Zero Future

In this chapter we will explore how the industrial sector can be re-engineered to run on a net-zero energy supply.

Industry Today: Energy Use and CO_2 Emissions Breakdown

Industry accounts for 33 trillion kWh of final energy consumption per year which represents 30% of global energy use. The associated CO_2e emissions are 16 billion tonnes or 32% of total emissions. The burning of fossil fuels to power industrial processes releases 11 billion tonnes and the remaining 5 billion tonnes of CO_2e are released directly from the chemical reactions used in the production of cement and ammonia, decomposing chemical feedstock, associated methane leaks, and NOx emissions from acid production.[1,2,3]

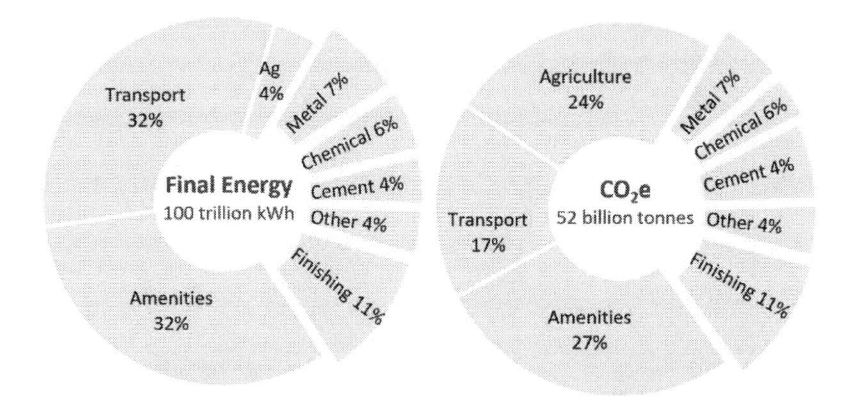

Industrial Energy Supply Mix: Coal (30%), natural gas (20%), and electricity (27%) are the main components of supply into industrial processes with oil (11%), waste heat from electricity generation (5%), and solid biofuels (7%) making up the total 33 trillion kWh. In addition, a further 10 trillion kWh of fuels are used as feedstock – with over 7 trillion kWh of oil turned into chemicals and plastics and 2 trillion kWh of hydrogen (from natural gas) turned into nitrogen fertiliser and used in oil refining.

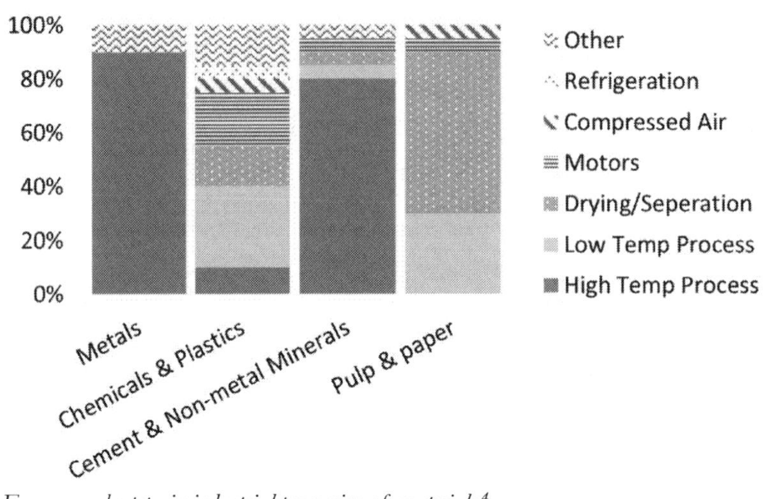

Energy use by type in industrial processing of materials.[4]

Tonnage Consumption: Today the world consumes 60 billion tonnes of industrial output every year which is an average of 7.5 tonnes of stuff per person. This includes all the raw materials used in construction, car production, machinery, electronic goods, white goods, clothing, paper and packaging, furniture, and general consumer items. This number does not include the tonnage of fossil fuels or food consumed, which would double the total again.

Most governments measure domestic material consumption to gauge tonnage use by the population – this number includes the production and importation of all raw materials minus the exports. However, the domestic material consumption doesn't include materials used in finished goods and artificially lowers consumption figures for developed economies which import much of their bulk goods from developing countries like China. This is the same accounting quirk which flatters the quantity of CO_2 emitted from developed countries. The idea that all goods and services have an associated energy cost or carbon cost of production is called embodied energy or embodied carbon – an idea which is key to understanding the true carbon footprint of consumption.

To compare the amount of material consumed in different parts of the world, it is best to adjust for the flow of material bound up in both raw materials and finished goods.[5] In their paper, 'The Material Footprint of Nations', Thomas Wiedmann and his colleagues calculated just this.[6] They showed that the US and UK consume 13-14 tonnes of industrial output per person every year. China consumes 9 tonnes and India is as low as 1.5 tonnes per person. Cleary there is a sizeable gap between developed and developing economies.

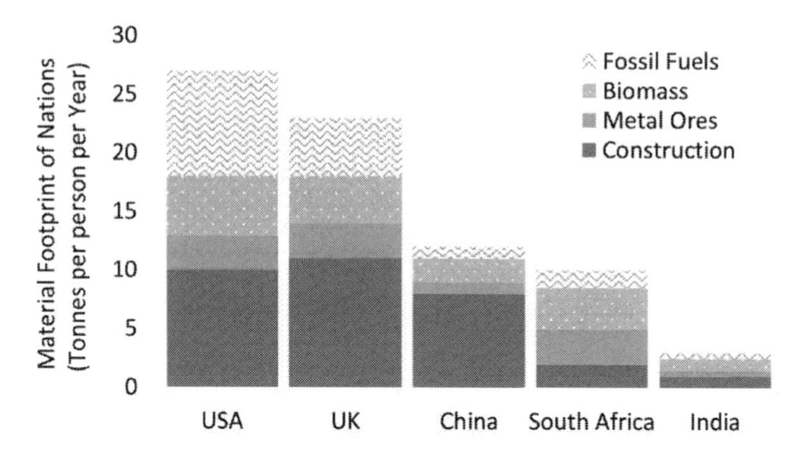

Material Footprint includes all domestic consumption and imported consumption minus exported product.[6]

Energy Consumption and Carbon Emissions by Material: Total tonnage consumption numbers can also be misleading because different raw materials and products require a significantly different amount of production energy (embodied energy).

Building aggregates like sand, crushed rock, and gravel account for 50 of the total 60 billion tonnes of global industrial output per year yet consume only 2% of all industrial energy use. Mining or quarrying is an efficient process, every tonne of aggregate consumes as little as 16 kWh of energy and emits just 6 kg of CO_2.

Metals (22%), chemicals and plastics (17%), and cement (14%) are by far the largest industrial energy users because these processes require large amounts of heat and pressure to drive the reactions.[7] Embodied energy is around 4,000 kWh per tonne for metal or chemical production with 1.5 tonnes of CO_2 emissions for every tonne of product. Cement uses 1,000 kWh energy per tonne and releases more CO_2 directly when calcium carbonate turns to calcium oxide (rock weathering in reverse) to emit a total of 1 tonne CO_2 for every tonne of cement.

Paper (6%), textiles (2%), wood products (1%), and nitrogen fertiliser (1%) round up the remaining raw materials which are used to produce the vast array of products in the modern world. Wood is the only industrial raw material with negative CO_2 emissions; every tonne of wood bound up in a permanent use such as construction will sequester 1.5 tonnes of CO_2.

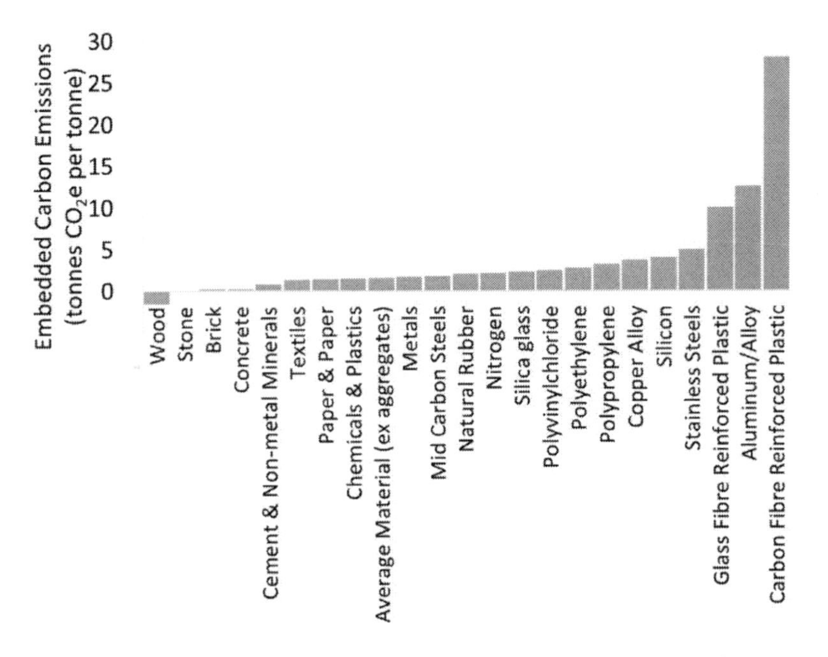

Embedded emissions of materials. Tonnes of CO_2 released during extraction and processing of one tonne of material.[8]

The remaining 35% of industrial energy use is spent turning these raw materials into finished goods. Excluding aggregates (which require little further processing) there are an annual 10 billion tonnes of raw materials which are machined, moulded, bent, chiselled, or printed into product. Processing takes an average of 1,000 kWh of energy per tonne and emits 1.5 tonnes of CO_2.

Industry Tomorrow – Planning for a 'Peak Stuff' Future

In our peak population world, we should budget for an industrial material footprint of 12 tonnes per person per year – a US and UK level. Assuming no change to the existing industrial system, this doubles the global material consumption to 120 billion tonnes, final energy use to 60 trillion kWh, and CO_2e emissions to 27 billion tonnes per year.

So how do we decarbonise industry in a peak population, peak stuff, world?[9]

Let's run through our options:

Metals: Net-Zero Technology Options

The production of metal is the largest use of industrial energy accounting for 7 trillion kWh final energy per year, 1,700 million tonnes of product and 2.7 billion tonnes of CO_2 emissions. Steel has the biggest share of production with 1,600 million tonnes metal per year: enough to fill Wembley Stadium 200 times over.[10]

Steel Production: Virgin steel production starts with the mining of iron ore which is then mixed with coking coal at 1200ºC in a blast furnace, converting the mix to purified metallic iron (pig iron) and carbon dioxide waste. The carbon rich pig iron is passed through a basic oxygen furnace to remove excess carbon and convert the iron into steel.[i] Two thirds of steel is produced in this way today. The remaining one third is recycled steel reformed in an electric arc furnace. The process uses an electrical current to heat the scrap metal to 1,800ºC and back into purified steel. This process uses just 1,200 kWh of energy per tonne and runs on electricity.[11,12,13]

Zero Carbon Steel: The electric arc furnace provides a natural route to decarbonising the steel industry at a comparable if not lower cost to the virgin process.[ii] Once powered by renewable electricity, it becomes zero carbon. The only hurdle is the availability of recycled steel which is already the most widely recycled material on Earth. Already, 85% of waste steel is reused, so there is little room to increase this percentage. However, steel producer Mittal estimates that based on the ageing of infrastructure (increased waste availability) and the projections for future construction growth, by the second half of the twenty-first century most steel can be produced from scrap.[14]

Replacing coking coal with hydrogen is another option for virgin steel production. Hydrogen can reduce iron ore to metallic iron with water as a by-product, not CO_2. However, given green hydrogen from electrolysis will cost at least three times the price of coal this will add around $200 per tonne to the production cost.

---

i The whole process uses 5,500 kWh energy from coking coal for every tonne of steel. With 0.35 kg of CO_2 emissions for every kWh of coal, this equals nearly two tonnes of embodied carbon dioxide for every tonne of steel produced.

ii Building a large virgin steel plant costs around $1,200 per tonne per year capacity, with a 25-year life. The energy cost is 5,500 kWh per tonne and the plant runs off mostly coal which costs $1-2c per kWh. Iron ore costs $100 per tonne of steel. Bringing the levelised cost to $300 per tonne before other operating expenses. Electric arc furnaces have a capital cost which is one third that of a virgin steel plant and consume one quarter of the energy. However, electricity costs at least five times more than thermal or coking coal and scrap steel is at least 50% more expensive than iron ore. Overall, the levelised cost of recycled steel works out similar to virgin steel.

Chemicals and Plastics: Net-Zero Technology Options

The production of chemicals and plastics is the second largest use of industrial energy accounting for over 5 trillion kWh final energy per year, over 1,200 million tonnes of product, and 2 billion tonnes of CO_2 emissions.[15] The chemicals industry spans millions of products based around either organic chemicals and plastics, which start life as fossil fuels run through a petrochemical cracker facility, or inorganic chemicals made from mineral resources.[iii] The oil and gas feedstock consumed in producing chemicals and plastics is another 10 trillion kWh where most of this carbon will be semi-permanently bound up in the plastics or chemicals which can take years to hundreds of years to decompose into CO_2 or other molecules.

Zero Carbon Chemicals: Switching petrochemical production to zero carbon requires shifting natural-gas-powered heating and compression to zero carbon electricity or adding carbon capture. High temperature fossil fuel heating processes can be replaced with resistance, induction, or electric arc heating. Switching from cheap gas to more expensive electrical heating will increase energy costs 2-3 times. Lower temperature processes which can use microwave heating may be more efficient. Decarbonising chemicals production will add $150 per tonne to production costs or 10% to the price of basic chemicals, even when using cheap solar electricity.[iv]

The cracker of the future consortium is a combination of some of the

---

iii Basic Petrochemicals Production: The basic building blocks of most organic chemicals and plastics are ethylene, propylene, and butadiene which have 2, 3, and 4 carbon atoms bound together in a chain surrounded by hydrogen. These products are produced from the steam cracking of longer chain oil molecules using high temperature and pressure. The process starts by diluting the oil feedstock in steam and heating the mixture very briefly under pressure to around 850ºC in the absence of oxygen (to prevent combustion). This transforms the feedstock into the basic chemical building blocks which are quickly cooled, separated, and compressed. The heating and compression processes require a large energy input which is usually fuelled by natural gas.

iv A world scale petrochemical cracker will cost around $3-4 billion to build and will produce around 2 million tonnes of product per year ($1,700 per tonne per year capacity). The facility will run for over 30 years giving a levelised capital cost of $100 per tonne. The embodied energy of petrochemicals is over 4,000 kWh per tonne and is fuelled mostly by natural gas costing $1-2c per kWh, a total of $150 per tonne cost. Natural gas or oil feedstock adds another $200-500 per tonne. So, the total production cost from a steam cracker is $450-750 per tonne product. [Costs from Plastics News "Ineos investing $3.4B in major European cracker project in Belgium", [online], plasticsnews.com, Jan 2019, accessed 2020.] [Energy use from T Ren et al., Olefins from conventional and heavy feedstocks: Energy use in steam cracking and alternative processes, Energy 31, 2006, 425-451].

largest global chemical companies, including BASF, Borealis, Lyondell Basel, SABIC, BP, and Total, who are attempting to electrify the cracking process used to manufacture the basic building blocks of the chemicals industry.[16]

Cement: Net-Zero Technology Options

Cement consumes over four trillion kWh of industrial energy to produce over 4,000 million tonnes of product every year.[17] The process energy use releases 1.5 billion tonnes of CO_2 and the CO_2 by-product of the reaction accounts for another two billion tonnes.[v] Cement plants are typically fuelled by coal or gas heating.[18,19]

Zero Carbon Cement: Removing the by-product CO_2 emissions from cement requires either carbon capture and storage or new chemistry[vi] and electrification of heating processes. The high carbon content of the exhaust gases makes the carbon capture process a little easier and cheaper than for electricity plants, but the shift to zero-carbon electricity for high temperature heating increases the fuel costs by 2-3 times. The basic cost of manufacturing cement is around $50-60 per tonne today. Carbon capture adds another $12 and switching to electrical heating adds another $25. The total costs add up to around $90 per tonne which would push the price up by 40%.[vii]

---

v Portland cement is the most common type of cement used in construction and concrete production (concrete is 3 parts aggregate and 1 part cement binder). Manufacturing starts with mining of limestone (calcium carbonate, CaCO3) and sand (SiO2) near the plant. These raw materials are then crushed and heated to 1,500°C before being rapidly cooled with bursts of air. This process decarbonises the calcium carbonate (CaCO3) and converts it to clinker (Ca3SiO5) – releasing CO_2 in the process. Every tonne of clinker releases half a tonne of direct CO_2 from the reaction, and another quarter to a half tonne of CO_2 from energy consumption in the process. The clinker is ground down and mixed with gypsum to produce Portland cement. Portland cement is known as a hydraulic cement because it sets when it contacts water..

vi Non-hydraulic cements offer an alternative chemistry to Portland cement, which is carbon neutral once set. If pure limestone (CaCO3) is decarbonised, then the clinker becomes pure Calcium Oxide or quick lime (CaO). CO_2 is still released in the production but now the setting process is driven by reabsorbing CO_2 (not water) and conversion back to calcium carbonate – creating a carbon neutral process. Major cement producer LafargeHolcim and Solidia technologies have launched a low carbon cement which uses a mix of chemistries to create concrete that emits 70% less CO_2. [LafargeHolcim, "Solidia low carbon solutions", [online], lafargeholcim.com, accessed 2020.]

vii Cement manufacturing is a high volume, low margin business which requires tight control of costs. A cement plant costs around $150 per tonne capacity per year giving a levelised capital cost of just $10 per tonne. Energy is provided by cheap coal or gas heating at $1-2c per kWh, so consuming 800 kWh per tonne of cement and energy adds $25 per tonne costs. Mining the raw materials adds another $20 per tonne. This brings the total basic production costs to $55 per tonne and with a retail price on cement of $100 per tonne there is little room for manoeuvre.

Wood Products: Net-Zero Technology Options

Every year, we use up to 1,000 million tonnes of dry wood products in construction, furniture, and consumer goods. Harvesting and processing the wood is relatively low energy at just 300 kWh per tonne and, if the wood is sourced from sustainably managed forests which can regrow, then the wood products will sequester CO_2 from the atmosphere.

Dry wood contains 50% carbon by weight so every tonne of wood has taken 1.8 tonnes of CO_2 from the atmosphere. Every tonne of wood product reduces atmospheric carbon dioxide by 1.5 tonnes including the processing emissions.

A greater proliferation of sustainably sourced wood products could add a significant offset to industrial emissions. If we replaced 30% of concrete with wood building products, we could drawdown 4.5 billion tonne of CO_2 each year which is nearly 10% of current emissions.[viii] Engineers have already designed and built 18 storey wooden skyscrapers in Norway, Canada, and the US.[20]

---

viii In our peak stuff world, 33 billion tonnes or 13 billion cubic metres of concrete will be used every year (double today). If we replace 30% of that concrete with wood building products, this reduces cement production by 2.5 billion tonnes, and increases wood consumption by 3 billion tonnes, sequestering an additional 4.5 billion tonnes of CO_2 from the atmosphere every year and locking the carbon away in long lived buildings. This would require an extra 3 million square km of sustainable forest which is 7% of global forest today.

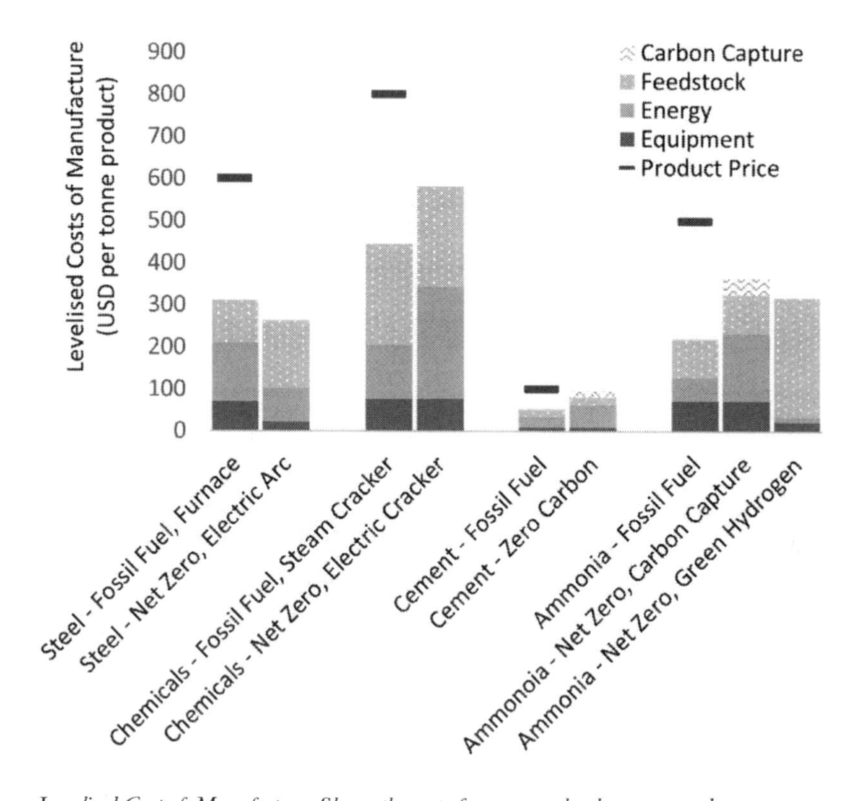

Levelised Cost of Manufacture. Shows the cost of current technology compared to a net-zero solution. Based on energy prices before tax and at wholesale rates, natural gas and coal at \$2c per kWh and net-zero electricity at scale \$6-7c per kWh. Author's standardised LCOE calculations, cost data referenced in preceding text.

Goods Finishing: Net-Zero Technology Options

Rounding off industrial energy, let's consider the further processing stages in the production of manufactured goods. Once the primary resources have been extracted and processed, the raw materials must be transformed into finished products. The many production routes and manufacturing techniques developed over the last 200 years are used to turn metals, chemicals, plastics, paper, textile, and wood into building materials, electronic goods, cars, packaging, clothes, furniture, and other consumer items.

The combination of these millions of different manufacturing processes uses roughly 11 trillion kWh of energy to produce 9 billion tonnes of stuff (excluding aggregates). That equals 1,300 kWh processing energy per tonne which uses half electricity and half fossil fuel and adds a further 5-6 billion tonnes of CO_2 into the atmosphere every year.

Decarbonising goods finishing has no single answer but broadly requires replacing fossil fuels with zero carbon electricity.

The last 200 years of industrial progress started with the invention of the steam engine in 1781, led to the development of mass production and specialisation of labour through the early twentieth century, and the integration of IT hardware in the 1980s. The next phase of industrial progress has been termed industry 4.0 and will be driven by the ongoing rapid developments in information technology software and ever-increasing penetration of hardware.

Manufacturing facilities, businesses, and homes will continue to add more digitally connected devices such as sensors, tags, GPS, and other monitoring tools to everything from machinery, transport equipment, and automated business control systems to home assistants, cars, fridges, and energy meters. The devices are collectively known as the internet of things (IoT) and they enable an increasing interconnectivity of us, the consumer, with industrial manufacturing.[ix]

Retail giants such as Amazon, Alibaba, and eBay have already disrupted "bricks and mortar" retail by integrating e-commerce and advanced logistics with finished goods wholesalers. Over the last 20 years, Amazon has gone from a small online bookseller to a trillion-dollar business, with over 50% of US online retail sales and more than 5% of all retail. Amazon are leading the charge on connecting the real world with the digital world and are already employing advanced machine learning algorithms to predict what you are going to buy so they can move the product to your nearest automated warehouse ready for one-day or one-hour delivery before you have even placed the order.

The next wave of digital integration will extend all the way up to basic industrial materials, energy, and labour. Once the entire industrial supply chain can digitally talk to one another, this will lead to greater levels of automation with advanced software, artificial intelligence, and big data left to control automated manufacturing processes, robots, and driverless vehicles, and drones for the day-to-day running of manufacturing and logistics.[21]

The digitisation of industry will improve efficiency through all manufacturing by streamlining resource management, creating less waste, allowing for more localised production, and fewer product miles (less outsourcing to cheap labour regions). It will build on a natural tendency towards more efficient and easier to automate electricity-

---

ix Progress will be boosted as 5G networks are rolled out globally enabling not only 100x faster data speeds but also importantly the ability to connect 100x more devices than 4G.

driven processes like injection moulding plastic (rather than shaping metal), offsite modular construction (rather than bespoke builds), and techniques such as 3D printing for small batch tailored products. Fundamental changes to the way goods are manufactured will naturally transition materials towards low embedded energy and low carbon.

Information about ethical sourcing, carbon emissions, and quality should be far easier to track and verify through each stage of the production process. This will allow online retailers such as Amazon to better inform us on the embedded carbon emissions of the products we buy.

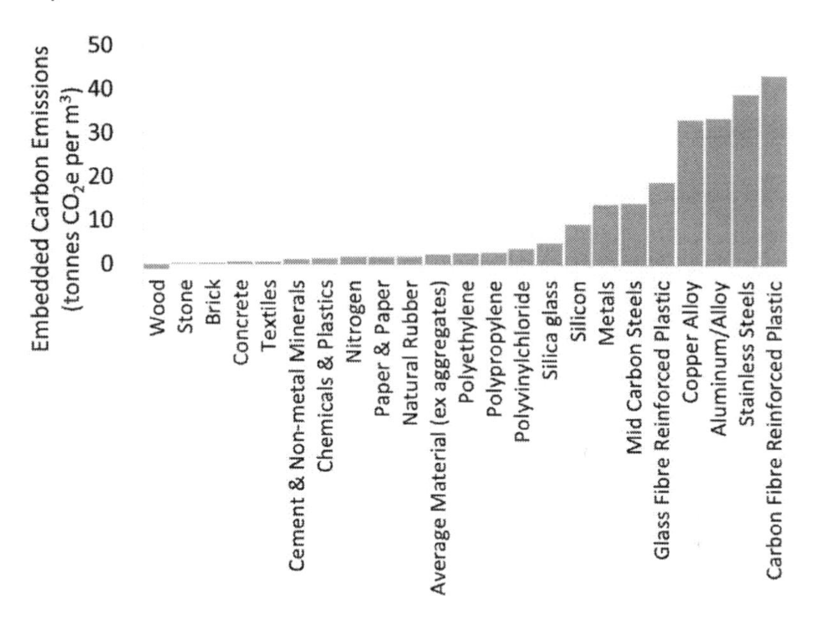

Carbon Emissions released in the extraction and processing of different materials. Tonnes CO_2 released per cubic metre of material produced.[22]

Re-engineering Industry for a Peak Stuff Net-Zero World

In a peak stuff world, all 10 billion people on the planet may consume a developed market quantity of consumer goods with an average 12 tonnes of stuff per person per year up from 7.5 tonnes today. Under the current industrial system that nearly doubles annual energy consumption to 60 trillion kWh and CO_2e emissions to 26 billion tonnes per year.

Re-engineering our industrial system for peak stuff and zero carbon will require greater electrification of heating processes, carbon capture for direct emissions, and greener chemistry.

Roughly one quarter of industrial emissions come from metals (20%), paper (4%), and textiles (2%), all of which should prove easier to decarbonise.[23,24] Electrification of metal and textile production when run on net-zero electricity should prove cost effective whilst bioenergy fuelled paper production is already leading the way in low carbon.

Another 40% of industrial emissions come from cement (25%) and chemicals (14%). These industrial processes will prove more challenging to de-carbonise.[25,26] Zero carbon cement production requires a switch to electric heating and carbon capture, increasing prices by 40%. The production of chemicals and plastics spans many millions of products, but at the heart of most processes are heat and pressure. Replacing natural gas heating with renewable electricity presents an economic hurdle given the lack of efficiency gains to offset. A zero carbon chemicals industry will require 10-20% product price increases.

A greater proliferation of sustainably sourced wood products could add a significant offset to industrial emissions. Replacing 30% of concrete volumes with wood products in long lived buildings requires an extra 7% of global forest management but could sequester and store 4-5 billion tonnes of CO_2 per year (nearly 10% of today's emissions).

The production of raw materials must be re-engineered to use zero carbon electricity, hydrogen, or recycled materials to reach zero carbon. When coupled with reduction and re-use of resources and a shift to wood products or materials with lower embedded energy this creates a route to decarbonising industry without changing consumption habits.

The digitisation of industry could connect the industrial supply chain, just as online shopping has already connected the consumer with retailers and distribution. Industry 4.0 can significantly cut waste, reduce energy consumption, lower the number of miles each product travels, and electrify industrial goods finishing. Information such as ethical sourcing, carbon emissions, and product quality should be far easier to track and verify through each stage of the production process, allowing online retailers to better inform us of embedded carbon emissions in products with no compromise on choice.

Switching all areas of our industrial processes to net-zero carbon using today's technology can be done at similar capital cost and using 40% less energy. However, average energy prices increase by 80% and feedstock costs increase by one quarter. The total annual cost of building out the industrial equipment and running the production increases by 10% compared to the existing fossil fuel system.

However, repeating the calculation for fully scaled renewables technology means the average energy price is just 30% more. The total net-zero industrial system costs 5% less than the fossil fuel alternative.

Assuming we can replace 30% of building products with sustainable wood, a share of metal moves to lower embodied energy products, and digitisation improves efficiency by one fifth, and total costs could be another 20% lower.

Re-engineering the industrial system for peak stuff and zero carbon would cost up to 25% less than remaining bound to fossil fuels. It allows for high levels of consumption for all, makes better use of mineral resources, and improves consumer experience. One third of industry already has zero carbon enabled technology and the remaining two thirds pose only incremental technology challenge. The biggest hurdle will be realigning the economics of industries such as cement or chemicals to incentivise change.

Global Industrial System (per year)		Today	Peak Industry (beyond the year 2060)			
			No System Change	Net Zero Today's Tech	Net Zero At Scale Tech	Net Zero At Scale + Efficiency
Energy Consumption	k billion kWh	33	63	37	37	28
Blended Energy Price	$c per kWh	5	5	9	7	7
Levelised Energy Costs	$ trillion	2	3	3	3	2
Goods	tonnes/capita	8	12	12	12	9
Feedstock Costs	$ trillion	1	2	2	2	2
Equipment Costs	$ trillion	2	3	3	3	2
Total Costs	**$ trillion**	**4**	**8**	**9**	**8**	**6**
CO2e emissions	billion tonnes	14	27	-3	-3	-7

Table showing the energy use, output, emissions, and total costs of our industrial system now and in different futures. All estimates are based on the detailed estimates and solutions given through this section and using the energy supply estimates given in chapters 7 and 8.

Chapter 11: Amenities in a Net-Zero Future

In this chapter we will explore how basic amenities and modern living can be re-engineered to run on a net-zero energy supply.

Basic Amenities: Energy Use and CO_2 Emissions Breakdown

Residential and commercial basic amenities account for 33 trillion kWh of annual final energy consumption or 33% of global energy use. Residential uses two thirds and commercial buildings the remaining one third. Providing basic amenities releases 14 billion tonnes of CO_2e per year.

Electricity, fossil fuels, and solid biofuels (fuel wood) account for one third each of the final energy supply which emits a total of 11 billion tonnes of CO_2 per year. Another 3 billion tonnes of CO_2e comes from associated methane leaks and halo-carbon emissions from refrigerants.[1,2]

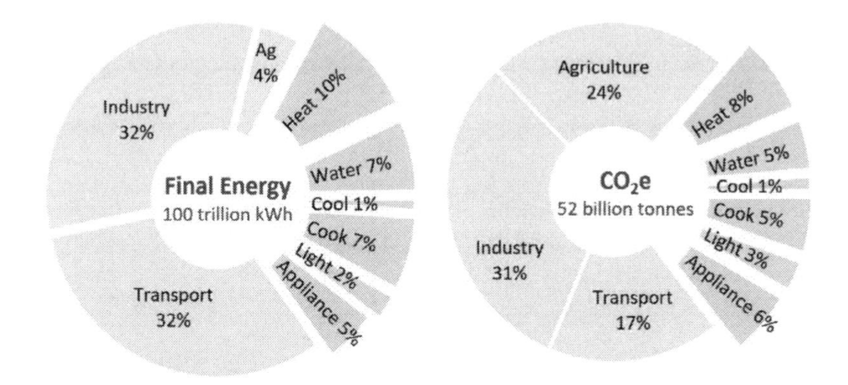

Residential and Commercial Sectors Tomorrow – Planning for a 'Peak Comfort' Future

In our peak world the global population will increase to 10 billion people and everyone on the planet should enjoy a high standard of living. In our peak consumption scenario, we will add heating and cooling to all buildings where needed and maintain an indoor temperature around 20ºC. This will increase space heating by 40% to 5.5 kWh per person per day and space cooling by 400% to 4 kWh per person per day. We will budget for electricity access for all and for enough energy to deliver 55 litres of hot water, 55 hours of artificial light, two hours of appliance use, and three hot meals per person per day.[i]

Assuming no change to the existing residential and commercial energy systems, this nearly doubles required energy consumption to 56 trillion kWh and increases CO_2 emissions to nearly 19 billion tonnes per year. So, let's look at our options to decarbonise residential and commercial energy use in a peak population, peak comfort world.

Space Heating and Cooling: Estimating Future Demand

Humans can endure everything from sub-zero temperatures right up to 40ºC with no help from heating or cooling sources, however 18-22ºC is the sweet spot for regulating our bodies, for comfort, and for productivity.

If a house or office space were perfectly insulated, then the temperature would never change and would require heating or cooling just once. Unfortunately, heat is lost (or enters) through conduction in solids, convection through air, and radiation of electromagnetic energy.

We can estimate the insulating ability of a space based on the construction materials and design. U-values measure the number of watts energy transferred per square metre for every degree-C difference between the inside and outside temperature. A roof with thick loft insulation will have a U-value of 0.6 compared to a poorly insulated roof with a U-value of 2.[3]

The average floor space of a house ranges from 500 square feet in China, India, and Russia to over 2,000 square feet in the US, Canada, and Australia. The global average is 700 square feet or 65 square metres with an average occupancy of four people.[4]

---

i An average person across the world today uses 4 kWh of space heating, 1kWh of space cooling, 45 litres of hot water, 50 hours of artificial light, enough heating for 2.5 hot meals and 2 kWh of electricity to run appliances each day.

Adding together the various U-values our average house will exchange 330 watts power with the outside for every 1ºC difference in temperature.[ii] Let's see how our house fares in different climates.

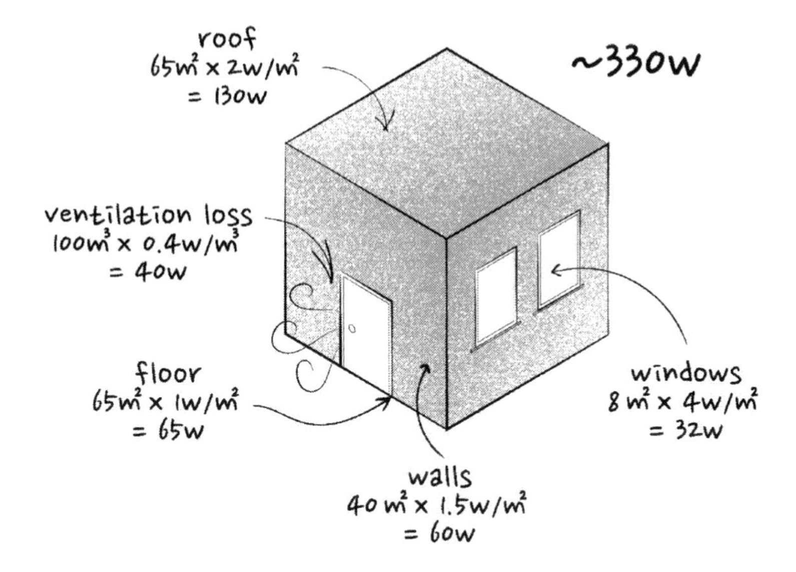

Diagram of the thermal energy leakage from an average house. Roof, wall, and window area is multiplied by the U-value and ventilation losses depend on the volume of the house. The total energy exchange adds up to 330 watts per 1ºC temperature difference between inside and outside.

Most people on Earth live between 20 and 40 degrees north of the equator where the climate is highly suitable for agriculture and where the greatest land mass resides. So, first imagine our house sits in Indianapolis, exactly 40 degrees north and landlocked in the middle of the US. Temperatures in Indianapolis average minus 4ºC in January, and up to 24ºC in August. If we want to keep our house between 18 and 22ºC for 12 hours per day we require 35,000 degree-hours of heating per year. This is the difference between our ideal temperature and the outside temperature for every hour in the house. Multiply degree-hours by our U-value and our house requires 11,000 kWh of heat energy to stay at an optimal temperature. A modern gas boiler is around 90% efficient, meaning that every one-kWh of heating requires just over one kWh of natural gas.

---

ii Let's assume our average house is single level so has a roof of 65 square metres with a U-value of 2, a floor of 65 square metres with a U-value of 1, then 40 square metres of walls with a U-value of 1.5, and 7 square metres of single glazed windows with a U-value of 4. This brings the total U-value for our house to 280 watt per degree C. We should add another 40 for ventilation losses (100 cubic metres volume x 0.4 ventilation value).

There are nearly 2 billion residential dwellings on Earth. Assume 1 billion require heating and at present about 75% are adequately heated this adds up to a total of 8,000 billion kWh of heating which is exactly the energy used on residential heating every year according to the IEA.[iii]

Next, let's move our imaginary house 20 degrees south and halfway round the world to Saudi Arabia. Saudi sits 20 degrees north and temperatures in January average 14ºC but by August the heat reaches 35ºC (day and night average). To keep our house between 18 and 22ºC for 12 hours a day, we need 25,000 degree-hours of cooling which, when multiplied by our U-value, requires 8,000 kWh of heat energy moved from the inside to the outside. Luckily, cooling with air con is more efficient than heating with natural gas. Air con is 300-400% efficient so every 3-4kWh of cooling requires only 1 kWh of electricity – the ratio is called the Coefficient of Performance (COP).

There are over 1 billion residential dwellings which could benefit from cooling. The US, Japan, and South Korea have >90% penetration of aircon already. China is about 40% but India, Africa, Latin America, and Southern Europe have less than 10% penetration.[5] The global average is around 20%. So, at present over 0.2 billion dwellings use aircon at home. Multiply by 8,000 kWh and divide by the COP and the annual final energy used for residential aircon comes out at about 500 billion kWh – again in line with IEA estimates for the world.[6,7,8]

Space Heating and Cooling in a Peak Comfort World: Adding commercial heating and cooling to our residential numbers bring today's total space heating to 11 trillion kWh and space cooling to over 1 trillion kWh. Take the global population from 8 to 10 billion people and bring heating and cooling to all residential and commercial dwellings which need it, and the energy required for space heating increases by 80% to 19 trillion kWh and energy for cooling increases over 4 times to 5 trillion kWh. Global air conditioner units increase from 1.3 to over 5 billion installed units. CO_2 emissions will climb from 4 billion to over 7 billion tonnes per year. So how do we decarbonise peak comfort heating and cooling?

---

iii 10% of Europe and many developing countries are in fuel poverty. 1 billion houses x 75% x 11,000 kWh = 8,000 billion kWh.

Space Heating and Cooling: Net-Zero Technology Options

Natural Gas Boiler to Hydrogen Heating: One third of heating is done using natural gas boilers. They are up to 90% efficient, relatively cheap ($1,500 installed), and will last for 10-15 years. However, gas boilers emit CO_2 and are too small to fit with carbon capture. Boilers could be adapted to run on hydrogen gas which produces only wastewater. The cost of the boiler would not differ much, but the cost of the fuel would. Retail priced natural gas costs about $3c per kWh and, as we calculated earlier, full scale hydrogen from renewable electricity would cost closer to $7c per kWh. The levelised cost of a hydrogen powered boiler is $9c per kWh heat compared to $5c using natural gas.

Air Source Heat Pump: Heat pumps are basically an aircon unit run in reverse.[9] They move heat from the outside to the inside using a fan and compressor. Just like air con they have an average efficiency or COP of 3-4 times. Every one-kWh of electricity provides 3-4 kWh of heat. The exact COP will range from 2 to 7 times depending on the difference between outside and the inside temperatures – they are more efficient at heating radiators or air to 40ºC rather than 60ºC and are more efficient in milder outdoor temperatures (but will work even in sub-zero conditions).[10,11] Air source heat pumps cost around $5,000 to install but, because of the high efficiency, require only one quarter of the energy to run. The levelised cost of buying and running a heat pump is comparable to a gas boiler at $5-6c per kWh of heat. Run on renewable electricity and you have zero-carbon heating.

Ground Source Heat Pumps (GSHP): operate on the same principle as air source heat pumps but use long lengths of plastic piping underground to circulate water and extract heat from the Earth. This provides a slightly higher average COP (3.5-4.5) because the ground stays warmer through winter. However, GSHP cost at least $20,000 due to the large-scale excavation or drilling work required and every job is different. The total levelised cost is an expensive $11c per kWh heat.

BioFuel Heating: One third of global heating still uses biomass such as wood on basic stoves or fireplaces to heat residential dwellings (mostly developing countries). This is one of the cheapest forms of space heating but is also damaging to health. Fuels such as wood contain water and impurities which stop the carbon burning cleanly and help produce harmful levels of particulate matter or toxic emissions, especially when used in an enclosed space. Moving away from biomass heating in developing countries is needed to improve global health.

District Heating: We ran through district heating in the seasonal storage section as a way to store cheap summer energy for use in winter heating or for water heating all year round. This can also be coupled with waste heat output from industrial processes or electricity generation (with carbon capture). This route has the potential to supply heating in the local vicinity at comparable costs to gas boilers if effectively utilised.

Better Insulation: Better insulating homes is one of the cheapest ways to lower heating (and cooling) requirements:

- Insulating an average roof costs just $100 and will last for 50 years. It can reduce the U-value from 2 to 0.5 watts per square metre per ºC which for an average house will save 3,500 kWh of heat energy per year. That works out at $0.1c per kWh equivalent heating, which is over 50 times cheaper than running your boiler.

- Smart Heating Systems are another interesting development. Zoned heating control, which turns on the heat only in areas of the house you are using through the day and night, can reduce energy consumption by 30% (3,000 kWh per year). It costs around $600 one off. Assuming the equipment lasts ten years, that equates to $2c per kWh – cheaper than firing up your boiler.

- Double glazed windows reduce energy loss but are not cost effective unless also installed for aesthetics and sound proofing. The cost of double glazing is around $800 per square metre. Our average house has seven square metres of windows bringing the total cost to $6,000. The U-value of this area declines from 5 to 2.5 watts per square metre per ºC, saving 635 kWh of heat energy per year. A levelised cost of $65c per kWh, over ten times more than a natural gas boiler.

- Passive house standards represent the highest standard of insulation in new building. The costs of implementing passive house standards in commercial building come at little extra cost and can reduce building energy consumption by up to five times, an immediate payback. Residential buildings can save between 50-90% of energy use at an extra cost of $150 per square metre on a new build so the extra investment pays back over 40 years.[12]

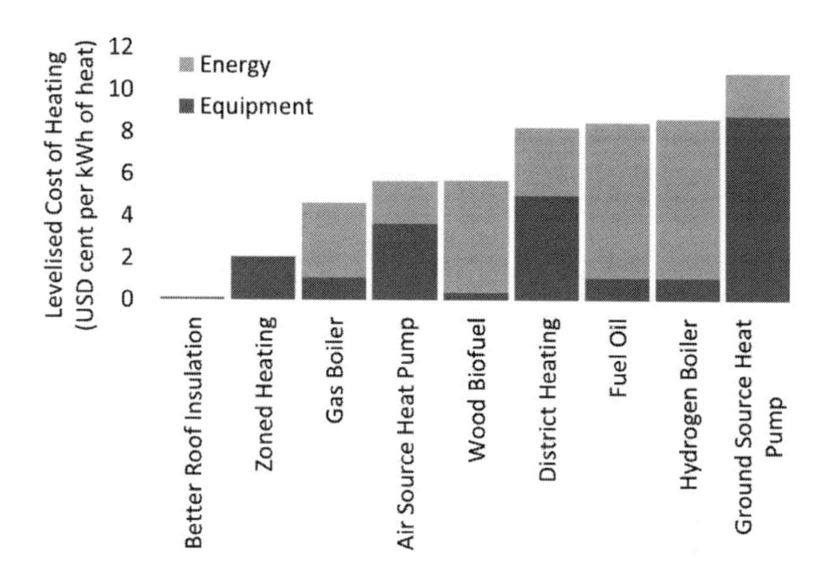

Levelised Cost of Heating. Based on fuel costs including T&D but before government taxes, natural gas $3c per kWh, wood $4.5c per kWh, green hydrogen $7c per kWh (at scale), and net-zero electricity $8c per kWh (at scale). Author's standardised LCOE calculations based on cost data referenced in preceding text.

Air Conditioning: Air conditioning can be run as zero carbon when using zero carbon electricity. Solar electricity generation will generally balance well with air conditioning as it is usually sunnier when it is hottest. Improving the COP or efficiency of air conditioning is especially important as more and more of the world installs air con.

Moving towards a zero-carbon space heating and cooling system requires replacing as many gas boilers, oil, and biomass burners with electric heat pumps as possible whilst increasing the level of insulation in all homes. For modern builds and good insulation retrofits this poses little problem. For older housing stock which cannot be so well insulated or is situated in an extreme cold environment, hydrogen boilers are likely the next best solution because they can run to higher operating temperatures than heat pumps.

Air conditioning will undoubtedly continue to increase with growing frequency of temperature extremes and will become essential across larger areas of the world. The focus on air con should be towards maximising efficiency standards and smart control to integrate with renewable grids.

Water Heating: Net-Zero Technology Options

Hot water remains a luxury of the privileged few. The United Nations estimates that 4 billion people around the world don't have plumbing or running water, 2.4 billion have no proper sanitation, and 0.8 billion don't have access to safe drinking water. Yet the world still consumes 7 trillion kWh of energy every year on hot water – that is an average 46 litres per person per day which emits 2 billion tonnes of CO_2.[iv] So how do we go zero carbon for hot water:

Hydrogen Boilers: These remain an expensive option given the higher cost of hydrogen compared to natural gas or wood. Hydrogen boilers have a levelised cost of $0.5c per litre (at scale economics) compared to natural gas boilers at $0.25c per litre.

Air Source Heat Pumps: A good option for space heating but much less efficient at getting water up to 60°C (COP of 2 at higher temperatures) so total levelised costs are around $0.5c per litre.

Solar Thermal Panels: Operate like Solar PV panels but produce hot water rather than electricity. The efficiency is high at 70% of the sun's energy, but to produce the 3,000 kWh of annual heating for a four-person house requires three square metres of panels at a cost of $5,000. This works out at $0.5c per litre but will deliver too much heating in summer and not enough in winter and requires back up.

Resistance Heater: A simple element heater could use zero carbon electricity. They are cheap to fit ($300 installed) and close to 100% efficient. However, scaled up renewable electricity will still cost three times more than natural gas heating so again the levelised cost will be around $0.5c per litre hot water.

Water heating in the future will need to move from today's natural gas boilers, fuel oil, and biomass to the use of resistance heating, resistance/heat pump combined heating, hydrogen boilers, and increased penetration of district heating.

iv Water has a specific heat capacity of 4,200 Joules per 1°C per kg. It takes 0.001 kWh heat energy to raise one litre or one kg of water by 1°C. The average global temperature is 15°C and hot water must be heated to 60°C to kill off any bugs. That means every litre of hot water requires 0.05 kWh of heating energy.

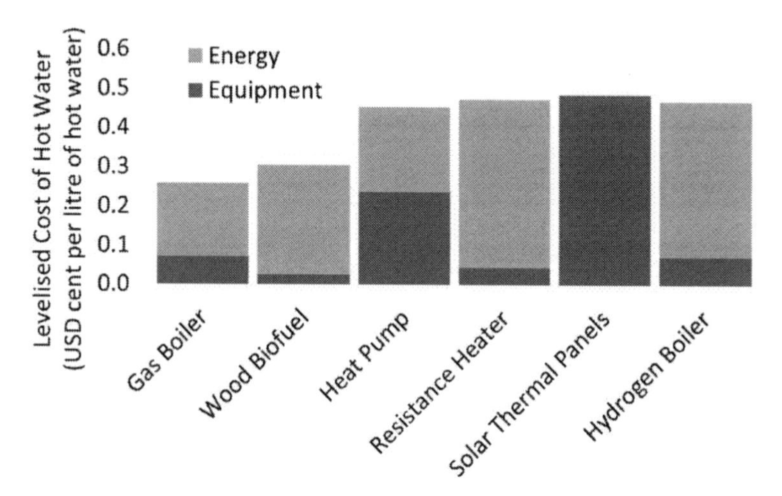

Levelised Cost of Heating Water. Based on fuel costs including T&D but before government taxes, natural gas $3c per kWh, wood $4.5c per kWh, green hydrogen $7c per kWh (at scale) and net-zero electricity $8c per kWh (at scale). Author's standardised LCOE calculations based on cost data referenced in preceding text.

Lighting: Net-Zero Technology Options

Artificial lighting accounts for 2 trillion kWh of final energy (mostly electricity) and is responsible for 1.2 billion tonnes of CO_2 emissions per year. There are 55 billion light bulbs installed globally (seven per person) which are used for an average of six hours per day. There are three basic types of bulb:[13]

Incandescent Bulbs pass electricity through a thin metal filament. The resistance to the flow of electricity turns the electrical energy into heat and light energy. A 60-watt light bulb will produce 800 Lumens of light or 13 lumens per watt. The theoretical best efficiency of lighting is 683 lumens per watt, making incandescent bulbs just 2% efficient. The rest of the energy is waste heat or light outside of the visible range. Incandescent bulbs cost just $50c but only last for 2,000 hours. The levelised cost of buying and running an 800-lumen incandescent light is around $0.5c per hour.

Fluorescent Lighting uses an electrical current to excite a gas inside a glass enclosure. This causes the emission of UV light which hits a phosphor coating and converts the non-visible higher energy UV light into visible light. The whole process produces far less heat than incandescent lighting and is about 8% efficient, producing 50 lumens per watt. Fluorescent bulbs cost $1 each and last for 30,000 hours. The levelised cost of an 800-lumen fluorescent light is around $0.12c per hour.

Light Emitting Diodes (LED) are the most efficient form of lighting, effectively a solar PV panel run in reverse. Electricity is passed through a semiconductor chip and forces electrons in the material to emit light. LED lights can produce over 100 lumens per watt which is over 15% efficient (similar to a solar panel). LED lights give off very little waste heat. LED bulbs cost $3 per unit but last for 50,000 hours and are very efficient. The levelised cost of an 800 lumen LED light bulb is around $0.07c per hour – seven times cheaper than an incandescent bulb.[14]

Switching to zero carbon electricity creates zero carbon lighting and as the share of LED bulbs increases from 27% to 100% this will halve energy consumption from existing lighting.[v]

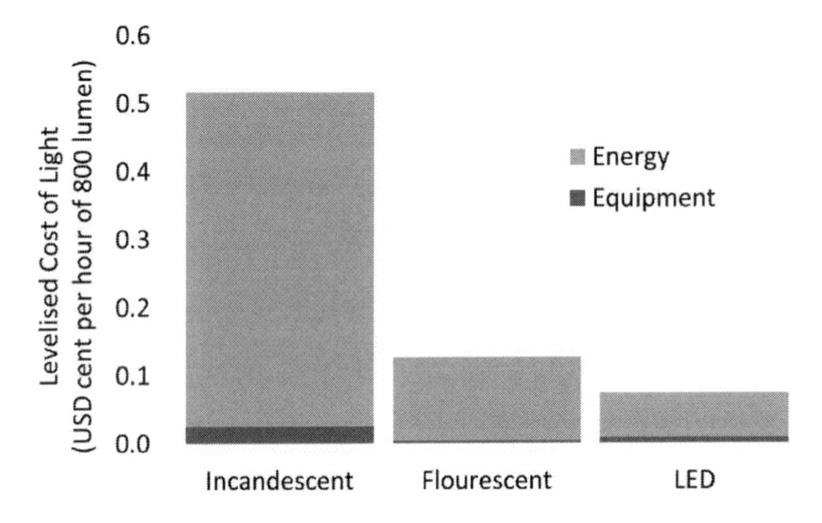

Levelised Cost of Lighting. Based on fuel costs including T&D but before government taxes, net-zero electricity $8c per kWh (at scale). Author's standardised LCOE calculations based on cost data referenced in preceding text.

v Incandescent lighting accounts for 9% of installed light bulbs, fluorescent bulbs account for 64%, and LED the remaining 27%.

Cooking: Net-Zero Technology Options

Cooking consumes another 7 trillion kWh of final energy and emits 2 billion tonnes of CO_2 emissions per year. Energy is supplied by a mix of predominantly natural gas, biofuels, and electricity:

Natural Gas Cooker: Natural gas cookers are commonplace in developed countries using piped natural gas to heat an oven or hob. The cooker costs $500 and retail natural gas is supplied at $3c per kWh. A natural gas oven is rated around 2 kW power. So, use your gas oven to cook a pizza for 30 mins and it will consume $3c of natural gas (1 kWh) and will cost you another $5c capital cost (assuming you cook three pizzas per day for the life of the oven). A total levelised cost of $8c per pizza.

Electric Cooker: This operates just like a gas cooker but uses electrical resistance to heat the oven and hob. Electric ovens retail for the same price and are rated at a similar 2kW power. The total levelised cost for an electric cooker comes out closer to $14c per pizza because retail electricity is more expensive than gas.

Biofuel Stove: Much of the world still cooks on basic wood fired stoves. These are inexpensive with relatively cheap fuel so have a levelised cost of cooking which is less than gas or electric. However, once again this type of fuel burning in an enclosed space can be damaging to respiratory health.

Microwaves: Another form of electrical cooking but this time electricity is used to generate electromagnetic microwave energy. The microwaves are just the right frequency of energy to vibrate water molecules in food and heat it. Microwaves are a direct form of cooking which wastes minimal energy. An 800-watt microwave costs just $50 and can cook a pizza in five minutes. The levelised cost is the cheapest of all forms of cooking at just $1-2c per pizza.

Decarbonising global cooking requires moving away from gas cookers in developed markets towards electric ovens and microwaves. Those who cannot give up gas hobs may have the option of hydrogen burners. Developing regions which use biofuel require greater electrification and to replace biofuel cooking with modern electric appliances for decarbonisation and health benefits.

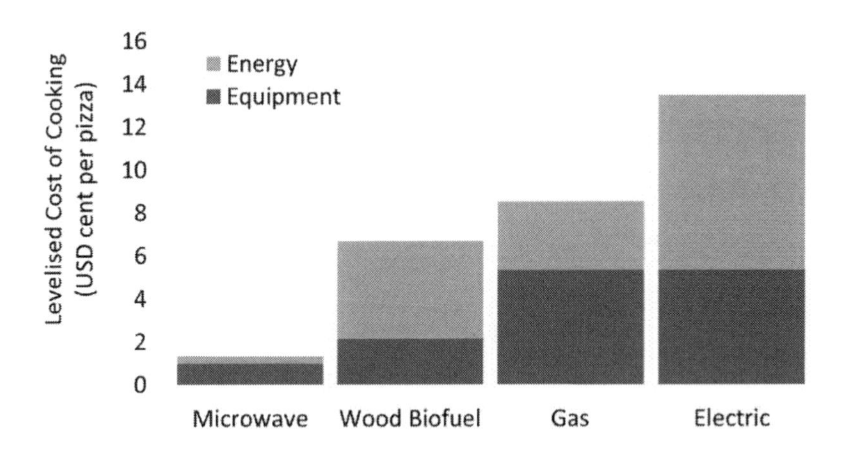

Levelised Cost of Cooking. Based on fuel costs including T&D but before government taxes, natural gas $3c per kWh, wood $4.5c per kWh, green hydrogen $7c per kWh (at scale), and net-zero electricity $8c per kWh (at scale). Author's standardised LCOE calculations based on cost data referenced in preceding text.

Re-engineering Residential and Commercial Spaces for a Peak Comfort Net-Zero World

In a peak comfort world, we should budget for adequately heated and cooled homes and workplaces plus access to basic amenities including 55 litres of hot water, 55 hours of artificial lighting, two hours of appliance use, and three hot meals per person per day. Enabling this transition will require the 0.8 billion people with no electricity access, 2.5 billion with no clean cooking solutions or sanitation, and 4 billion with no running water, are brought up to modern standards.[15,16]Assuming no change to our current system, energy consumption will double to 56 trillion kWh and CO_2e emissions will increase to 19 billion tonnes per year (from 11 today).

Re-engineering residential and commercial spaces for peak comfort and zero carbon will require electrification of heating, cooking, and hot water alongside greater penetration of district heating and efficiency improvements across buildings, cooling, and lighting. Replacing old appliances with new more efficient versions has the potential to save up to 50% of the energy used in fans, fridges, ovens, and TVs, and up to 70% savings in water use.[17]

Space and water heating accounts for 5 billion tonnes of CO_2 emissions per year. Developed countries use primarily natural gas or oil-fired boilers for heat, whilst developing regions (particularly rural areas) use basic wood or coal fired stoves or open fires. Moving space and water heating towards zero carbon requires replacing boilers and stoves with electric heat pumps or district heating (where possible) and hydrogen networks in extreme cold areas. Heat pumps cost three times more than gas boilers but consume 3-4 times less energy. On average, capital costs for heating will rise but will be broadly offset by energy savings.

Space cooling emits 0.6 billion tonnes of CO_2 per year from the electricity used in the 1.3 billion air conditioning units around the world. Penetration of air con remains relatively low (20-30%) but could increase up to 3-4x in a peak comfort world. Luckily, air conditioning is already relatively efficient and uses electricity which can be easily switched to zero carbon.

Better insulation will play a big part in more efficient houses and commercial spaces. Basic insulation measures cost up to 50 times less than the equivalent heating and allow for equipment such as heat pumps to run more efficiently at lower temperatures.

There are 55 billion lightbulbs in the world which emit over one billion tonnes of CO_2 from the electricity they consume. Shifting the remaining 9% of incandescent bulbs and 64% fluorescent lighting to more efficient LED bulbs will halve the total cost for artificial lighting.

Global cooking uses 7 trillion kWh of energy and emits 2 billion tonnes of CO_2. Switching the gas cookers and biomass (wood) stoves towards renewable supplied electric cookers and highly efficient microwaves will cost 30% more but brings residential emissions to net-zero and eliminates health damaging pollutants from indoor biomass burning.

Switching residential and commercial basic amenities to net-zero using ***today's*** supply and demand technology would reduce energy consumption by 30%, increase energy prices by 50%, and require 30% higher capital cost of equipment. Overall, the total cost of buying and running net-zero basic amenities using today's technologies would be 15% more expensive than sticking with fossil fuels.

However, re-running the calculations with ***fully scaled zero carbon*** technology reduces energy consumption by 35%, increases average energy prices by just 5%, and capital costs by 30%. The total cost of buying and running a fully scaled net-zero system of basic amenities will be 5% cheaper than sticking with fossil fuels.

Fully implementing zero carbon solutions at scale and with a shift to other energy efficiency solutions including all LED lighting, and moving to best efficiency aircon and better insulation across half of buildings, could reduce costs by another 5%.

Re-engineering residential and commercial spaces for a peak comfort, zero carbon world will prove 10% cheaper than sticking with the existing fossil fuel system. The transition creates a route towards a modern standard of living for all and significant health benefits for half of the world's population currently with no access to clean cooking, sanitation, or electricity. Most of the technology is already commercially available and half is cost comparable today. The biggest hurdle to decarbonising residential and commercial amenities is the roll out of cheap zero carbon electricity, public awareness of alternative heating appliances in developed markets, and technology availability and financing in developing markets.

Global Basic Amenities (per year)		Today	Peak Amenities (beyond the year 2060) No System Change	Net Zero Today's Tech	Net Zero At Scale Tech	Net Zero At Scale + Efficiency
Energy Consumption	k billion kWh	33	56	38	37	33
Blended Energy Price	$c per kWh	5	5	8	6	6
Levelised Energy Costs	$ trillion	2	3	3	2	2
Amenities	Util/capita/day	106	125	125	125	125
Equipment Costs	$ trillion	1	2	2	2	2
Total Costs	**$ trillion**	**3**	**5**	**6**	**5**	**4**
CO2e emissions	billion tonnes	11	19	0	0	0

Table showing the energy use, output, emissions, and total costs of our amenities system now and in different futures. All estimates are based on the detailed estimates and solutions given through this section and using the energy supply estimates given in chapters 7 and 8.

Chapter 12: Agriculture in a Net-Zero Future

In this chapter we will explore how the agricultural sector could be re-engineered to reach net-zero.

Agriculture Today: CO_2 Emissions Breakdown

The agricultural sector emits nearly 13 billion tonnes of CO_2e per year which is 24% of global annual emissions. Unlike the rest of the economy, agriculture uses relatively little energy. Agricultural energy use contributes just 1.5 billion tonnes of annual CO_2. Another 6 billion tonnes of CO_2 equivalents (CO_2e) are released directly from agricultural processes. The final 5.5 billion tonnes CO_2e come from the expansion of agricultural land at the expense of natural forest or peatland.[1,2,3] Crop and pastureland use 33% of dryland on Earth to support nearly 8 billion humans who live on less than 2% of land: the expansion of agriculture is a far bigger space problem than urban sprawl.[4]

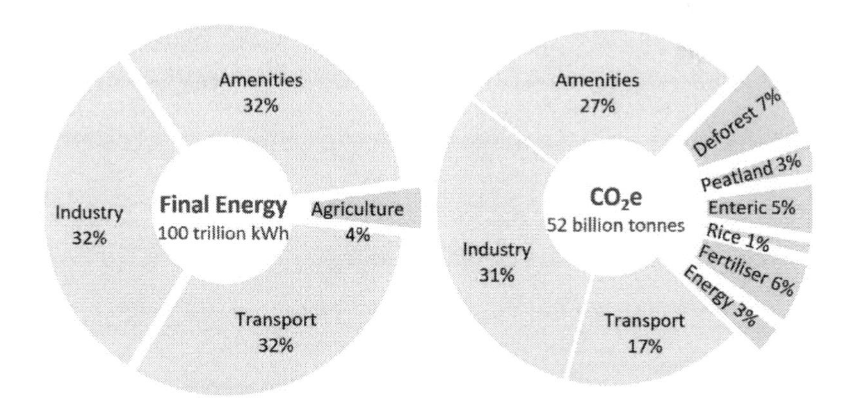

Energy Use: The agricultural sector uses 4 trillion kWh of final energy every year (1.5 billion tonnes CO_2, 3% global emissions). Oil products are used to power agricultural machinery (tractors etc.) and the remainder of the energy is mostly used for heating and equipment.

Enteric Fermentation: Refers to the methane released from ruminants. Animals with multi-chambered stomachs break down tough plant matter such as cellulose using bacteria. The process takes place without oxygen so instead of forming CO_2 they belch or fart methane which has 28 times greater warming potential compared to CO_2. The main culprits are cows and sheep which are responsible for around 2.5 billion tonnes of CO_2 equivalent methane emissions per year (5% global emissions). With 70 million tonnes of beef, 90 million tonnes milk solids, and 14 million tonnes of mutton produced every year, enteric fermentation emits around 15 tonnes of CO_2 per tonne of product.

Fertiliser: Both synthetic fertiliser and natural manures increase the amount of nitrogen in soils and nitrogen run-off from fields. This increases the rate at which the solid or liquid nitrogen is oxidised into nitrous oxide gases in the atmosphere which have a warming potential of 265 times that of CO_2. Fertiliser and manure use accounts for 3 billion tonnes of CO_2 equivalent NOx per year (6% global emissions).

Rice Paddy Flooding: Large areas of otherwise dry soil are flooded to grow rice. The water prevents oxygen in the air getting to the soil so dead plant matter is broken down into methane rather than CO_2. Nearly 800 million tonnes of paddy rice are produced each year emitting 0.7 billion tonnes of CO_2e.

Deforestation: Every year approximately 150,000 square km of forest is cleared to make way for agricultural land (mostly in the tropics). Roughly 70,000 square km of forest regrows. So, a net total of 80,000 square km or 0.2% of forest is lost per year. Forests have over 50,000 tonnes of CO_2 per square km bound up as carbon in the biomass and soils which is released back into the atmosphere through burning, rotting, and cultivating. This emits 4 billion tonnes of CO_2 per year.

Peatland Draining: Wetland areas where dead plant matter has been buried and transformed into peat (the first step to coal). Peatland covers three million square km of Earth or 3% of land and holds 550 billion tonnes of carbon. Peatland draining or burning releases 1.5 billion tonnes of CO_2 into the atmosphere every year.

Agriculture Consumption Today: Tonnes, Calories, and Land Use

The agricultural industry produces nine billion tonnes of crop products every year.[5] Half is used directly as food, 45% is fed to animals for meat and dairy produce and the remaining 5% used for biofuel production. An average of 1.2 tonnes of crop per person per year.

The average tonne of crop product contains nearly 2 million kcal (food calories) of energy. This means we produce around 6,000 kcal[i] of edible crop product per person per day. The US consumes over 15,000, Europe consumes 8,000, and Asia and Africa less than 5,000 kcal per person per day.

Berners-Lee, Kennelly, and Watson calculated global available calories and waste in their 2018 *Elementa* paper.[6] In addition to the 6,000 kcal of edible crop energy grown per person per day, they estimated we use another 4,000 kcal of energy from pastureland to rear animals. This brings the total energy use to nearly 10,000 kcal per person per day. As the authors point out the recommended calorie intake for humans is 2,350 kcal per day. So where are these extra calories going?

Animal Products: Over 5,500 kcal per person per day are lost in meat and dairy production. More than 6,100 kcal (1/3 crop and 2/3 pastureland) are fed to animals and turned into just 600 kcal of meat and dairy. Animal rearing is less than 10% efficient due mostly to waste heat and excrement during the animal's life. As scientist Hope Jahren writes in 'The Story of More', "When an animal eats grain, its body uses the grain for many things, but mostly not building flesh".[7]

Biofuels: Every year 21 billion gallons of starch ethanol are produced mainly from US corn and eight billion gallons of sugar ethanol are produced from Brazilian sugarcane.[8] Corn produces 110 gallons ethanol per tonne and sugarcane produces 20 gallons ethanol per tonne.[9] This adds up to a combined 600 million tonnes of crop use (5% of total crop tonnes) or 300 kcal of edible crop energy per person per day.

Waste: A total of 1,500 kcal are lost along the agricultural production chain. After harvesting losses and seed replanting only 90% of starting crop calories are left (600 kcal loss). Storage and trading results in another 5% of losses (270 kcal). Processing and distributing crop and

---

i Food calories or kcal are a measure of energy contained in food. There are 860 kcal per kWh.

meat products wastes another 10% (310 kcal) then household waste accounts for a further 10% loss (280 kcal).

Overconsumption: The final 150 kcal per person per day losses are due to overconsumption – we eat more than we should, put on weight and have to consume more calories to sustain our heavier (less healthy) bodies.

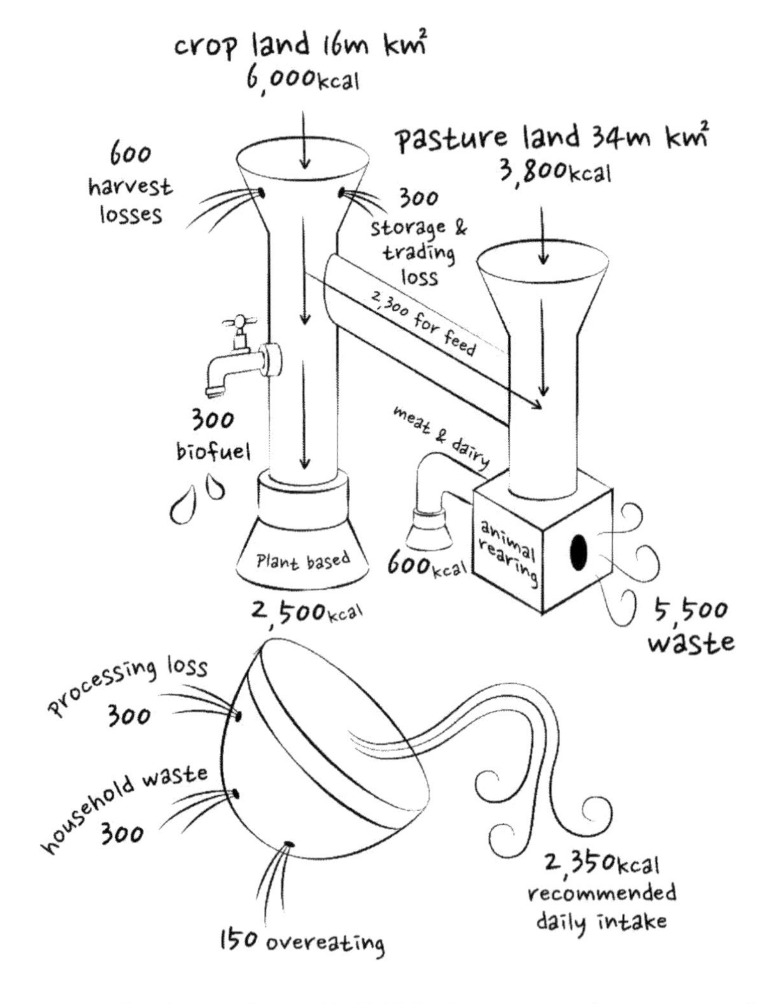

Today's agricultural system. Starts with 6,000 kcal per person per day grown on cropland and 3,800 kcal per person per day from pastureland. Finishes with 2,350 calories per person per day recommended intake. The rest is lost along the way. Author's calculations adapted from Berners-Lee et al. [10]

Due to the inefficiencies of our current system, agricultural land covers 49 million square km or 33% of dry land on Earth. Pastureland used for animal grazing is 34 million square km and crop land is 16 million square km. So, for each person on the planet, we use 6,000 square metres or about one football field of agricultural land for growing and rearing food. If you eat mostly animal produce you use four football pitches. If you eat a vegan diet you require just 700 square metres of land: the equivalent area of the penalty box.

Agriculture Tomorrow: Feeding the World Without Heating the Planet

The challenge for future agriculture is to produce enough food for the growing global population, to distribute those calories more evenly, and move carbon emissions to net-zero.

Agriculture today already provides more than enough calories to feed everyone on the planet despite the losses from meat, dairy, biofuels, and waste. Yet there are an estimated 800 million people undernourished or 10% of the global population (mostly in developing countries) and 1.9 billion overweight adults, including 650 million obese (mostly in developed countries).[11,12] Rebalancing the global allocation of calories will improve health in both developing and developed countries and is a key challenge for global food systems.

We can understand the impact of agriculture on the environment by modelling different future scenarios, from a system that doesn't change with growing population to a world where everyone is vegan. Here are the outcomes:

Current Agriculture System at Peak Population: First, let's imagine a planet with 10 billion people each consuming an average European diet. That means 30% more mouths to feed and each person increases meat and dairy consumption from 66 kg to 100 kg per year. Peak population, peak meat. We would need to increase annual crop production from 9 to 15 billion tonnes and animal products from 500 to 1,000 million tonnes per year. Assuming no change to average crop yields, waste, and overconsumption means that land use would increase from 49 to 90 million square km (to 60% of dryland). Ongoing CO_2e emissions excluding land use change would grow from 8 to 14 billion tonnes per year. CO_2 emissions from forest clearing and peat land would increase more than ten-fold from 5.5 billion tonnes per annum to 80 billion tonnes per annum, whilst the population was still expanding (land change emissions stop once peak population, peak meat consumption is reached). The total cumulative CO_2 emissions from land use change until peak population would add over 2,500 billion tonnes of CO_2 into the atmosphere, which is the same as all greenhouse gas emissions in human history and nearly five times more than the remaining carbon budget to keep us under 1.5°C.

We don't reflect this worst case in our baseline of inaction because the limits to agricultural yield have decent headroom (unlike fossil fuel energy efficiencies),[ii] but this simple maths shows how important agricultural yield improvements are.

Agriculture System with Best in Class Yields at Peak Population: Agricultural yield is the number of tonnes of crop or calories of foodstuff which an area of land can produce per year. Sugarcane, corn, wheat, rice, potatoes, and soybeans make up over 60% of total crop production. The yield of each is different[iii], but if we compare the same crop in different regions of the world such as the US, Europe, China, and Latin America that employ modern farming techniques, these areas have over double the yield compared to India or Africa, where traditional farming techniques are more common. Modern farming uses 40% of global crop land but produce 60% of all crops. India and Africa have 35% of global crop land but produce only 20% of food.

Let's assume that we can boost yields to today's best regional average in every part of the world. That means increasing yields to North American levels for most crops apart from wheat where China yields the most. It also means everywhere becomes as efficient at rearing animals as Latin America. In total, this would increase crop tonnage by 90% and animal products by 80%, using the same land area. Our peak population still uses 15 billion tonnes of crops and 1 billion tonnes of animal product and the associated emissions still increase to 11 billion tonnes per annum. However, land use for agriculture remains largely unchanged – avoiding any further CO_2 emissions from deforestation and peatland draining.

If yields are raised to best in class levels, this prevents the need for further agricultural land expansion and can provide 10 billion people with a

ii At present, population is growing at 0.9% per year, food consumption is growing at 1.6% per year, but agricultural land is only expanding at 0.3% per year, because of continuous crop yield improvements. If yields hit limits and we move to peak population, peak meat consumption this creates huge pressure on land availability.

iii This varies by type of plant. Sugarcane is the world's highest yielding crop by weight at 7,000 tonnes per year per square km, but sugarcane contains only 0.3 million kcal per tonne – a total of 2 billion kcal per square km per year. Corn on the other hand yields just 600 tonnes but each tonne contains 350 million kcal putting it on par with sugarcane in total.

European standard of food consumption. Achieving continued yield improvements, however, will require advances to technology and farming practices in developed areas, technology sharing and education for more traditional farming, strengthening of global trade, and reconfiguring agricultural policies around the world. Agricultural subsidies often create adverse economic signals, leading to the wrong crops being grown in the wrong areas of the world and holding back global yields.

A Vegan Planet at Peak Population: The production of meat and dairy is a very inefficient use of calories. We saw earlier that 6,000 calories per person per day are used to produce just 600 calories of meat and dairy. Add to this the 15 tonnes of CO_2e methane released per tonne of beef, dairy solids, and mutton, thanks to enteric fermentation, and you can see why animal products are high on emissions.

For these reasons going vegan is greatly beneficial for the planet.[iv] Let's assume in our peak population, peak consumption scenario that yields, waste, biofuels, and calorie overconsumption remain unchanged but everyone on the planet became vegan, this would drastically reduce the need for agricultural land. Total crop production could hold steady despite 30% more mouths to feed because the wasted crop calories going into rearing animals would be consumed directly. The other 3,500 kcal per person per day from pastureland are no longer required, so up to 34 million square km of forest could be planted instead.

The net result is ongoing emissions of CO_2e (excluding land use change) drop to 3.5 billion tonnes per year (from eight billion). More significantly, the reforestation and restoration of peatland could cumulatively sequester over 2,000 billion tonnes of carbon dioxide, the equivalent of all human CO_2 emissions since the start of the industrial revolution, and would more than quadruple the remaining carbon budget to keep us under 1.5°C warming.

Veganism is already on the increase, driven by traditional animal welfare and ethics but increasingly also driven by health and environmental concerns. Many consumers are adopting a more vegan or vegetarian diet, replacing meat and dairy with plant-based alternatives or substitutes.

A company called Beyond Meat has been the bell weather for the rising interest in substitute meat over the last ten years. The company was founded by Ethan Brown in 2009. Ethan had been working the clean energy sector but became increasingly interested in ways to cut the

iv The USDA and Weber & Matthews estimate that beef and lamb emit 14g of CO_2e per kcal. Chicken, pork, fish, dairy and fruit emit around 4g per kcal. Most other products emit 0.5-2g CO_2e per kcal.

environmental impact of meat. The company uses heating, cooling, and pressure, to create the fibrous texture of meat from plant-based proteins mixed with plant-based fats, minerals, colours, and carbohydrates, to replicate the appearance, flavour, and juiciness of meat.[13] In 2019, the company raised $240 million from investors to fund manufacturing expansion due to rising demand. Shares were listed on the stock exchange for the first time (an IPO or initial public offering). On the first day of public trading, when the newly minted shares can be bought and sold on the exchange, the share price increased by 160%, making Beyond Meat the best performing first-day IPO in nearly two decades.[14] As I write this, the business is valued at over $9 billion (with a brief stint at $14 billion on the way).

Beyond Meat's main competitor is another California based company called Impossible Foods and goes one step further by adding an iron rich compound called heme to their plant-based meats. Heme gives meat its distinctive taste and blood-red colour. Impossible Foods use their heme, produced via fermentation, to create a bleeding burger made from plants.[15]

Further out, cultured meat could soon be hitting the supermarket shelves. As an environmental and animal friendly alternative to vegan or plant-based options, several start-up companies are trying to scale up techniques which use stem cells (cells which can form any type of body tissue) from animals to grow muscle fibres in reactors. These artificial muscles can be ground up, mixed with plant-based fats, and turned into burgers or sausages or anything else you can think of. The stem cells can even be grown directly into structured muscle and fat cells to create cuts of meat like chicken breast. The look, texture, and taste of meat without the need for all that pastureland and crop feed. California based Memphis Meats is a prominent start-up company which has attracted investment from Richard Branson, Bill Gates, and food giants Cargill and Tyson. Memphis Meats use bioreactors to produce the world's first cell-based meatball in 2016 and first synthetic poultry in 2017.[16] Cultured meat is still in development and costs need to come down before it can compete with farm reared meat.

Developments in the animal feed space are also accelerating. One prominent early stage example is Robert Downey Junior's investment into Ynsect through his Footprint Coalition Fund. Ynsect is a company pioneering the large-scale production of mealworm protein from food waste, as an alternative fish food. Mealworms produce the same amount of protein as traditional fishmeal with 98% less land and 50% fewer natural resources.[17]

A Zero Waste World at Peak Population: The next improvement we can look at is waste. What if we continued to grow crops at the same yield, eat the same types of food, and continue using biofuels, but we did it without waste?

Efficiency improvements of 30% would offset the 30% growth in the global population but not the rise in meat consumption per person. Underlying CO_2 emissions increase from eight to 12 billion tonnes per year and agricultural land expands from 49 to 68 square km emitting over 1,000 billion tones CO_2 in the process (double the remaining 1.5°C budget). Better than doing nothing but not as effective as yield improvements or veganism.

The rise of digital agriculture could prove a large part of the solution to reducing losses by incorporating more technology into the farming process. Big data, artificial intelligence, the internet of things (IoT), e-commerce, payments technology, sensors, GPS, drones, weather prediction, automation, and advanced imaging technologies improve efficiency across all aspects of the agricultural supply chain. Essentially, digital agriculture builds on the amount of information available to farmers and automates the collection, processing, and interpretation of that data to improve yields and reduce waste. Beyond the farm, digital integration of trading, storage, and processing creates a more efficient system.

Improvements could be made on-farm, enabled by better monitoring, data collection, and interpretation that allows for better timing of growing seasons, optimum crop selection, and more precise fertiliser, herbicide, and pesticide application to improve yields and minimise waste. Off-farm, buyers and sellers of agricultural products can be better matched to avoid food miles, length of storage, and number of transactions before the crop hits our table – this helps reduce waste. Digital payments, availability of credit, and insurance should become more widely available in developing countries via smartphones, helping to modernise traditional farming practices in areas such as India or Africa. All of this improves yields and reduces waste.

Lowering Carbon Intensity of Agricultural Practices: The final set of tools towards lowering the agricultural environmental footprint are low carbon farming practices. Let's look at a few examples:

Better Soil Management: Soils contain 2,500 billion tonnes of CO_2 equivalent, more than three times the CO_2 stored in the atmosphere and four times the amount of carbon in all plants and animals. Many agricultural practices disturb the soils and release carbon. Better management of soils could lead to net uptake of CO_2 if done well. Techniques include:

- ***Cover Crops:*** rather than leaving soil bare after harvest, clover or legumes are planted which add nitrogen to the soil, supress weeds, prevent soil erosion, and take in CO_2 from the air as they grow.
- ***Minimising Tillage:*** minimises the exposure of soil to the air which reduces the amount of carbon dioxide release.
- ***Rotational grazing:*** helps to keep carbon in the soil by preventing over-grazing and death of pastureland.
- ***Fertiliser application:*** to prevent excess soil nitrogen or nitrogen run-off.
- ***Agroforestry:*** planting trees or shrubs in and around pastureland or cropland to prevent soil erosion and to capture CO_2 from the atmosphere.
- ***Biochar:*** take trees from sustainable forests (regrown capturing more carbon) and heat in the absence of oxygen to create biochar, which is spread on farmer's fields, boosting crop production by 15% and leaving the carbon in the soil.[18]

Enteric Fermentation: It is possible to supress or capture some of the methane emissions from cows and other ruminants. Dutch company DSM has launched a feed additive under their clean cow platform. The additive is mixed with the daily feed and acts to supress the enzyme which triggers methane production and can reduce emissions by 30% in cows and other ruminants. Alternatively, if cattle are housed indoors the methane can be captured directly.

Rice Paddy Methane: Using methods such as a mid-season paddy dry down or planting the rice on raised beds and flooding only the furrow can halve the CO_2 emissions without impacting yield.

Zero Carbon Energy: Switching agricultural machinery and vehicles from combustion to electric and switching coal and gas heating to electric. The switch would cut energy use by four-fold.

Organic Farming: Using no artificial fertiliser, pesticides, or genetic modification provides a route to reduce emissions of NOx and leaching of chemicals into the surrounding environment whilst also encouraging biological diversity. But caution must be used. A study of studies by Ponisio et al. shows that organic crop yields are 20% lower than conventional yields for major crops.[19] Offsetting the yield losses would require half the world to go vegan. For certain products, organic farming practices can compete with conventional methods but should be used selectively.

The net impact of low carbon technology options if applied to full effect in our peak population, peak meat world, would help to reduce ongoing CO_2 emissions (excluding land use change) by half to four billion tonnes per year (from eight billion). However, the impact of increasing land use would still climb with the growing global population and meat consumption.

Net-Zero Carbon Agriculture: Finding the Path of Least Resistance

Reviewing the different improvements available to reduce the environmental impact of agriculture, veganism (or at least vegetarianism) has the largest beneficial impact on both land use change and ongoing agricultural emissions. Improving agricultural yields follows closely behind. The next best options are reducing waste to help minimise the need for deforestation and then low carbon technologies for reducing ongoing emissions.

Realistically, a combination of all options implemented to a lesser degree will prove the path of least resistance to decarbonise agriculture. Here is just one scenario where agriculture gets to net-zero carbon emissions:

- **Meat consumption per person holds steady:** If fairly distributed, that requires more than 50% cuts for big meat-eating areas such as the US, Europe, and Brazil. Americans consume an average of 100 kg of meat per year, the equivalent of 2.5 quarter pounders per day. Reducing this level of meat consumption towards the global average of 40 kg would allow the 3+ billion people living in areas such as India or Africa to enjoy a 3-to-10-fold increase in meat consumption.

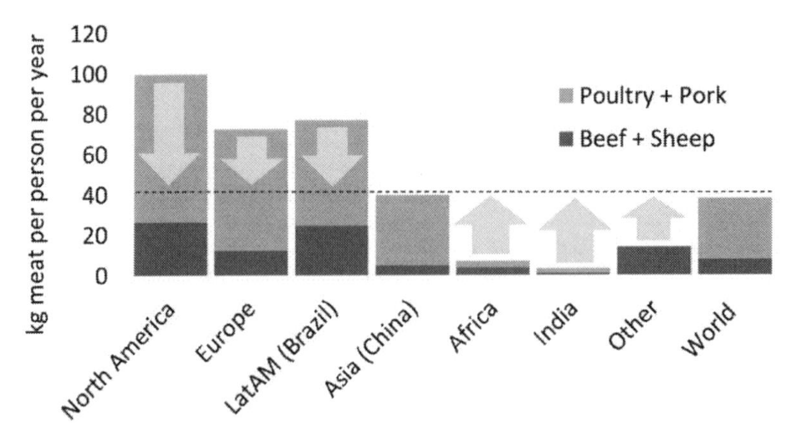

Meat consumption (excluding dairy) in kilograms per capita per year by region. Data from OECD [20]

- **One third less Beef and Sheep consumption:** Replaced by pork, fish, or poultry. This limits the amount of enteric fermentation or methane emissions.
- **One third less waste:** Takes crop to table efficiency to 80% (from 70%) and crop to meat to table efficiency to 8.5% (from 7.5%).
- **Yield improvement by one third to best:** Moving all agricultural land to current best in class yields whilst keeping the same share of different crops would increase production to two billion kcal per square km (from one billion). However, not all agricultural land is of the same quality and it depends on weather, soil, temperature, water availability, and nutrients. So, let's assume that average global crop yields can reach one third of the way to best and increase to 1.4 billion kcal per square km per year. Importantly, though, this relies on moving India and Africa towards more modern agricultural practices, it doesn't require further breakthroughs in seed science (breeding or genetic modification), crop protection (pesticides/herbicides), or fertiliser – but these of course may also help.
- **One third deployment of low carbon technology:** This can reduce carbon intensity of ruminants by 10% by deploying methane reducing feed over one third of farms. We can reduce methane emissions from rice paddies by 15% by deploying best practice over one third of fields. And we can reduce the release of nitrous oxide from soil by 33% if we deploy best practice soil management across one third of agricultural land.

- **Zero Carbon Energy:** We can decarbonise the energy used in farming by switching the 40% oil related energy use to electric agricultural machinery. We can switch the 20% gas and coal-based heating energy to electric using heat pumps. This brings annual energy use down to 1trillion kWh per year (from 4) and allows us to power agriculture with renewable energy.

The outcome of all of these actions is to bring ongoing agricultural emissions down to five billion tonnes of CO_2 per year (from eight billion tonnes CO_2e) from enteric fermentation, paddy field flooding, and excess fertiliser. However, these emissions can be offset by a five million square km lower land requirement (versus today) allowing for carbon capture through reforestation to offset.

Land used for agriculture today, and under each of our peak food scenarios (as % of all land on Earth).

Re-engineering Agriculture for a Peak Food Net-Zero World

The global agricultural system is responsible for 13 billion tonnes of annual CO_2e emissions. Ongoing emissions add up to a total of nearly eight billion tonnes from a combination of energy use (1.5b), enteric fermentation (2.5b), excess fertiliser application (3b), and rice paddy flooding (0.7b). Land use change adds another 5.5 billion tonnes of CO_2e emissions per year from deforestation (4b) and peatland draining (1.5b).

The world produces a total of 1.2 tonnes of crop product per person per year. This amounts to a total of 6,000 kcal per person per day and another 4,000 kcal are sourced from pastureland used to rear animals. The average human being requires only 2,000-2,500 kcal per day to maintain an optimum weight. This means our current agricultural system is just 30% efficient from field to fork. In other words, we already produce enough food to feed the world three times over.

Assuming no change to the current agricultural system including crop yield, waste, and farming practices, then a planet with 10 billion people eating European levels of meat per person will emit 14 billion tonnes of ongoing emissions (from eight). Agricultural land will double to 60% of dryland on Earth, forest clearing would emit a total of more than 2,500 billion tonnes of CO_2, five times more than the remaining 1.5ºC carbon budget. This illustrates why continued crop yield improvements are so important. Luckily, we have plenty of headroom to improve agricultural practices around the world and boost food production yield by up to 90% if best in class yields could be achieved everywhere.

Fundamental change to the agricultural system is required to support a net-zero carbon transition. One of the most powerful changes is diet. A total of 6,000 kcal per person per day are used for rearing animals with only 600 kcal of meat and dairy to show for it. Eliminating this wasted energy in an all-vegan world would increase agricultural field to food efficiency from 30% to 70% and not only halve ongoing agricultural CO_2 emissions, but also allow for reforestation of farmland. This could sequester 50 years' worth of current CO_2 emissions and quadruple the remaining 1.5ºC carbon budget.

Waste is the next biggest improvement helping limit agricultural emissions. Eliminating waste across harvests (10%), storage and trading (5%), processing (10%), and households (10%) would improve field to food efficiency from 30% to 40% and would hold ongoing emissions flat as the population grows whilst eliminating any further emissions from land use change.

Low carbon techniques and technologies offer a route to further decarbonising ongoing emissions by reducing carbon intensity. Better soil management, crop cover, and agroforestry, coupled with methane reduction from cows, rice paddy dry downs, and zero carbon energy and equipment, could reduce ongoing emissions, but this has no impact on land use.

Farming will have to adopt a sensible mix of all options. We can reach a peak food, net-zero agricultural system holding meat and dairy consumption steady per person (this implies over 50% cuts for big meat eaters) and switching one third of the remaining beef and lamb consumption to fish, chicken, or pork. We will need to reduce waste by one third, improve crop yields by one third (by modernising developing market farming techniques), and we will need to deploy low carbon technology over one third of all farming.

Switching agriculture to net-zero carbon requires implementing established best practice and already available commercial technology worldwide. Net costs to the economy are negligible because the higher capital costs should be offset by better crop yields and energy efficiency. The most difficult part of decarbonising agriculture will be improving availability of education, credit, and access to markets for emerging market farmers, and convincing the western world to eat half as much meat.

Agriculture is the one area of the economy where it is near impossible to get to zero carbon without a fundamental change in lifestyle for developed countries. Some would consider less meat in the diet as a positive and healthy life choice, others would consider it an insult to liberal free choice. Either way, meat consumption represents an immoveable hurdle to zero carbon unless lab grown meat products become cost effective and accepted. Agriculture is the one area where changing consumption habits probably represents the lowest friction road to net-zero.

Global Agriculture (billion tonnes CO_2e per year)		Peak Food World					
	Agriculture Today	No System Change	Best Yield	All Vegan	Zero Waste	Low Carbon Intensity	Mixed Approach
Energy Use	1	2	2	1	2	0	0
Enteric Fermentation	3	5	5	0	5	3	2
Rice	1	1	1	1	1	0	1
Fertiliser	3	6	3	1	4	0	2
Ongoing Emissions	**8**	**14**	**11**	**3**	**12**	**4**	**5**
Net Deforestation	4	48	0	-45	21	48	-5
Net Peatland Change	2	18	0	-17	8	18	-2
Annual Land Change (to Peak)	**6**	**66**	**0**	**-61**	**28**	**66**	**-7**
Total CO_2e per Year	**13**	**80**	**12**	**-58**	**40**	**70**	**-3**
Land Change Total Impact		**2,653**	**16**	**-2,452**	**1,139**	**2,653**	**-293**
multiple of 1.5C budget (x)		4.8 x	0 x	-4.4 x	2 x	4.8 x	-0.5 x

Table showing the carbon dioxide emissions of the agricultural system today and the scenarios we laid out for different possible futures. Mixed approach is our blended net-zero solution.

Chapter 13: The Net-Zero Economy

Social and Economic Benefits

The last six chapters ran through detailed supply and demand options for switching the global economy to net-zero carbon. We compared capital costs, running costs, safety, reliability, and technical limitations of all options. We used all tools at our disposal to build a viable global solution to climate change and air pollution which could provide a sustainable, high quality of life for a peak population planet. Drawing heavily on technology to reduce the carbon intensity of electricity supply whilst electrifying demand and bringing substantial efficiency gains in the process. Hydrogen, biofuels, and carbon capture supplement areas of the economy which are hard to electrify. Consumption reductions play a role by reducing meat production and freeing up valuable agricultural land for reforestation. Negative emission technology is a last resort for the most stubborn emissions. We have pieced together a cornucopian vision of plenty, whilst recognising our neo-Malthusian planetary boundaries.

In this chapter we will compare the total social and economic outcomes of a net-zero future. We will explore the benefits to health, wealth, jobs, resource availability, and energy security in addition to the damages we avoid from limiting climate change. We will meet our final dial of doom and boom: the *experience curve*, which defines the rate at which net-zero technology costs decline. And finally, we will tweak all three dials of doom and boom to show why a rapid transition is optimal.

Recapping Energy Supply in a Zero Carbon World

We re-engineered the use of energy in each four sectors of the economy to match demand and net-zero supply, making sure that we deliver enough energy and in the right form to power a net-zero economy. The following energy supply is one of many possible routes to provide a technically and economically viable route to decarbonising the economy in a peak population, peak consumption world.

Electricity use becomes more than 70% of final energy. Solar photovoltaic and wind make up the bulk of generation with a small baseload contribution from hydroelectric and, if needed, next generation nuclear or natural gas with carbon capture plus storage. Every 1 kWh of wind or solar electricity uses 0.5 kWh of short-term energy storage, mostly from lithium-ion batteries and pumped hydro.

Fossil fuels decline to less than 20% shrinking from 80% of final energy today. Natural gas and carbon capture (7%) is used for hard to electrify high temperature industrial heating processes and oil (8%) is used as petrochemical feedstock.

Hydrogen provides 10% of final energy produced from water splitting using renewable electricity and used mostly in difficult to electrify transport applications and nitrogen fertiliser production. Hydrogen provides a seasonal storage solution for renewables and adds flexibility to short term energy storage capabilities.

Biofuels remain 10% of final energy. The use of solid (wood) biomass declines to 4% (from 10%) as developing regions modernise heating and cooking. Liquid biofuel grows share from 1% to 4% of final energy shifting from first generation ethanol to third generation bio-algae for long haul air travel and long-distance shipping.

Over 80% of the technology is already commercialised and will continue to improve cost and efficiency with scale of production. The remaining 20% of technologies, including hydrogen (from water splitting), next generation nuclear, and third generation biofuels, work today but require further development before hitting the commercial experience curve.[1]

Once each of the technologies is deployed at full scale production, large enough to serve a fully decarbonised economy, the experience curve and technology roadmaps tell us the average price of fully loaded zero carbon energy will drop to $7c per kWh, just 30% more than today's mixed fossil-fuel-based system.

Recapping Energy Demand in a Zero Carbon World

Transport: Public transport expands. Cars, vans, and trucks electrify. Air travel, haulage, and shipping use a mix of electric, hydrogen, and biofuel energy for short, mid, and long-distance routes. Better integration of public and private travel systems allows for greater penetration of ride sharing, shared ownership, and efficient integrated journeys. Digitisation of industry creates more localised manufacturing, reducing product miles. Two thirds of the technology is already commercialised, one third requires further development. A fully scaled net-zero transport system will use nearly two thirds less energy than the fossil fuel equivalent and total costs will be nearly 40% cheaper.

Industry: High temperature heating processes switch from natural gas or coal to net-zero electricity. Carbon capture and storage is used to sequester direct emissions before they hit the atmosphere. Steel, paper, and textiles can be decarbonised at similar manufacturing cost or less. Cement and chemicals will require price increases. Incorporating more negative emissions wood into construction, efficiency improvements from smart manufacturing, and a switch to low embodied energy materials will halve total industrial energy consumption compared to fossil fuels. The total cost of net-zero industrial production will fall by up to 25% compared to the fossil fuel equivalent. Most of the technology is already commercialised but requires up front economic incentive to become more widely deployed.

Basic Amenities: Convenience living is re-engineered for net-zero and widely deployed to provide a modern standard of living for all. Natural gas or wood burning is replaced by district heat, electricity, or hydrogen for space heating, cooking, and hot water. LED lighting and low-cost insulation improve efficiencies. A net-zero residential and commercial sector will consume one third less energy with total costs up to 10% lower than the fossil fuel equivalent. All technology is commercially available.

Agriculture: Big meat-eating countries halve meat and dairy consumption and switch one third of beef, to pork or poultry. Digital agriculture and shared best practice cuts waste and improves yields by one third. Low carbon technology including soil care, clean cows, and rice paddy management are deployed across one third of farms. Demand for agricultural land declines by 5%, despite 30% more mouths to feed, which eliminates the need for deforestation and lets us reforest up to 100 thousand square km per year – enough to bring agricultural emissions to net-zero.

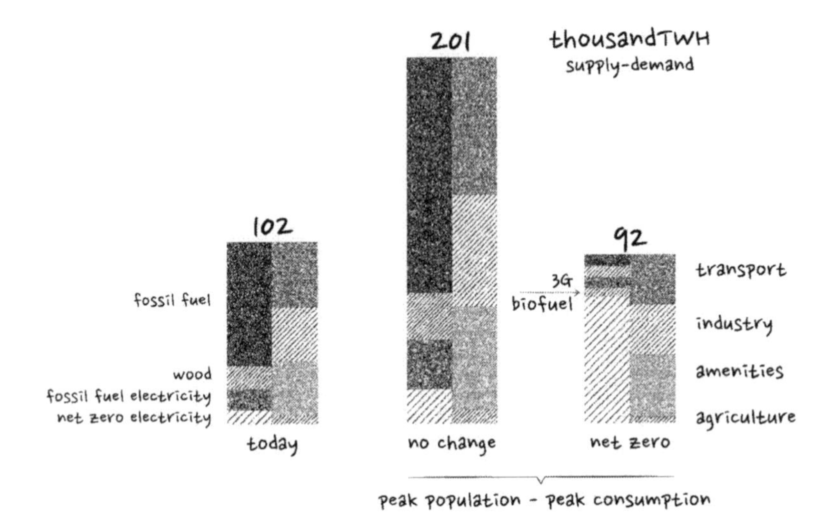

Final energy supply and demand for the global economy today, and peak population (10 billion), peak consumption world either remaining bound to fossil fuels or under a net-zero system. Left side of each column set is supply, right side is demand by use. Labelled shares remain in order across the three diagrams (3G biofuels added in net-zero).

Economic Benefits of a Zero Carbon Economy

Let's compare the total cost of re-engineering the global economy for peak consumption, peak population, net-zero carbon to a scenario where the fossil fuel systems don't change (our baseline of inaction).

Today's Economy: Across transport, industry, basic amenities, and agriculture the world consumes just over 100 trillion kWh of final energy (mostly from fossil fuels) and emits over 50 billion tonnes of CO_2e per year. The average cost for one unit of wholesale energy is $5.5-6c per kWh which brings the total energy cost to $6 trillion per year. Another $6 trillion per year is spent on the equipment used to convert this energy into a useful work such as motor vehicles, industrial machinery, and gas boilers. A total of $12 trillion per year or 13% of global GDP is spent on energy and energy equipment supporting the lives of nearly eight billion people on the planet but providing a high standard of living for only 10%.

Peak Population, Peak Consumption based on the Current System: In a world with ten billion people enjoying a high standard of living and with no change to current fossil-fuel-based systems, energy use doubles to 200 trillion kWh per year and annual CO_2e emissions climb to over 100 billion tonnes. The total energy ($11 trillion) and equipment costs ($12 trillion) increase to $23 trillion per year to support 30% more people on the planet with each person averaging 50% greater consumption.

Peak Population, Peak Consumption, Zero Carbon at Current Technology Costs: Re-engineering the global economy for net-zero using the cost and efficiency of today's technology would reduce peak energy use by 45% to 110 trillion kWh, with an average energy price of $10c per kWh. Total energy costs are $11 trillion dollars, but capital costs increase to $16 trillion per year. The total cost of peak consumption at zero carbon using today's technology is $27 trillion or 15% more than the fossil fuel alternative.

However, we know today's technology costs will decline as they scale up...

Peak Population, Peak Consumption, Zero Carbon at Full Scale: Once all key technologies are deployed at full scale, production costs decline, and efficiencies improve. All ten billion people on the planet can enjoy a high standard of living whilst consuming just 90 trillion kWh of energy per year, which is 55% less than the fossil fuel equivalent and is also 10% lower than today. The average price of energy will equal $7c per kWh and total energy costs add up to $6 trillion per year, 40% below the fossil fuel equivalent. Adding equipment costs of $11 trillion takes the total cost to $17 trillion dollars per annum. **This scenario represents the final state of our net-zero energy transition where everyone on the planet can enjoy a high standard of living at 25% lower cost compared to remaining bound to the fossil fuel alternative.**

			Peak Consumption (beyond year 2060)			
Global Energy System (per year)		Today	No System Change	Net Zero Today's Tech	Net Zero At Scale Tech	Net Zero At Scale + Efficiency
Energy Consumption	k billion kWh	102	200	113	110	92
Blended Energy Price	$c per kWh	6	6	10	7	7
Levelised Energy Costs	$ trillion	6	11	11	7	6
Capital Costs	$ trillion	5	10	14	11	9
Other Costs	$ trillion	1	2	2	2	2
Total Costs	**$ trillion**	**12**	**23**	**27**	**21**	**17**
CO2e emissions	**billion tonnes**	**52**	**106**	**-4**	**-6**	**-10**

Table showing the energy use, weighted energy price, levelised cost breakdown, total costs, and emissions of our energy system, both now and in different futures. All estimates are based on the detailed estimates and solutions given through chapters 7 to 12.

Contrary to popular belief, we can switch the global economy to net-zero carbon with no compromise on the highest standards of living. The final levelised cost of the switch is 25% cheaper than sticking with our existing fossil fuel system. A saving of $6 trillion per year or $600 per year for each person on the planet.

The switch does require higher upfront costs whilst certain net-zero technologies ride down the experience curve. The cost of new technology such as solar, wind, and batteries decline by 10-30% every time installed capacity doubles, but most of the capacity is installed at the cheaper end of the curve.

The cost of the energy system is less than 1% more through the first quarter of the transition and 25% cheaper once complete.

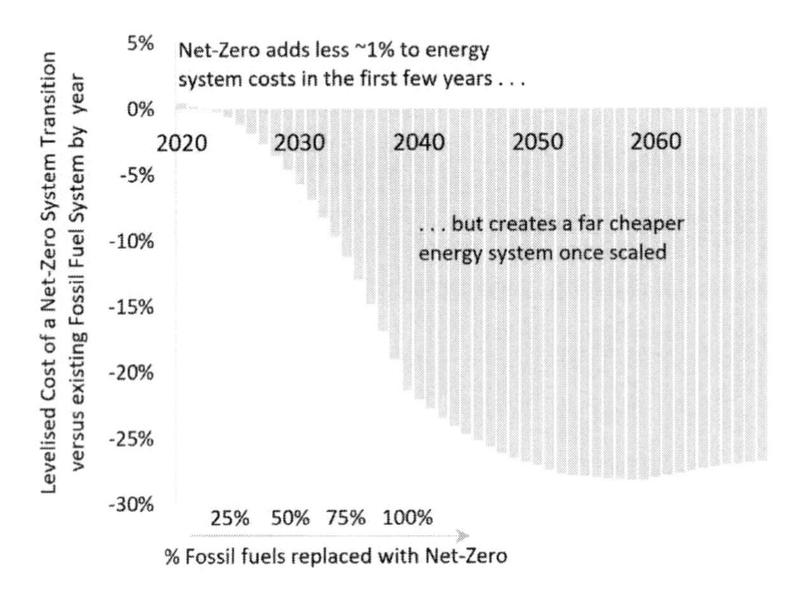

Annual levelised system cost difference between a net-zero transition (over 20 years) and sticking with our fossil fuel system. Includes energy supply costs and demand side capital costs, feedstock changes, and added labour costs. The early years are <1% more expensive but the experience curve quickly drives down net-zero technology costs so once 25% of the transition is complete net-zero becomes the cheaper pathway. Author's calculations based on the aggregated assumptions and cost analysis from chapters 6 to 12.

Societal Benefits of a Zero Carbon Economy

Switching to a net-zero economy is twice as efficient and 25% cheaper than continuing on with our existing fossil fuel system. Moving to net-zero carbon also brings with it health benefits, employment benefits, and better use of the Earth's natural resources.

Health Implications of Net-Zero Carbon

Most health risk is associated with respiratory disease and heart failure caused by harmful emissions such as particulate matter, nitrous oxides, ozone, and sulphur oxides from the burning of fossil fuels and wood. Statistically, the safest forms of energy generation are nuclear, solar, wind, and hydroelectric, with less than two deaths per billion kWh of energy supply. Coal, oil, and wood biofuel are the worst offenders with 50-250 deaths per billion kWh.

Today's energy generation is responsible for nearly eight million deaths per year and 2-3 times as much serious illness and suffering. To improve air pollution whilst still burning fossil fuels is possible but would cost around 0.5% of GDP to upgrade to the best performing technology.[2] Switching the economy to net-zero will clean the air we breathe to an even higher standard, it will reduce the mortality rate by 90%, preventing seven million premature deaths per year, and do it with a cheaper energy system.

Energy Employment in a Net-Zero Carbon Economy

Renewable technologies have higher capital costs and higher labour costs than fossil fuel alternatives yet can deliver electricity at a similar, if not cheaper, price point. The reason is economic rent on fossil fuels. Effectively, the extortionate profits extracted by resource owners. For example, the cost of oil extraction in Saudi Arabia is less than $10 per barrel and yet oil sells on international markets for an average of $65 per barrel.[3] Saudi makes $55 of profit on each barrel of oil, not because it has a better product or a better extraction process, but because it just happens to own easy to extract oil deposits. The world pays a total of $1.5 trillion per year (2% GDP) over and above the cost of extraction to access and use these finite deposits (mostly oil) owned by resource rich countries (the equivalent of up to $2.3c per kWh). As much as 25-50% of GDP comes from oil rent in many petrostates. On the other hand, wind and sun cost nothing and are owned by nobody. In a net-zero world the annual $1.5 trillion transfer of wealth to international cartels such as OPEC can instead be paid to the workers on wind, solar, and hydroelectric projects.

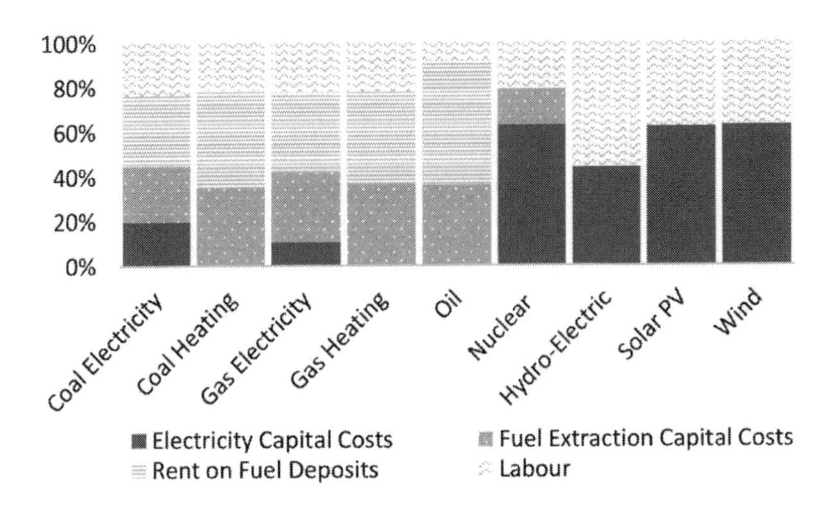

Share of final energy costs split. Rent data from World Bank.[4] Labour data from Wei et al. "Putting renewables and energy efficiency to work", capital cost data per chapter 7 modelling.

There are an estimated 13 million jobs in the global fuel energy and electricity sectors today and this represents about 0.3% of the global working age population. Fossil fuel energy sustains 30 to 120 full-time jobs for every one billion kWh of energy per year. Nuclear, hydro, wind, and biofuels require 140-240 jobs and solar 600 full-time jobs per billion kWh per year (though the average solar salary is one third of other technologies).

A peak population, peak consumption, fossil-fuel-powered world would require 25 million energy employees to deliver 200 trillion kWh per year. A zero-carbon economy requires over 30 million energy employees to deliver 90 trillion kWh per year plus more jobs upfront for deploying energy efficiency measures.

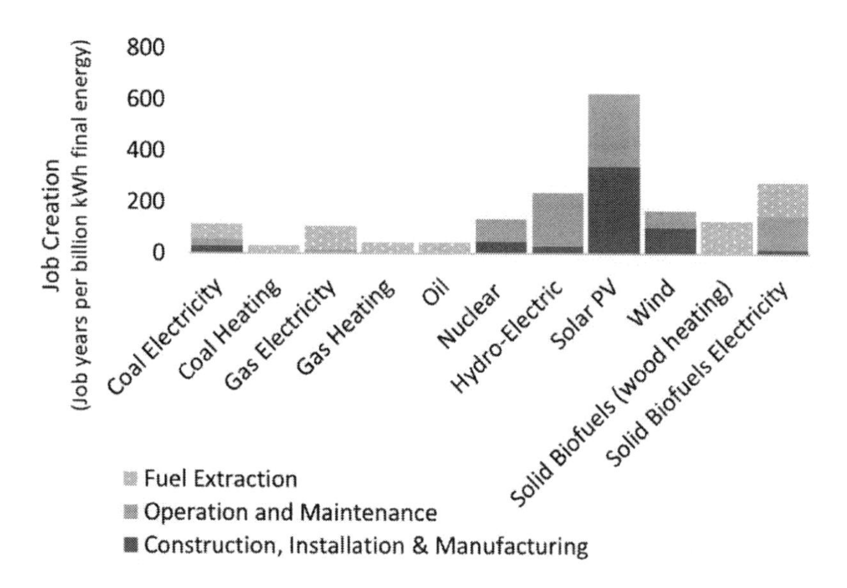

Split of labour by energy source. One job-year sustains one job for one year (multiply by final energy per year to calculate sustained job creation). Data from Wei et al. "Putting renewables and energy efficiency to work".[5]

Labour Inclusion in a Net-Zero Economy

Beyond direct employment, a fair transition to net-zero ensures the one billion plus people who have no access to the modern world and a sub-standard quality of life are connected to the economy and given the opportunity to prosper. The total cumulative cost of energy access and clean cooking and heating for all is around $2 trillion of our total net-zero spend or about 2% of global GDP.[6] This is some of the most effective transition money which can be spent – eliminating the four million indoor air pollution deaths every year, breaking poverty traps, and accelerating economic progress in developing countries. Driving a stronger and more equitable system for all.

Additionally, the transition to a net-zero energy system must bring everyone along and that includes the five million workers in fossil fuel industries across the world. Plans must include ways to transition lost fossil fuel jobs into net-zero employment and to direct newly created work into communities which need it most, an equitable transition. Building out our net-zero future will create a diverse pool of labour and require a huge range of skills. The engineering, geology, and operational experience in the oil and gas industry will prove invaluable for building out offshore wind, hydrogen infrastructure, biofuels, water desalinisation, storage, and carbon capture projects. Rapid expansion of the electrical grid will require engineers, electricians, and manual labour. Keeping the systems running requires physical, digital, and systems maintenance and the accompanying skills across each. Net-zero can boost jobs, improve working conditions, and build skills. Mark Carney writes, "For sustainable investment to go truly mainstream, it needs to do more than exclude incorrigibly brown industries and finance new, deep-green technologies. Sustainable investing must catalyse and support all companies that are working to shift from brown to green. We need approaches which capture Fifty Shades of Green."[7]

Sustainability of Energy Resources in a Net-Zero Economy

On today's rate of consumption, there are 50 years of gas, 60 years of oil, and 130 years of coal reserves left. Moving into higher cost deposits could at least quadruple those estimates. Continuing to burn fossil fuels is not only more expensive and more damaging to health, jobs, and the environment, but is also unsustainable beyond the end of this century without increasing cost of extraction. Transitioning to net-zero creates a sustainable energy system driven mostly by sun and wind, using a total of just ~1% of dry land on Earth.

Sustainability of Natural Resources in a Net-Zero Economy

With no change to our energy and agricultural systems, global water use will quickly breach sustainable limits and the world will be forced to desalinate sea water at a cost of at least $3 trillion per year. Instead, switch to net-zero and industrial water use declines by more than one third in 2100 thanks to less waste and a shift to wood products rather than cement and metals. Changes to diet and our agricultural system could halve the volume of water required for irrigation, with less demand for animal feed, less global warming, and fewer droughts. Uneven distribution of water resource will, however, still require many areas to improve access to water and storage or deploy desalination by the end of this century.

Fresh Water per Year (km^3 / billion tonnes)	Today	2100 (Inaction)	2100 (Net Zero)
Domestic use	475	608	608
Industry use	721	2,532	1,654
Agriculture use	2,767	8,997	4,555
Total Water Withdrawals	**3,963**	**12,138**	**6,817**
Thermal pollution from power	719	1,413	530
Total Water Used	**4,682**	**13,552**	**7,347**
Warming	1	3	2
Renewable fresh water supply	53,688	53,688	53,688
Additional capture from dams	3,500	3,500	3,500
Flood run-off	-40,000	-40,000	-40,000
Climate change decline	-417	-2,396	-974
Inaccessible water	-2,000	-2,000	-2,000
Net Water Available	**14,771**	**12,792**	**14,214**
Water Surplus (Shortage)	**10,090**	**-760**	**6,866**
Percent of Available Water used	32%	106%	52%

Water supply and water use today[8] and the year 2100 with either our baseline of inaction or net-zero. Domestic water use grows with population. Industrial water use is based on the mix and amount of power generation and industrial goods (water consumption ranges from 0.05 litres per kg for aggregates to 300 litres per kg for textiles). Agricultural consumption based on required crop acres/yield and prevalence of drought, 20% of crops are irrigated today, that rises to 60% in our baseline of inaction or 35% in our net-zero system. Author's calculations based on the aggregated assumptions and cost analysis from chapters 4-7.

Fertilisers represent another finite resource for the planet. Nitrogen, phosphate, and potassium (NPK) are the three key elements which support plant growth. Nitrogen is artificially synthesised from the air and is abundant so long as the energy used is clean. Phosphate and potassium are finite resources with no current substitute. However, there are enough reserves to last another 50-200 years and resources to last at least 800 years under all future scenarios.[9,10]

Energy Security in a Net-Zero Economy

Today's global energy supply relies on the extraction of finite fossil fuels which are unevenly distributed across the world, partially controlled by international cartels, and vulnerable to terrorism, political pressures, and unexpected disruption of supply. The way we use these precious resources is also very inefficient with at least a third of the energy wasted.

Net-zero will create a far more evenly distributed energy supply with no economic rent. Supply will be better distributed and markets more competitive. Final use will be far more efficient using half the energy of its fossil fuel equivalent for the same amount of utility. The IPCC notes that enhanced energy security afforded by renewables compared to fossil fuels is worth as much as 0.2% of GDP.

Aesthetics of a Net-Zero Economy

Farmland has come to be considered part of the natural world by many but remember those vineyards in Napa Valley or hedgerows criss-crossing England's green and pleasant land are a human creation. Two thirds of the Earth's surface was covered in forest not long ago but developed countries have already cut down half.

We currently use 33% of all dryland for crops and pasture, we inhabit less than 2%, and we use a negligible land area to generate fossil fuel power. Left unchecked and agriculture could grow to use more than half of all land on Earth to feed ten billion people consuming a meat rich diet.

Switch to a net-zero world and we limit agricultural land use to 30%. Wind, solar, and 3G biofuels claim less than 1% of land area, leaving more than 2% or four million square km for rewilding and restoration of forests, whilst limiting the destruction of natural habitats from climate change.

We must ensure though that we compensate those who haven't already decimated their natural forests and ensure a fair transition to a more sustainable planet. We must accept that the sight of solar panels or wind turbines, whilst offensive to some, represents a better use of land than unproductive agriculture. We must accept that no solution comes without small compromise.

Peak Population in a Net-Zero Economy

Add together the economic and social benefits from a cheaper energy system, better health, more energy jobs, and greater labour inclusion, and not only do we avoid climate change damages, but we also accelerate human progress and prosperity.

The Speed of the Transition and the Experience Curve: Our Third and Final Dial of Doom and Boom

Making no further changes to our fossil fuel powered economy and the planet warms by at least 3.3°C by 2100 with 0.5-1 metres of sea level rise. Climate change and air pollution could kill more than one billion people, throw over 2.5 billion people into malnutrition, force hundreds of millions to abandon their homes, and wipe out 5% of GDP by 2100. These estimates assume we don't breach any tipping points in the physical or socio-economic systems. If we did, this would significantly amplify the death, damage, and destruction. The Precautionary Principle tells us we should avoid potentially unlimited damages if the cost of the solution is limited (and ends up cheaper), but exactly how fast should we act? Clearly it makes no sense to ban fossil fuels overnight as this would destroy the economy and society, but how fast is best?

Economists, governments, and policy makers try to weigh up the best course of action on climate change using Integrated Assessment Models (IAM). IAMs project the damages from a warming planet through time, and balance this with the cost of a net-zero transition over time. A continuous benefit-cost analysis attempting to pick a set of actions that leads to the maximum utility. The speed at which these models transition the economy are optimised to reach a balance between limiting the damage from warming and limiting the added cost of moving too quickly.

We have built up all the information needed to perform the same calculation. However, our model has two fundamental differences:

We know the cost of net-zero carbon technology declines with scale of production. Prices of net-zero technology don't decline with time, they depend on the magnitude of production and the faster you move, the faster it gets cheaper - ***the experience curve***.

We know that at full scale a zero-carbon energy system is 25% cheaper than the fossil fuel equivalent. This is based not just on the price of energy, but the amount of energy required (efficiency) and the capital cost of all the equipment - ***the whole system***.

These two key pieces of information tell us there is little reason to delay the transition. We shouldn't sit on our hands waiting for technology to get cheaper: prices decline the faster we transition. Our net-zero solutions will quickly move from 15% more expensive today to 25% cheaper than fossil fuels at full scale, and we will reap the added benefits of cleaner air, fewer premature deaths, more jobs, and long-term sustainable energy.

We will cap global warming at the lowest possible limit with the least future damage and minimise the risk of hitting dangerous tipping points.

The limiting factors for the speed of transition are the higher upfront investment spend which displaces other investments in the economy, availability of labour, the difficulties of writing-off fossil fuel equipment before the end of its useful life, and the risk of moving too quickly for governments, businesses, and individuals to adjust. Factoring all of these limitations, the optimum speed of a net-zero transition is 20 years.

Enough time for workers to re-train and governments and businesses can refocus their investment spending. It means most fossil-fuel-based cars, boilers, and industrial equipment naturally reach the end of their useful lives and can be replaced with electric alternatives. Some recently built fossil fuel fired power stations are written off early and more than two thirds of identified oil, gas, and coal reserves are left in the ground ($20 trillion stranded assets by end of the transition). But even if the world pays to compensate these forced write downs, it will be more than offset by moving to a cheaper net-zero energy system. The co-benefits of clean air and avoiding climate damages come on top.

The biggest technical threat to a 20-year transition will be the challenge of developing the 20% of fall-back technologies not yet on the experience curve. These are the technologies which do need time before prices decline and include next generation nuclear, 3G algae biofuels, carbon capture, and hydrogen infrastructure. Not only does research and development take time, but it also takes 10-20 years to build up enough of a track record to deploy infrastructure on a massive scale. If development in these technologies proves too slow, then a similar cost-benefit outcome is 80% of the transition over the next 15 years and the remaining 20% of the transition in the following 15 years.[i] Alternatively, we may well reach a point where already established solar, wind, storage, and electrification become so advanced we no longer need other options.

The **experience curve** is our third fundamental dial of doom and boom, the technological sensitivity which defines how quickly net-zero solutions become available and cost competitive: dial it up and the market will quickly deploy a cost-effective solution, dial it down and solving climate change looks difficult and expensive.

---

i Once emissions are reduced by more than 80%, less than 10 billion tonnes of CO_2e will be released each year which will considerably slow warming.

Required Investment and the Impact on GDP: How Much Does the Solution Cost?

If technologies continue to follow their current trajectory, the total required investment for a 20-year net-zero transition is $70 trillion. This compares to the $24 trillion of investment required if we remain bound to fossil fuels. However, the final energy system costs of net-zero end up at $17 trillion compared to $23 trillion per year for fossil fuels. So, spending the extra $46 trillion upfront saves $5.5 trillion per year down the line. An 8-to-14-year payback followed by cost savings forever. A 7-12% return for the planet plus we avoid damages, tipping points, death, and suffering from climate change and air pollution.

Investment over 20 years ($ trillion)	Discounted into Today's Money Equivalent	Undiscounted Spending (excluding Inflation)
Net Zero Transition		
Net Zero Energy Generation	30	40
Net Zero Transmission and Distribution	10	13
Net Zero Extra Spending on Cars, Insulation etc	13	17
Net Zero Total Spend Over 20 Years	**53**	**70**
Remaining With Fossil Fuels		
Fossil Fuel Energy Generation	15	18
Fossil Fuel Transmission and Distribution	4	5.4
Fossil Fuel Extra Spending	0	0
Fossil Fuels Total Spend Over 20 Years	**19**	**24**
Additional Investment Required for Net Zero	**34**	**46**
Net Zero Energy System Savings (per year)	**2.4**	**5.4**
Rate of Return	7%	12%
Payback (years)	14	8

Required investment into energy generation, transmission, and distribution plus the extra upfront purchase cost of net-zero demand solutions compared to existing fossil fuel solutions (e.g. added cost of electric cars, heat pumps, and house insulation) for a net-zero transition. Compared to investment costs if we remain bound to fossil fuels. System savings are the annual saving from buying and running a net-zero economy once scaled, compared to a fossil fuel alternative. Author's calculations based on the aggregated assumptions and cost analysis from chapters 7-12.

The cumulative GDP gains over the rest of this century total $170 trillion in today's money. That's $50 trillion from moving to a cheaper energy system and $120 trillion of avoided climate change damages. On top of these market gains we know there will be significant co-benefits with seven million fewer premature deaths from air pollution each year and reduced suffering, more jobs, and greater energy security, we also avoid the worse impacts on the natural world and dangerous climate tipping points.

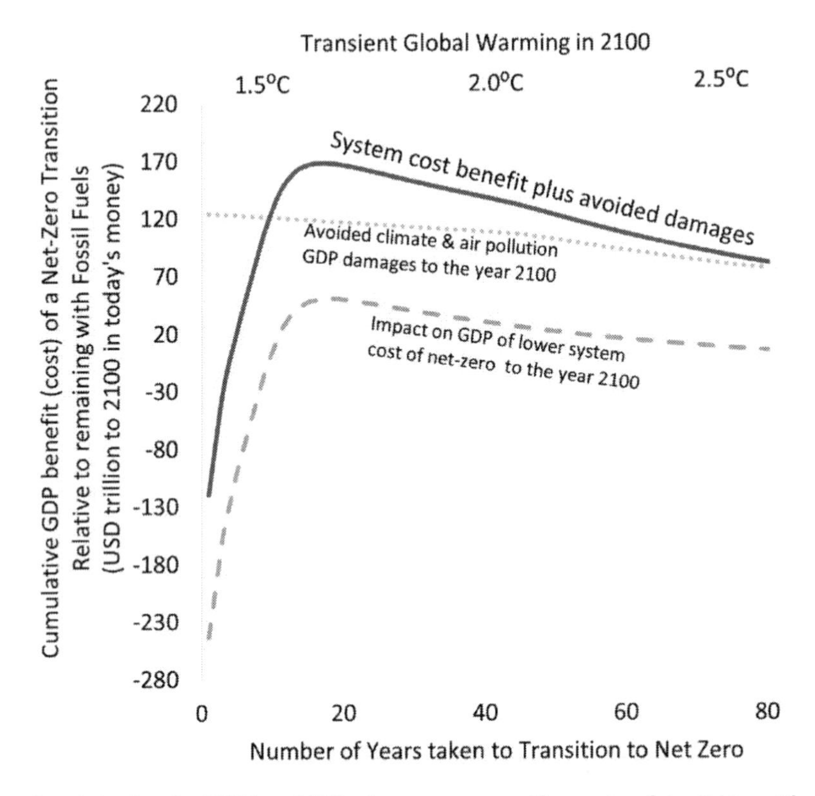

Cumulative benefit (NPV to 2100) of a net-zero transition compared to sticking with fossil fuels depending on the number of years taken to transition. The economic benefits of moving to net-zero are greatest with a transition between 15 and 20 years. This is the best balance of write-offs, labour availability, and technology costs versus the speed of getting to a cheaper energy system. Foregone climate damages save the economy more money the faster we transition and limit warming – do nothing and these costs are $180 trillion. GDP based on Cobbs-Douglas function. Author's calculations based on the aggregated assumptions and cost analysis from chapters 3 to 12.

A 20 year transition reduces GDP by at most 0.5% over the next two decades because the upfront spending for net-zero crowds out investment into other areas of the economy. But the early 0.5% losses are more than compensated by a cheaper energy system and the ongoing avoidance of climate damages into the future. A small price upfront for a materially better system down the line. Half of one percent of GDP equates to just $60 per person per year to solve climate change and the investment is quickly earnt back many times over.

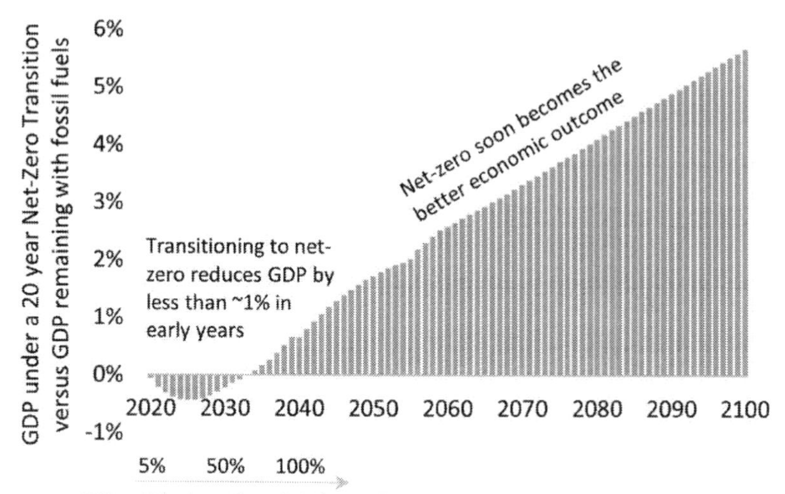

Annual benefit (cost) of a net-zero transition compared to remaining with our fossil fuel system. Over the 20 years of the first transition GDP is up to 0.5% lower each year due to higher upfront spending on net-zero generation, T&D, and demand side equipment (which crowds out other investments in the economy) plus fossil fuel write downs (mostly power plants towards the end of the transition). However, after the first 20 years net-zero provides a significantly better outcome as the energy system is cheaper and the world avoids the worst climate change damages. Non-market benefits (fewer deaths, less suffering, and preserving the natural world) are not included and come on top. GDP calculated based on Cobbs-Douglas function including market-based climate damages, higher net-zero investment spending, write downs, and relative ongoing system costs. Author's calculations based on the aggregated assumptions and cost analysis from chapters 3-12.

Stress Testing Our Vision of the Future: Tweaking the Dials of Doom and Boom

But let's pause for moment and make sure net-zero is the all-round better option. Let's revisit the scientific, economic, and technical uncertainty in our estimates. Let's flex the equilibrium climate sensitivity, the discount rate, and the experience curve to understand the limits of net-zero. Let's turn the dials of doom and boom.

First, the equilibrium climate sensitivity. Remember, this is the expected global warming when atmospheric CO_2 doubles and is the key scientific number which underpins the rate of physical change. The defined range of the climate sensitivity is 1.5ºC to 4.5ºC and through this book we used the consensus 3ºC. But what if we flexed this number? If we turn the equilibrium climate sensitivity to 1.5ºC, the avoided damages fall from $120 trillion to $15 trillion this century as physical changes are smaller. This is still a better outcome for net-zero and there are another $50 trillion of energy savings on top. Flex the climate sensitivity to the top of the range and the avoided damages triple to $400 trillion – an even more compelling case for net-zero.

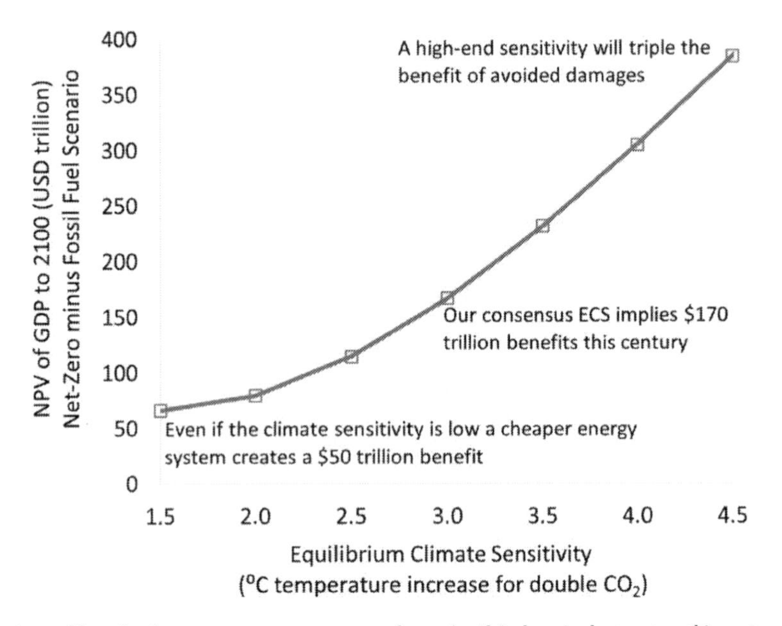

Annual benefit of a net-zero system compared to a fossil fuel equivalent system this century. Shows the difference in the net present value of GDP due to the different energy systems and damages using 3% discount rate across each equilibrium climate sensitivity. Assumes 20-year transition and GDP is calculated based on Cobbs-Douglas function including market-based climate damages, higher net-zero investment spending, write downs, and relative ongoing system costs. Author's calculations based on the aggregated assumptions and cost analysis from chapters 3-12.

So regardless of where the rate of warming falls, the outcome for net-zero is better than remaining bound to fossil fuels, all else being equal. But what about the discount rate? Remember we said this is one of the most important numbers in calculating the economic outcome. A higher discount rate reduces the value of future climate damages in today's money and disproportionately increases the cost of net-zero technology. Our 3% rate gives $170 trillion of benefits, and even high-end estimates such as Nordhaus' 4-5% create around $50 trillion of greater wealth this century. In fact, fossil fuels need a (real) discount rate of more than 7.5% to prove the better economic option this century, and that's only if we ignore non-market impacts like air pollution deaths, and damage to the natural world.

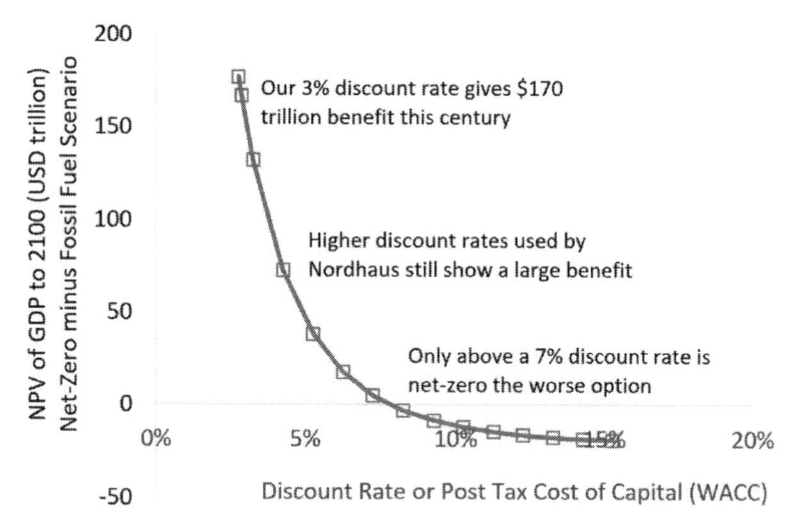

Annual benefit (cost) of a net-zero system compared to a fossil fuel equivalent system this century. Shows the difference in the net present value of GDP due to the different energy systems and damages across each discounting rate/post tax cost of capital. Assumes a 20-year transition and GDP is calculated based on Cobbs-Douglas function including market-based climate damages, higher net-zero investment spending, write downs, and relative ongoing system costs. Author's calculations based on the aggregated assumptions and cost analysis from chapters 3 to 12.

Okay, so even with unreasonably high discount rates, net-zero still stacks up as the better option. But what about the rate of price declines for technologies like wind, solar, and batteries? We can re-run our calculations once again and change the experience curve or rate of price declines with scale. This analysis shows the 20-year transition to a net-zero energy system creates more wealth this century unless the rate of price declines for net-zero technologies are more than four times slower than they have trended in the past (and ignoring the damages from climate change).

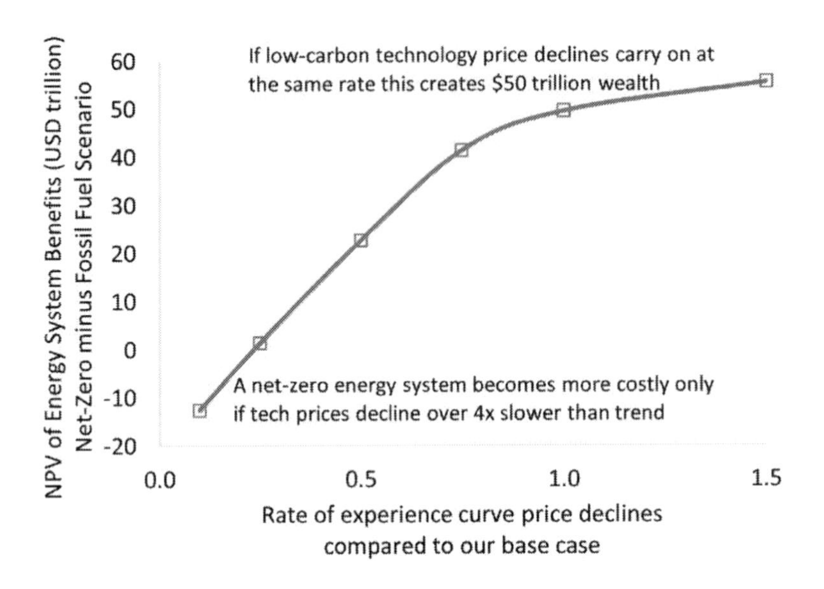

Annual benefit (cost) of a net-zero energy system compared to a fossil fuel equivalent system this century. Shows the difference in the net present value of GDP due to the different energy systems and using 3% discount rate for post-tax cost of capital used for calculating levelised energy technology costs. Assumes a 20-year transition and GDP is calculated based on Cobbs-Douglas function including market-based climate damages, higher net-zero investment spending, write downs, and relative ongoing system costs. Author's calculations based on the aggregated assumptions and cost analysis from chapters 3 to 12.

Net-zero solutions stand up as the better option compared to fossil fuels under all reasonable scientific, economic, and technical assumptions. Net-zero will create a cheaper energy system, is better for GDP growth, avoids major climate damage, and brings many co-benefits. The only way to make an argument for continuing to burn fossil fuels is to use an unreasonably high discount rate, assume we deploy best case adaptation measures, hope we don't hit any dangerous tipping points, ignore non-market death, suffering, and destruction of the natural world, whilst assuming net-zero technology price declines suddenly become four times slower than trend.

Russel L. Ackoff, Wharton Professor of Management Science, remarked, "The more efficient you are at doing the wrong thing, the wronger you become. It is much better to do the right thing wronger than the wrong thing righter. If you do the right thing wrong and correct it, you get better".[11] He may have been reflecting on social policy, but the principle applies to energy systems. Net-zero is the right solution and even if executed badly will still prove the better outcome compared to fossil fuels.

Net-Zero and Competition for Resources: Are There Better Ways to Use Our Capital?

Net-zero solves both climate change and air pollution problems plus it's cheaper, more sustainable, more reliable, and creates more jobs. Surely implementing net-zero is a no-brainer, right? Well, not so fast. We wanted to find the optimal way to allocate our finite resources and net-zero seems to tick all the boxes for our problem. But remember there are many other problems out there in the world, with many different solutions. Net-zero is a tool we have added to our workbox, but that doesn't mean we should try to fix every problem with it.[12] Maybe there are better ways to save lives, end suffering, and improve human prosperity.[ii] Let's try and compare some different tools.

We can use quality adjusted life years (QALY) to compare the cost of saving lives. QALYs are used in healthcare systems to assess the value of medical interventions and to best allocate resources.[13,14] One QALY represents one year in perfect health. So, if a heart transplant can extend the life of a patient by 30 years and that person will live in two thirds of perfect health it represents 20 QALYs. A heart transplant costs $1.4 million, so the upfront cost per QALY is $70,000. Now imagine that person is well enough to go back to work and let's say they can earn an average developed market wage of $45,000 per year: it takes 30 years to cover the medical costs with recouped income. Heart transplants are not such an efficient means of allocating resources to save life or to boost the economy.

Next let's extend the analysis to a broader problem, malaria. Each year malaria kills 600,000 Africans, a loss of 46 million QALY according to the World Health Organisation. The world currently spends $4 billion per year fighting malaria but the *Lancet* estimates that by boosting spending a further $2 billion, malaria could be completely eradicated by 2050. Discounting that spending in today's money adds up to $126 billion; divide by the discounted life years saved into the next century and this gives an upfront cost per QALY of significantly less than $100. A very cheap way of saving lives and as malaria is a burden to the African economy –amounting to $12 billion dollars per year – so eradicating malaria has a ten year payback in USD and far faster in equivalent utility: a much more efficient means of saving life and boosting the economy compared to heart transplants.

But how do medical interventions and developing market aid programs compare to our net-zero plan? Well, with a discounted extra upfront cost

---

ii Schelling's conjecture asks whether development might be the best defence against climate change impacts.

of $34 trillion[iii] to save seven billion life years from air pollution, natural catastrophes, and starvation, the cost per QALY is $5,000. A more expensive upfront cost of saving lives compared to action on malaria, AIDS, or clean water, but cheaper than most medical interventions. However, the speed at which net-zero will payback is fast. The energy savings payback takes about 14 years, but include the damages avoided from climate change and the payback is five years. Deploying net-zero boosts the economy and accelerates human progress through this century.

So, should we stop funding malaria prevention and channel all money into net-zero? Of course not. If we ask our global planner how best to maximise utility for all humans, then likely a huge transfer of wealth from rich to poor countries is optimal.[15] Yet that transfer of wealth is happening too slowly. Development support, aid, and knowledge transfer are essential to support better resilience in poorer areas of the world and this money must be ring fenced and increased. Net-zero is better financed through mostly private investment seeking a competitive market rate of return and should be facilitated but not bankrolled by public funds: net-zero investment must not displace development aid.

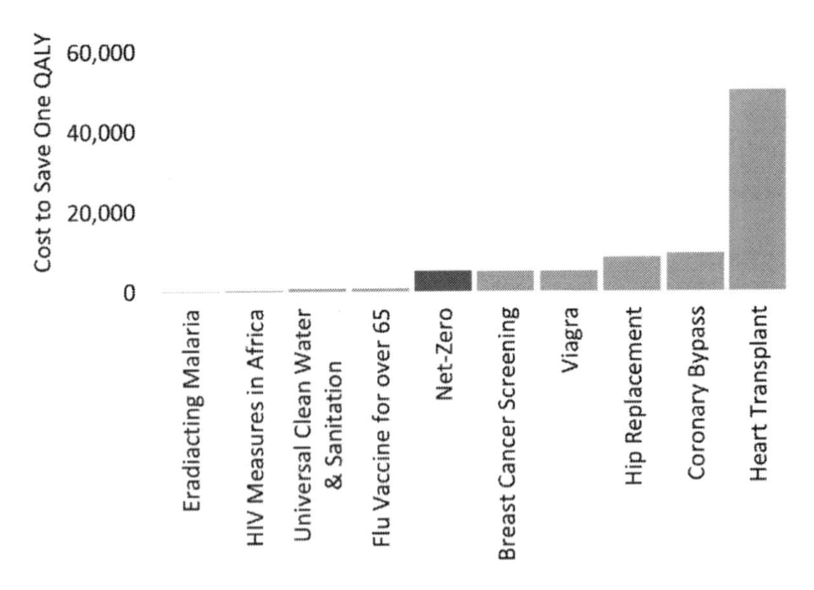

Cost to save one Quality Adjusted Life Year by intervention. USD cost and future life years saved both discounted at 3% into today's equivalent. Malaria based on World Bank, Malaria

iii The undiscounted extra cost is $46 trillion ($70 trillion minus $24 trillion). The discounted extra cost is $34 trillion ($53 trillion minus $19 trillion). We use the discounted value of investment ($34 trillion) and discounted future energy savings ($2.4 trillion) and avoided climate damages ($4.2 trillion) so we can compare to other problems today.

Free Future, and Lancet estimates.[16,17,18] *Clean water and Undernutrition based on UN sustainable development goals and estimates of cost-benefit from World Bank, Lancet, and World Health Organisation.*[19,20,21] *Other interventions from QALY databases or reports.*[22,23,24,25,26]

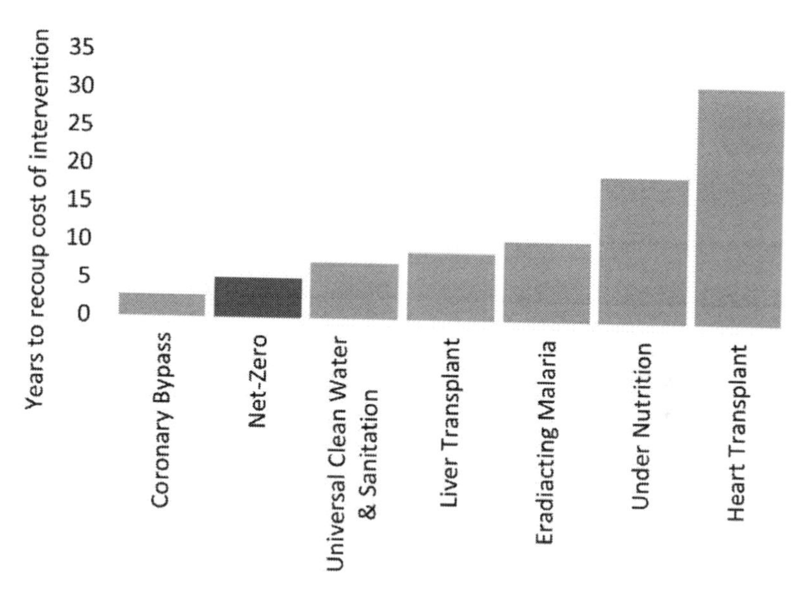

Years to recoup the cost of each intervention from lower costs, avoided damages, and recouped income (in today's money).

Taking the Meter Reading: Other Evidence for the Economic Benefits of Net-Zero

In 1976, energy policy analyst and disenchanted nuclear technologist, Amory Lovins published 'The Road Not Taken', calling for a "soft path" of development away from fossil fuels or nuclear and towards wind, solar, and efficiency gains: "a largely or wholly solar economy can be constructed in the United States with straightforward soft technologies that are now demonstrated and now economic or nearly economic".[27] Fifty years later and we still haven't taken that soft road but things are about to change.

Hopefully, I have convinced you that a net-zero system will be cheaper, cleaner, safer, more reliable, and provide more jobs than remaining bound to our existing fossil fuel system. We estimated the transition requires a total of $70 trillion of upfront investment (up to $2.5 trillion on supply and $1 trillion on demand per year) which is $46 trillion more than sticking with fossil fuels but creates annual savings of $5.5 trillion per year – a ten-year payback.

Sounds too good to be true? Well, I'm not the only one who thinks so. For the last decade there have been many major global reports which reach the same conclusion: net-zero is possible and affordable.

- In 2012 the 'Global Energy Assessment – Towards a Sustainable Future' showed 41 possible energy scenarios which would limit global warming to less than 2ºC and cost around $1.3-1.8 trillion per year.[28]
- Johan Rockstrom and Jeffrey Sachs described their sustainable development trajectory in 2013 as part of the 'Sustainable Development and Planetary Boundaries' report. The plan is to transform energy through electrifying everything, decarbonising electricity, efficiency gains, selective adaptation of biofuels, and land use change and to improve agricultural productivity by 70% and cut food waste to hold land use steady. They would create more sustainable cities, encourage biodiversity, increase accountability and governance, and complete the demographic transition to peak population.[29]
- In 2015, the investment bank Citi published 'Energy Darwinism: Why a Low Carbon Future Doesn't Have to Cost the Earth' and concluded that moving to a low carbon system would require at most just 1% extra GDP investment upfront, and at most just 0.1% GDP extra energy costs to the consumer. They estimated a $2 trillion lower cost to build out a low carbon energy system over 45 years compared to fossil fuels (including system costs).[30]
- In 2016, Christopher Clack and team reported in *Nature* that the US electricity sector could be 80% decarbonised using wind, solar, and high voltage connection with no increases to the cost of electricity.[31]
- In 2017, the 'Renewables Global Futures Report – Great debates towards 100% renewable energy', interviewed 114 energy experts from around the world. Two thirds thought 100% renewables both realistic and affordable.[32]
- In 2018, a review conducted by Jesse Jenkins and team concluded that the lynchpin of reaching zero carbon is the decarbonisation and expansion of the electricity sector. The authors suggest all technology options including nuclear, biofuels, and carbon capture must stay in play to ensure an affordable transition. Noting a renewables only transition would require grid expansion, flexible demand, very low cost wind and solar, and seasonal storage, which the authors believe has just 50:50 chance

of happening, "effectively a coin flip".[33] They expanded on this work in 'Net-Zero America' in 2020 estimating the US needs a 2-4 times increase in electricity and transmission, whilst overall energy demand will decline by one third, and net-zero can be achieved with lower than historical energy system costs (as a % GDP) whilst creating greater employment and significant health benefits.[34]

- 'Achieving the Paris Climate Agreement Goals', published in 2019, estimated energy demand in 2050 of 75 trillion kWh with two thirds renewables, creating 15 million more jobs for a total cost of $49 trillion.[35]
- The IPCC 1.5ºC special report in 2019 highlighted a transition to net-zero by the middle of this century gives a good chance of limiting warming to 1.5ºC. The aggregated solutions generally agree on about two thirds of energy provided by renewable electricity, negative emissions technology such as BECCS, and five million square km of pastureland converted to forest. With an estimated supply side investment cost of about $3 trillion for 30 years.[36]
- The IEA Sustainable Development Scenario puts the world on a path to limit warming below 2ºC, net-zero by 2070, at a cost of $2 trillion per year on energy supply and $1.6 trillion per year spent on energy efficiency (upgrading buildings, transport etc). A cost of 25% more than the current pathway but "partially counterbalanced by reduced fuel costs, which mitigates the impact on the energy bills paid by consumers".[37]
- The IRENA Global Energy Transformation Roadmap, 2020, estimates 98 trillion kWh energy use per year by 2050 with half electric and mostly renewables. Wind and Solar PV reaches $4c per kWh by 2040. They estimate a total cost of $110 trillion to 2050 (2% of GDP, $3 trillion per year) or $75 trillion in today's money. The added cost is $19 trillion but creates externality savings from air pollution and climate damages of $50-$142 trillion (3-8 times the investment). They estimate 42 million jobs in the energy sector and new energy job creation will exceed job losses in every region.
- Stanford Professor Mark Z. Jacobson backs 100% wind, water, and solar. His detailed study from 2019 shows 100% renewables scenarios for 143 countries around the world are possible. Concluding the transition will cost $73 trillion in today's money

at an average energy price of $9c per kWh using 0.65% of land cover. His 100% WWS world uses 76 trillion kWh energy per year at an energy cost of $7 trillion per year – a seven-year payback compared to fossil fuels.[38]

- Comparing the output from some of the major integrated assessment models which go by some highly creative names such as REMIND, WITCH, MESSAGE, IMAGE, AIM, or GCAM, the consensus is that 50-70% of demand should be electrified, fossil fuels should shrink to less than 20% of energy, and renewables come to account for 50-100% of electricity. The total energy demand reduced by an average of 40%.[39]
- The Bloomberg NEF New Energy Outlook Climate scenario estimates 100 trillion kWh renewables power is required by 2050 with two thirds direct use and one third for hydrogen and a total transition cost of $78 to $130 trillion.[40,41]

Despite a growing body of work highlighting the direct cost benefits of net-zero, the mainstream economist view asserts that solving climate change comes at a long-term economic cost. The last IPCC report from 2014 suggested an average GDP or consumption loss of 1-4% by 2030 and to 3-11% by 2100 is required – this equates to reducing growth from a range of 1.6%-3% to 1.5%-2.9% per year.

Yet these estimates are based on economic models using old cost data for renewables and consistently underestimating the rate at which the price of net-zero technology declines as deployment is scaled.[42] Historically, cost reductions and the 'rate of learning' in the energy sector has been slow: extracting fossil fuels relies on geology so projects are one off, bespoke, expensive and high risk, plus the more fuel you extract, the harder it becomes to extract more. On the other hand, renewable technologies like solar, wind, or batteries are modular, so the more units you produce, the greater economies of scale and learning pushes down the cost. Understand these differences and you understand why net-zero gets so cheap so quickly. Forecasting agencies and analysts continue to use outdated cost estimates, underestimate the rate of decline, and understate the potential capacity additions of renewables.[43] They also fail to build technology experience curves into many of the models.[44] This means economists and governments using these models and forecasts continue to overestimate the cost of net-zero and delay decisive action. [45] As Al Gore puts it in his call to action, 'An Inconvenient Sequel', "The Sustainability revolution… combines the scale of the industrial revolution with the speed of the digital revolution."

Collaborating to Drive Change: Putting Our Natural Bias to One Side

Hans Rosling wrote in the book *Factfulness*, "If you are good with a tool, you may want to use it too often. If you have analyzed a problem in depth, you can end up exaggerating the importance of that problem or of your solution. Remember that no one tool is good for everything. If your favourite idea is a hammer, look for colleagues with screwdrivers, wrenches, and tape measures."

Whilst much debate still surrounds the optimum mix of technologies to achieve net-zero, and experts all have their particular 'hammer', the general pathway is becoming clear. Push as hard as possible with wind, solar, and storage technologies which are clean, cheap, and can be deployed very quickly. As the electricity grid fills up with variable renewables, it may prove more effective to install a baseload from nuclear or fossil fuels/biofuel with carbon capture and storage. However, these are large complex builds which require extensive supply chains and waste management. Nuclear facilities can take ten years or more to complete and carry high risk, usually requiring backing from governments. The long build times require a decision is made about halfway through the transition as to whether nuclear or CCS baseload is required or not. Either way, we shouldn't delay the first 80% of the transition because we can't agree on the last 20%. Nearly all experts agree on what needs to be done now: deploy renewables as fast as possible, there is no reason to wait.

Too little work has been publicised widely enough on the costs and benefits of the transition and so the idea that zero carbon is the more expensive option is still pervasive. Economists and policy experts continue to argue over the details of their integrated assessment models but continue to underestimate the rate at which net-zero technology costs have declined.

In his essay, 'The Use and Misuse of Models for Climate Policy', Robert Pindyck highlights the flaws of integrated assessment models including uncertainty on the discount rate, uncertainty on the climate sensitivity, damage functions which are based on arbitrary mathematical functions, and the inability to deal with tail risk or tipping points. He asserts that despite the increasing sophistication of the modelling, there are so many unknowns and expert guesses going in that, inevitably, junk is coming out, no matter how many Monte Carlo statistical simulations are run.[46]

However, these problems are blown away once we realise how close we are to net-zero becoming the far cheaper solution. Arguments over discounting, equity adjustments, tipping points, and the social cost of carbon become irrelevant because we are no longer trying to balance avoided damages against higher mitigation costs – with net-zero solutions we can avoid the worst impacts of climate change *and* we can move to a cheaper energy system. Too much time has been spent trying to quantify the problem and we have missed the solution staring us in the face. Thankfully, there is now a growing body of work showing that we can cut air pollution, avoid the worst impacts of climate change, and we can do it at a lower system cost. The economy can decarbonise and never look back.

As economist and former director of the Earth Institute, Jeffrey Sachs puts it, "We must still make the transitions to renewable energy, sustainable agriculture, and a circular economy that safely recycles its wastes. Until those transitions are accomplished, Malthus's spectre will continue to loom large."[47]

We need to break down the engrained idea that fossil fuels are cheap, and zero carbon is expensive – this is not the case. Net-zero will prove cheaper, safer, and more reliable. We should be moving as fast as we can, pushing the 80% of commercialised technology down in cost, and developing a range of options for the last 20% to diversify and de-risk the final push. A fast transition will give the best possible economic and employment outcome before even considering the socio-economic benefits from reducing air pollution and avoiding climate damage.

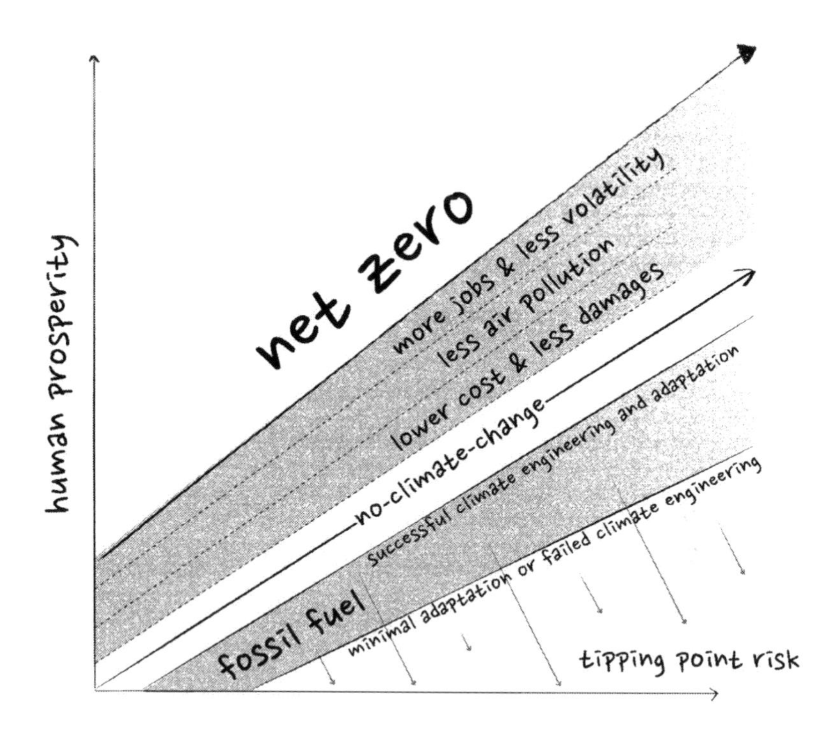

A growing body of work shows the benefits of net-zero and yet we have barely started on our journey. Why has progress been so slow? What is blocking decisive action? What needs to be done?

The next chapter delves into the economics and politics of change.

Chapter 14: A World Trying to Transition

Progress and Hurdles to Action

In this chapter we will outline the required rate of investment and technological change required to drive a rapid transition to net-zero. We will recap on what has been done, what needs to be done in the future, and the hurdles which are delaying decisive action.

A Brief History of Progress on Climate Change

Thomas Robert Malthus published his book *An Essay on the Principle of Population* in 1798, claiming that humanity is destined to remain bound to the same level of prosperity forever. Then, just seven years later, James Watt's steam engine patents expired and ushered in a wave of mechanisation across agriculture and industry. The fossil fuel powered industrial revolution took hold, and humanity hasn't looked back since. Just as Malthus predicted, the global population swelled from less than one to nearly eight billion people. However, prosperity increased faster, with the average wage rising from $1,000 to $18,000 per year. Life expectancy has doubled, child mortality and extreme poverty are 10 times lower, and deaths on the battlefield have declined 20-fold. We are living through a period of unprecedented wealth, quality of life, and peace: it seems Malthus picked the worst possible moment in human history to unleash his theory.

However, another all-the-more subtle story has been playing out through the last 200 years. Since 1824, when Joseph Fourier first proposed the "greenhouse effect", we have been unravelling the complex science of global warming and the unforeseen damage that burning fossil fuels is having on the planet.[1] We are also coming to understand the damaging consequences of inaction, the risk of dangerous tipping points, and the possibility that humanity could be dragged back into destitution. Will Thomas Malthus have the last laugh after all?

The story of global warming has many chapters and many prominent actors, from the pioneering scientific work by Foote, Tyndall, Arrhenius, and Callandar in the early twentieth century, to the definitive understanding and scientific basis for climate change pieced together by Revelle, Keeling, Manabe and Wetherald (amongst many others) in the fifties and sixties.[2,3,4,5]

By the early 1970s, John Sawyer (correctly) predicted 0.6ºC temperature increases in the twentieth century, the CIA secretly acknowledged global warming was already happening, and Wallace Broecker coined the term climate change.[6,7,8] In the 1980s, attention was temporarily diverted from the climate to the hole in the ozone and the Chernobyl disaster, until James Hansen delivered his famous speech to US congress in 1988 stating his team were "99% sure global warming is upon us".[9,10,11]

The intergovernmental panel on climate change, the IPCC, was established in 1988, stating two years later, in its first report, that the planet was likely to warm by 0.3ºC per decade under a business-as-usual scenario.[12,13] In 1992, the Earth Summit in Rio saw the signing of the United Nations framework convention on climate change (UNFCCC), an international treaty based on the pillars of stabilising the climate at safe levels, whilst recognising differing responsibilities across the world, and later extended into the Kyoto protocol[i] (1997) which agreed developed countries would cut emissions by about 5% relative to 1990 levels.[14,15]

The third IPCC report, in 2001, projected temperatures in 2100 reaching 1.4-5.8ºC warmer unless something was done.[16] The fourth IPCC report, in 2007, started to flag the risk of climate tipping points.[17] By 2008, the world was realising we needed to move faster in order to effectively combat climate change and the fifteenth conference of parties (COP15) in Copenhagen, 2009, was the platform for change.[18] However, the global Credit Crunch, in late 2008, had already diverted attention and, in the weeks leading up to the meeting, a network of climate deniers leaked partial extracts of hacked emails from a number of climate scientists at the University of East Anglia in the UK, misrepresenting the content, seeding doubt, and creating a media fuelled smear campaign to disrupt the talks.[19] Copenhagen resulted in no binding agreement and very little progress.

---

i Kyoto is Tokyo's former name.

It wasn't until 2015 and COP21 in Paris when a binding global agreement to limit emissions was reached between over 190 countries.[20] It may have taken nearly one quarter century of talks, but Paris represented a level of international cooperation not achieved since the nuclear non-proliferation treaty in 1968.

The long-term objective of the Paris Accord is to make sure global warming stays "well below" 2ºC and to "pursue efforts" to limit the temperature rise to 1.5ºC. Nationally defined contributions (NDCs) on agreed emissions limits by 2030 are submitted by each country and include roadmaps as to how the reductions will be made. Developed countries also committed to at least $100 billion of funding to developing countries each year to support the transition. National emissions stock takes are recorded every five years and reporting is a legal obligation but, unlike Kyoto, the emissions targets have no legal binding or penalty for non-compliance.

Paris represented a real political breakthrough in climate change action and yet since 2015 emissions have continued to increase and concerns have continued to grow louder. Major developed economies are unconditionally committed to reducing emissions by 25-40% by 2030. Developing economies are committed to either absolute reductions or relative limits which allow some countries to increase emissions over the next decade. More than half of developing countries have nationally determined contributions conditional upon receiving technology transfer, financing, and capacity building (skills) from developed countries which are not being paid in full.[21]

As the Paris commitments stand, global CO_2e emissions will remain at roughly the same level for the next ten years. In just one decade we will add another 500 billion tonnes of CO_2e to the atmosphere and blow through most of the remaining carbon budget to limit warming to 1.5ºC or one third of the remaining 2ºC budget. Individual country emissions reductions targets (NDCs) must continue to ratchet tighter if we are to avoid potentially dangerous tipping point temperature increases.

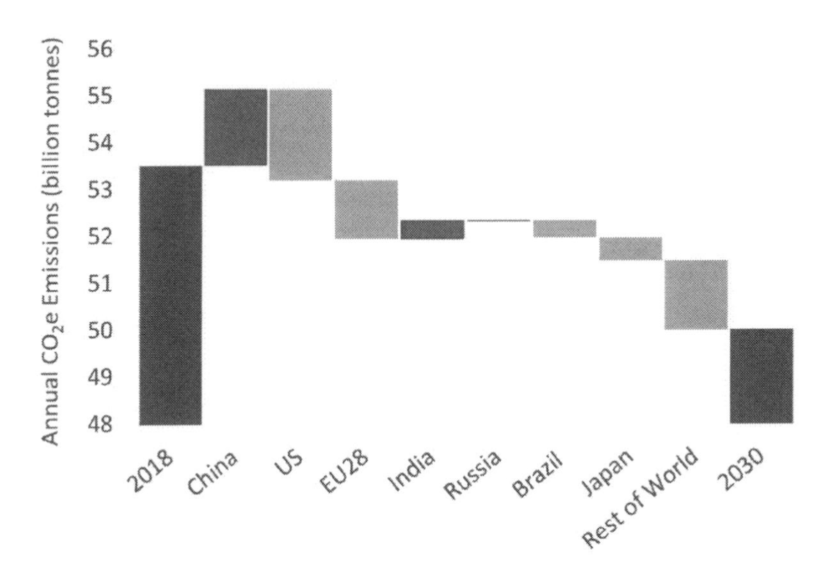

Annual CO_2 equivalent GHG emissions change from 2018 levels of 53.5 billion tonnes to 50 billion tonnes annual emissions in 2030 based on meeting all unconditional and conditional NDCs. China and India should emit 2 billion tonnes more CO_2e by 2030. The rest of the world should emit 5.5 billion tonnes less than in 2018. Author's calculations based on NDCs and cross referenced with UN gap report.[22]

Despite 200 years of ground-breaking science, unprecedented technological progress, and a quarter century of international talks, global emissions have continued to rise. If we don't accelerate action the planet will be at least 3ºC warmer by 2100.

Rate of Technological Change: What is Required for an Optimal Transition?

Reaching net-zero is not trivial. Our back-of-the-envelope calculations show it requires a complete overhaul of the world's energy system and fundamental changes to industry, transport, agriculture, and everyday living. But done well, the transition will barely register for the top 10% already accustomed to the highest quality of life in developed countries. Meanwhile, the remainder of the world can quickly raise living standards towards the top bracket, and the poorest 10% will for the first time connect to the modern world with electricity access, clean cooking, and basic amenities. Executed over two decades and the transition will accumulate $50 trillion of benefits from lower energy system costs and more than $120 trillion of avoided climate damages this century whilst minimising death, suffering, and destruction of the natural world.

Twenty years sounds fast, but nations have achieved rapid transitions in the past. France increased nuclear power generation from 2% to 80% of electricity between 1975 and 2000, Brazil took biofuels from 1% to 25% of road fuel between 1975 and 2002, China has covered an extra 5% of their land area in forest since the 1990s, and the UK has effectively phased out coal electricity in the last five years. These isolated actions show just how quickly change can be accomplished, but a handful of measures is not enough: the countries of the world must now act together across all areas of the economy.

A twenty-year transition requires annual capacity additions of at least 500 GW of solar PV and 250 GW of wind; that's over 1.5 billion solar panels and 50 thousand wind turbines installed each year. Coal, oil, and gas are pushed out of the energy system so the production of new combustion vehicles, gas boilers, and fossil fuel equipment with up to 15 years installed life, needs to end in the next five years. As renewables become a greater share of electricity generation, over 3 billion kWh of short-term energy storage needs to be built annually, the equivalent of 150 million residential batteries plus 100 pumped hydro systems per year. Longer term storage and transport needs require annual installation of 15 thousand hydrogen electrolysers and 50 district heat systems.

Offsetting the stubborn emissions from agriculture, petrochemicals, and industry requires ten billion trees planted and over 400 industrial carbon capture devices installed per year. We must also figure out whether we need a baseload from nuclear and/or combustion with carbon capture and, if so, we will need to build enough capacity to pump up to 24 million cubic metres of CO_2 into the ground each day: around two thirds the rate at which we currently extract oil and gas.

Solar and wind capacity additions are headed in the right direction but need installing five times quicker. The use of coal is shrinking, but the rate of decline needs to double. Gas heating and electricity cannot be touted as a transition fuel, capacity must turn course and decline. The ever-growing consumption of oil products must reverse.

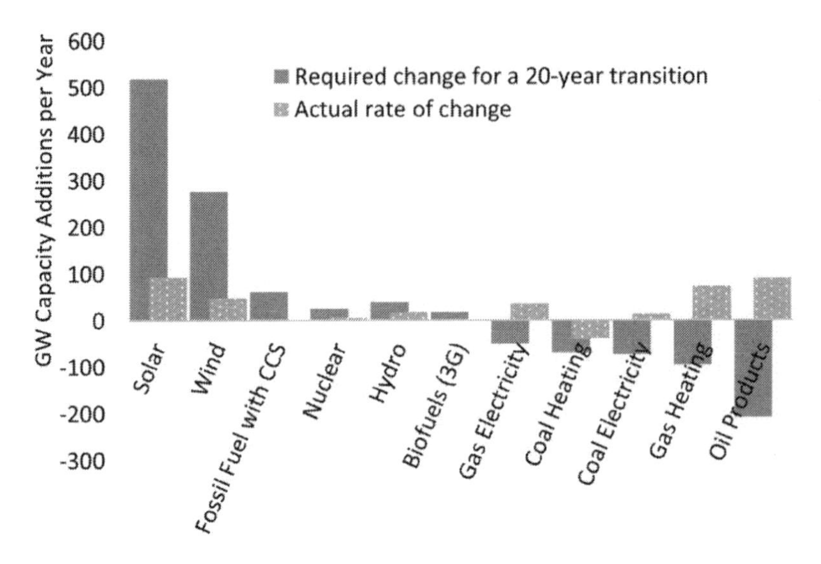

Solid: Required annual average rate of change in capacity for 20-year net-zero transition (assumes growing population and consumption). Dotted: Average rate of change over the last five years (wind, solar, and hydro equal two years average) - numbers based on energy consumption change using average load factors (and 100% for fossil fuel products for comparison). Author's calculations based on modelling in chapter 7-12, data from BP *statistical energy review 2019. GW = million kW.*[23,24]

Electricity makes up 21% of final energy today and is growing at 0.2% per year. Reaching two thirds of final energy in two decades requires 2% annual increases, ten times faster than the ongoing rate of change. Electric vehicle production ramps from 2 to 60 million cars per year, residential heat pump production increases from 3 to 25 million units per year and new net-zero technologies scale.

Whilst the ambitions of Paris need to be tripled, the actual rate of supply side change must move five times faster and demand must electrify ten times faster.

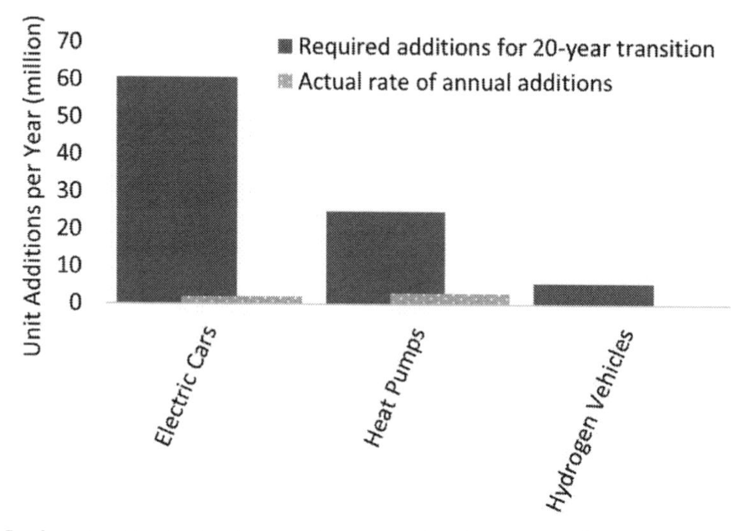

Number of units added each year today compared to the rate of addition to transition to net-zero in 20 years. Author's calculations based on modelling in chapter 7-12, current data sourced from world heat pump market study.[25]

The headline numbers sound daunting but break the transition down over the global population and it seems far more manageable. Over the next two decades, we install just four solar panels, plant 27 trees and develop 16 square metres of biofuel area for each person on the planet. We add the equivalent of one (radiator sized) battery in every house and one wind turbine between every 7,000 people. One quarter of car owners shift to ride sharing and the remaining three quarters replace their old vehicles with electric. Half of broken boilers are replaced with heat pumps. It starts to feel far more achievable.

Magnitude of Financial Change: How do we Source and Spend the Money?

Industries, technologies, and the way we live our lives change very quickly: think vinyl to CDs to streaming, or landlines to mobiles to smart phones. Once a better solution comes along it very quickly displaces the old. As Kingsmill Bond argues in his '2020 vision report' for the Carbon Tracker Initiative, we are past the net-zero innovation phase, and moving into the second and third phase of the energy transition where fossil fuels peak this decade, and renewables will rapidly come to dominate the energy system.[26] Once we understand net-zero is win-win, this exponential change will follow. But demand for change is only effective if financing solutions can keep pace: five to ten times more money must be readied for net-zero solutions.

A Very Brief History of Money & Finance

For most of human history, humans ran a sort of social co-operative where food, fire, and tools were gifted to other members of the tribe as a social security net. But with the advent of the agricultural revolution, and specialisation of production, came the need to barter. One farmer could exchange his excess grain for another farmer's excess livestock in a mutually beneficial exchange. As populations increased and empires expanded, bartering became more difficult: the greater number of goods and services made it harder to find the right trade, it became less practical to store accumulated wealth, and farmers wanted to barter for goods before their harvest was ready. Debt, credit, and money was born.

At first goods would be exchanged for clay tokens at the local temple and redeemed when needed, but as trading with foreign merchants evolved, a unit of exchange not tied to the local priest was needed. Precious metals like gold and silver first served this purpose and by 3,000 BC the Mesopotamians had developed a large-scale economy based on shekels, a fixed weight silver coin representing a fixed quantity of barley as a measure of value.

Roll forward to the twelfth century and the modern banking system was taking shape, not as many suppose by the merchants of Venice, but it was the Knights Templar that ran the world's first banking corporation. The Knights Templar were a military order who pledged themselves to a life of devotion to the church, and with it a special obligation to protect access to the Holy Land. They soon discovered that whilst guarding the highways and by-ways of Europe they could provide not just safe passage for pilgrims, but also for the money, gold, and valuables of wealthy land owners, kings, queens, and popes – all for a nice fee. The Knights Templar were the premium banking service in Europe for over two centuries, until King Philip IV of France, after exhausting higher tax collection and money printing, turned his attention to nationalising the huge wealth of the knights' order. He started with a smear campaign, accusing the Templars of having sex with corpses, eating dead knights, and draining the bodies of infants to render the fat into sacred oils. He then turned to good old arrest, torture, and killings until the Knights Templar had given up their riches and the order was abolished.[27]

The dissolution of the Templars left a void in the banking system, and over the following years families in the merchant cities of Italy and Flanders would slowly take up their business. The new merchant banking system was supported by more robust legal systems which protected property rights, and accountancy which helped enforce trading contracts. Merchant banks evolved to connect the wealth of lenders with the funding needs of borrowers through commercial loans used to build out the factories and ships needed for the production and trading of commodities. Merchant banks would control most access to financing for much of the second millennium.[28]

Through the twentieth century merchant banks outgrew their family-owned origins, broadened their scope of activities, and have become the corporate investment banks we know today. Similarly, access to capital has evolved with a rapid expansion of bond and stock (shares) markets allowing corporates and governments to circumvent merchant banks and raise money directly from lenders. More recently, new types of investors such as business angels, venture capital, and private equity companies have brought significant funding to companies or projects in the start-up, early, or latter stages of the business life cycle. Access to money has become increasingly democratised and the growing wealth of the world is increasingly managed by professional institutional investors at mutual, pension, insurance, and sovereign wealth funds plus the hedge funds, private banks, and family offices managing the money of the super-wealthy.[ii]

Sourcing the Money: Finding $3.5 Trillion Per Year

Now we understand net-zero is cheaper, cleaner, and more reliable, but also requires more upfront investment to yield big savings down the line. Spending on the global energy supply system such as generation, storage, and transmission will have to increase to about $2.5 trillion per year for the next 20 years. The extra upfront cost of demand side equipment like the batteries for electric cars, extra cost of heat pumps compared to gas boilers, and insulation refits will add another $1 trillion per year. So where do we find $3.5 trillion every year for the next 20 years?

The first step is to push the ongoing $1.5 trillion of annual energy investment away from fossil fuels and into net-zero solutions. Next, we

---

ii Shares and stocks mean the same thing, a share of ownership in a project or company. Mutual funds are pools of invested money from individuals, pensions are pools of invested money from individuals (usually with tax benefits) which pay out only at retirement age, insurance funds are pools of money from policy payments which are invested to meet future insurance pay outs, sovereign funds are excess government money held as investments. Hedge funds are pools of money sourced from accredited investors with a large minimum buy-in and are managed with either a greater degree of risk taking or scope of investments than institutional money. Family offices are a team of investors that manage the money of a wealthy family.

must reduce the annual $500+ billion[iii] of fossil fuel subsidies which only serve to transfer taxpayer's money to the profits of fossil fuel companies.[29,30] The world already spends around $250 billion on low carbon transport solutions and efficiency so this bringing our total to $2.25 trillion per year.

So, we only need an extra $1.25 trillion of investment which can come from either reducing consumption or displacing other investments. This represents 1.5% of the $90 trillion of current global GDP.

Of that $90 trillion GDP we spend $68 trillion on the consumption of short-lived goods and services like food, healthcare, and entertainment where we could cut back and direct the increased investment into net-zero. Most likely this would require increased taxation with the money directed into investments by the government. A transfer of private household wealth/capital to the public coffers.

The other option is to displace part of the $22 trillion of GDP spent on long term investments such as land improvements, industrial machinery, roads, rail, and buildings. We could divert $1.25 trillion of this spending into net-zero infrastructure investment at a 3% real rate of return. This would require a mix of government debt raising and the re-allocation of investments by corporates and savings by individuals. A transfer of wealth/capital from the government to private households.

---

iii Fossil fuel subsidies are hard to measure in total, consumption subsidies (which are used to reduce the price of fossil fuel energy) are estimated to have averaged $300-$500 billion per year over the last decade according to the IEA. Other subsidies include tax breaks, favourable rates for land or government services and helping fund investment projects which could bring the total towards $1 trillion per year.

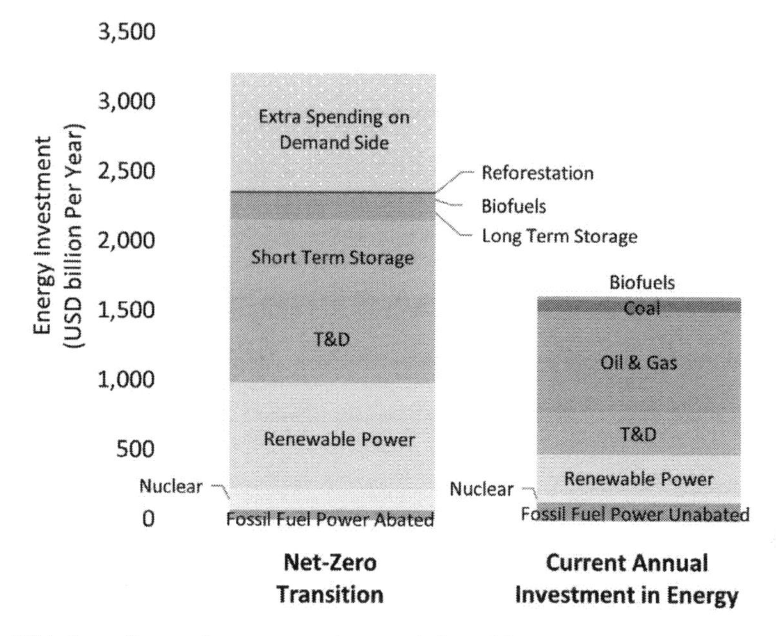

Global spending on the energy supply network for a 20-year net-zero transition (annual average) and the spending on the energy supply network (2018).[31] *(T&D stands for Transmission and Distribution). Extra spending on demand side is extra (not total) cost of buying net-zero appliances and solutions like the battery in electric cars, heat pump versus boiler or insulation refit etc. Author's calculations for net-zero transition based on modelling in chapters 7-12.*

Deploying the Cash: Who Holds the Purse Strings?

The world is already spending over $300 billion per year on renewable energy generation and over $250 billion on low carbon transport infrastructure, energy efficiency, and adaptation measures.[32]

Government grants (paid for by taxes) account for 5% of the financing and are non-repayable funds used for technology development and commercial demonstration projects, to encourage technology adoption. Examples are electric vehicle grants, insulation refits, and international green development.

Green banks and development banks account for 30% of today's funding. These financial institutions are usually run or directed by the government for a specific mission, in this case the low carbon transition. They are guaranteed and capitalised (given equity money) by the government which allows the bank to issue debt or bonds to the market and raise much larger sums of money from private investors at very low interest rates.[33] They lend this money out to project developers who build out green infrastructure such as solar or wind farms. The loan is paid back by the future cash generated from the project.[34]

Commercial banks, infrastructure funds, and institutional investors have contributed another 10% of loans to project financing of net-zero infrastructure. But whilst green banks mostly use debt (loans), there are corporates, institutional investors, and infrastructure funds which may want to hold an equity stake (ownership) in the project. Like debt, equity gets paid from the cash flows of the project, but as this is allocated payment after the debt, it is therefore higher risk and expects a higher return. This provides another 10% of funding.

Finally, companies can develop low carbon projects as an integral part of their own business called "on balance sheet" financing. The company itself is funded by debt and equity from investors but usually at a higher rate of return because companies generally take on more risk. On balance sheet financing is best suited to high risk but potentially more lucrative ventures.

Source	Type of Financing	Terms of Financing	Annual Investment (USD billion)
Government	Grant	No return	30
Green Banks & Development Banks	Project Finance	Below Market Rate Debt	65
		Market Rate Debt	110
Commercial Banks	Project Finance	Market Rate Debt	75
Infrastructure Funds and Institutional Investors	Project Finance	Market Rate Debt & Equity	15
	Project Finance	Market Rate Equity	45
Corporates	Balance Sheet	Corporate Rate Debt	95
	Balance Sheet	Corporate Rate Equity	75
Individuals	Balance Sheet	Opportunity Cost Equity	50
Total			**560**

Annual financing by source, type, and terms adapted from Global Landscape of Climate Finance, Climate Policy Initiative, 2019.[35]

The $0.56 trillion of investment each year needs to increase nearly seven-fold to $3.5 trillion to achieve an optimal transition but is increasing by just $0.05 trillion per year. The change is simply too slow. We must ensure the financing is ready to go.

This requires using all sources of funding by better matching projects, reducing project risk, and creating accommodating policy.

Government grants or loans must incentivise households to electrify demand whilst also providing support for industrial demonstration projects in highly capital intensive[iv] research and development areas like hydrogen, next generation nuclear, and carbon capture, all of which may or may not need to be deployed at scale later in the transition.

Green banks should be used for aggregating smaller projects or for funding capital intensive projects which need to build longer track records. This creates sizeable, low risk investment opportunities which will eventually attract more corporate loans and big institutional money[v] into new technology areas. Offshore wind and battery storage are the next big opportunity.

Private financing from commercial banks, infrastructure funds, and institutional investors must go green to further accelerate the deployment of already de-risked technologies like onshore wind and solar. Governments should ensure regulations make green financing as easy as possible with standardised contracts, quick permitting, and validating trustworthy contractors to reduce the uncertainties and further drive down financing and transaction costs in all global markets.

Individuals place their money with net-zero aligned retail banks and invest their savings in sustainable funds. Over the last decade the cumulative money invested into Environmental Social Governance (ESG) funds is $1 trillion and another $2 trillion of green and sustainable bonds have been issued by governments and corporates: representing less than 3% of the $200 trillion bond and share markets, still plenty of room for increases.[36,37]

Early stage but high risk-reward technologies like 3G biofuels, hydrogen transport, and new energy storage technologies still need funding from angel investors, venture capital, and support from government-academic-corporate collaboration.

---

iv Capital intensive means each project requires large investment.

v Institutional investors are the managers of the money from pensions, mutual funds, insurance funds, and sovereign funds.

Global Capital/Wealth of the World is $400 trillion

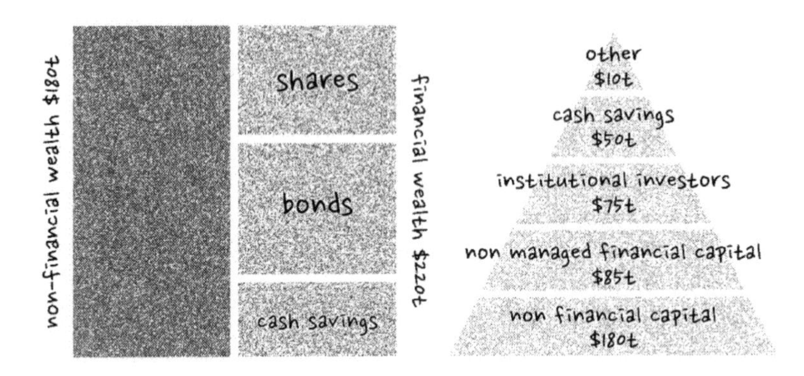

Left Chart: Global net capital made up of ~90% private wealth and ~10% public wealth (China, Norway and Russia). Non-financial wealth is mostly residential property and to a lesser extent commercial property and equipment, agricultural land, and precious metals. Financial wealth is mostly the value of equity and debt invested in companies (or government organisations) and represents the non-physical value of organisation, information, reputation, trademarks, patents, branding, and know-how which allows the company to generate income. Right: Who manages the capital? Other represents hedge funds, venture capital, private equity, and family offices who are willing to take higher risk investments. Cash savings are typically in bank deposits. Institutional investors are pension, mutual, sovereign, and insurance funds which professionally invest money but take lower risk. Non-managed is all other financial capital. [38,39,40,41,42,43]

Why Has Progress Been So Slow?

We are 30 years on from the formation of the IPCC and we have increased the share of zero carbon technology by just 3%, to reach 15% of final energy use today.[23,24] Why has change been so slow? The problems range from the Chernobyl and Fukushima nuclear disasters slowing and ultimately reversing nuclear development[vi], to new production of cheap oil and gas from an increasingly global unified market, the rise of developing economies demanding the rapid scale up of cheap energy, and protectionist policies unwilling to enact change.

The list of problems is long, but in the end it's the balance of fear which has stopped humanity from transitioning away from fossil fuels. Ultimately the perceived threat of climate change and damage from air pollution hasn't proved big enough to take on the political, social, and economic risk of transitioning the energy system. But the longer we wait,

---

vi Nuclear energy supply peaked at 17% of all electricity globally in 1993. Nuclear now accounts for just 10% with many more plants set to retire over the next decade.

the bigger and more immediate the threat becomes, the more people die from air pollution and, although the cost benefits of a zero-carbon system will become increasingly clear, our window of opportunity for the best possible outcomes inevitably narrows.

Hurdles to Action: The Problems Holding Back Change

Our cornucopian economist would argue if net-zero is so cheap, why is it not already being deployed? The answer: it is, but hurdles to action are slowing progress.

Renewables have already reached cost parity with fossil fuel electricity across most of the world and electric vehicles are a few years away from competing with combustion engines. Wind, solar, and electric cars are growing exponentially from a small starting point but will soon create disruptive change. The result will be cheaper electricity, cheaper storage, cheaper driving, and further economic incentive to electrify everything. The transition to a cheaper, better system is happening and is the very definition of cornucopian substitution and human innovation, but change is too slow. In the last chapter we examined why a 20-year transition provides an optimal outcome, yet inefficiencies and hurdles are holding back change. So, let's take look at how apathy, misaligned economic interests, inequality, freeriding, political wrangling, and deliberate sabotage are slowing progress.

Hurdles to Action – Attribution and Understanding: Chapter four illustrated why climate change is at least a top five global problem with the potential to take the undesirable top-spot on deaths, suffering, and economic loss. Air pollution, hidden hunger, productivity losses, and natural catastrophes are already taking their toll on the world, but attribution of these impacts is tough. We have evolved to pass on information through story telling, stories grounded in specific events we have seen, heard, or felt at a specific time and specific place. We store them in our long-term memory, share and aggregate them to take a view or form an opinion. Climate change, in much the same way, is the statistical variation in weather patterns which when aggregated create a trend. But individuals cannot gauge one-degree average temperature change over multiple decades. Humans are predominantly sight animals – seeing is believing – but CO_2 and temperature are invisible, and the weather seems unpredictable. How do we experience climate change? Scientists can calculate the increased likelihood of flooding, fire, or drought due to a warmer atmosphere, but cannot definitively assign a single weather event to global warming because the system is too complex, with much internal variability. Climate change is measured in decades and over the entire planet and is neither specific in time nor specific in space. Individuals, governments, and corporates operate on shorter timescales and in smaller regions. From next month's pay cheque to five-year election cycles and ten-year investment plans, we are not used to thinking and planning for long term change. When farmers in Illinois were asked to recall growing conditions over the last seven seasons, their memories were more influenced by their pre-existing beliefs than the actual weather: those that believed in global warming thought it got drier, those that didn't remembered no change.[44]

The impact of global warming seems distant. The risk is that we reset expectations each year for more extreme weather, drought, and fire, and therefore normalise the increasing threat. We win the ongoing fight against our future selves and future generations but at what cost? Broad-based understanding of the evidence and actions of climate change is needed to give individuals, corporates, and governments the confidence to enact behavioural change.

Hurdles to Action – Market Failures and the Tragedy of the Commons: In his essay written in 1833, British economist William Forster Lloyd used the hypothetical example of farmers over-grazing cattle on common land to describe how a system designed for individual gain can create a detrimental outcome for all.[45] In 1968, US ecologist Garret Hardin further explored this type of social dilemma and coined the term "the tragedy of the commons". Hardin believed family planning (as a human right) and the welfare state were to blame for overbreeding and "no technical solution can rescue us from the misery of overpopulation", going on to say, "the only way we can preserve and nurture other and more precious freedoms is by relinquishing the freedom to breed".[46] Hardin was a firm neo-Malthusian. The idea of the tragedy of the commons stuck and helps explain why the market, left to its own devices, is unlikely to create a perfect solution to greenhouse gas emissions.

The market can be thought of as trillions of two-way trades, the mutual exchange of goods and services between two parties, where both parties end up better off (or else why would they make the exchange?). In a perfect market, these trades continue until they reach the Pareto optimum, the point at which no more mutually beneficial trades can be made. However, real markets are not perfect, we have already come across externalities, the idea that an unwitting third party is impacted as a result of the two-way trade. Think about each time we exchange our paid labour to use a service which burns coal, oil, or gas; we make a two-way trade, but we also emit pollution which damages a third party (the planet and everyone on it). Our individual gain negatively impacts the common good, so without intervention the harm from emitting greenhouse gases goes unchecked. Simply, we can pollute the atmosphere, but we don't pay for the damage.

But even if we could bring the damage inflicted by greenhouse gases inside the market system, there are many other imperfections which get in the way. For example, hyperbolic discounting is the aversion to spending money upfront, even if you save money long term. Spending an extra $5,000 buying an electric car may work out cheaper through the life of the vehicle, but for many consumers it feels more expensive and they don't want to part with the cash upfront: we mentally use a very large discount rate.

Next let's think about heat pumps which have a more similar upfront cost to gas boilers, but with less environmental impact; surely uptake here should be easy? Except there is a lack of information and a preference for the status quo. If consumers don't have the information on how they work, or recommendations from friends and family, we tend to wait for others to test the waters. Better the devil you know.

Well, what about insulating the roof, surely that's a no-brainer? It's cheap, pays itself back in a year, and saves money on heating for another 50. But only if you live in the house. Landlords have no incentive to insulate their rental properties because the tenants receive the benefit.

Switching to LED light? Too small a saving to bother. Building a nuclear plant? Too big to finance. A new electric cooker? But I only bought the gas one two years ago or I'm thinking of moving soon, so why bother? The list goes on.

Hurdles to Action – Regional Inequality: Warmer temperatures mean more rain, cloud, melting ice, and fire, faster growth on land, decay in the oceans, and slowly rising seas. We saw in chapter four how, overall, this change will cause death, suffering, and economic loss. But it will not be evenly distributed. India, Bangladesh, and Pakistan are home to nearly two billion people and will be hit hardest by most of the worst impacts of climate change. Storms, floods, productivity losses, water shortages, declining agricultural output, and unbearable heat. The geography, agriculture-led economy, and dense population in these areas leaves them incredibly vulnerable to a warmer climate. Compare this to more northerly, colder climates such as Canada, Russia, and parts of Northern Europe, where a few degrees of warming may initially prove beneficial, assuming the world doesn't hit a catastrophic tipping point.

For the north, small warming means milder winters, better crop growth, and new trade routes through the melting Arctic ice. The challenge in bringing the world to action will be mobilising those countries less affected or even set to benefit from a warmer climate.

CO_2 emissions per person tend to be higher in countries which are less impacted by climate change. So, although nearly 25% of the global population live in India, Bangladesh, and Pakistan, these countries contribute less than 10% of annual global CO_2 emissions. The three billion people on Earth (40% of the population) worst affected by climate change are responsible for just 14% of emissions. The 1.4 billion people (18% population) least impacted are responsible for over 40%. Aligning the interests of this regional imbalance remains a hurdle blocking decisive action.

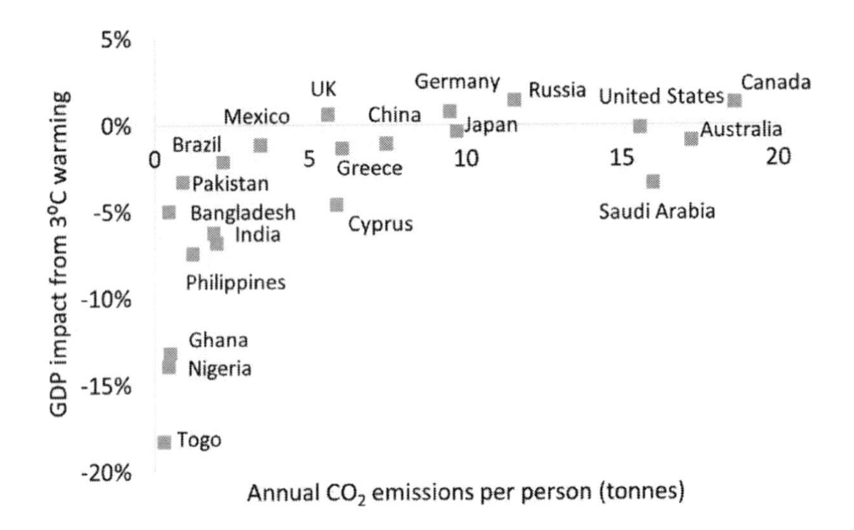

Annual average CO_2 emissions per person (tonnes) in each country versus the estimated GDP impact of that country from 3°C warming. CO_2 per person per region from Janssens-Maenhout et al.[47] *GDP impact from Roberto Roson, Martina Sartori et al. and includes impacts from labour and agricultural productivity, health, tourism, energy demand, and sea level rise.*[48]

Hurdles to Action – Participation: Net-zero is slightly more expensive today but becomes significantly cheaper once scaled, but if the world doesn't transition at the same rate this creates hurdles. Those who move first bear more of the experience costs and install a greater share of zero carbon technology at higher prices. A first mover disadvantage. Furthermore, should a portion of the world refuse to decarbonise, then costs increase for everyone else because the optimal mix of technologies cannot be deployed.

Europe has been an early adopter of renewables technology and played a large role in driving wind and solar costs down the early part of the experience curve. In 2004, Germany was the first country in the world to implement a generation subsidy which paid €46c per kWh of solar electricity (four times the global retail price). The levels of installation were uncapped and the generous pay-out created the beginnings of the solar boom. However, as Jenny Chase (head of Solar at BloombergNEF) points out in her book *Solar Power Finance* those subsidies have left Germans paying €6c per kWh or 20% of their electricity rate to cover past renewables incentives.[49] Compare that to the US, the richest country in the world with highest per capita emissions, and the take-up of net-zero technologies has lagged Europe by at least five years: this is learning by watching or freeriding.

Game theory, which is the science of strategic decision making, is often used to try to understand the likely outcome of climate negotiations and the extent of global participation. Individual actions by countries across the world are often compared to the "prisoner's dilemma", a paradox where two bank robbers are held as suspects in police custody but with no evidence. Both are offered the option to testify against the other for a reduced sentence. The best outcome is for both suspects to co-operate with each other, keep quiet and go free, but neither robber is sure whether the other will talk or not. Game theory suggests both will pursue their own self-interest, opt for a reduced sentence, and each of the suspects will do time: a less than optimal outcome.[50]

Climate negotiations are likened to the "prisoner's dilemma": why would individual countries take action to reduce emissions when they cannot be sure other countries will act? Self-interest surely suggests taking no action and waiting for others to pay the higher upfront technology cost whilst enjoying the free-rider benefit. We are back to the tragedy of the commons.

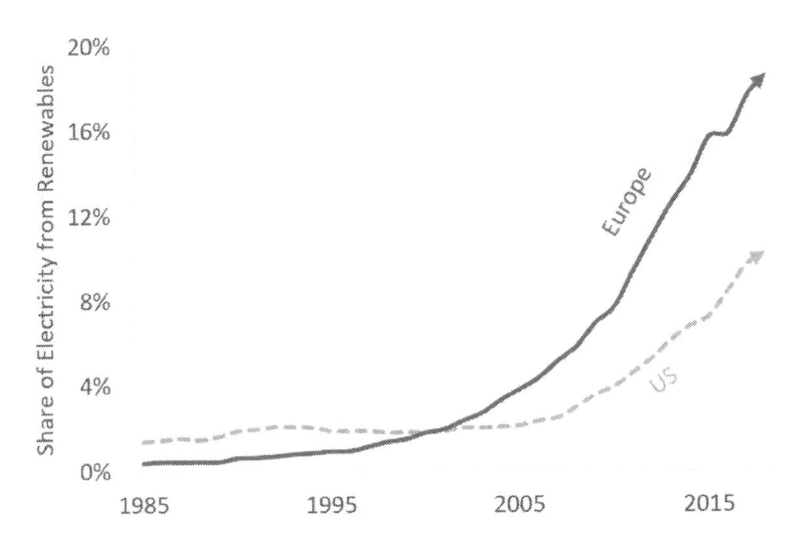

Share of electricity generated from renewables in Europe and US. Data from BP statistical review, based on gross output.[23]

Hurdles to Action – Wealth Inequality and Views on Climate Justice: The ability to adapt to a changing climate, and deal with the impacts of warming, are highly dependent on wealth. Building sea walls, air conditioning, irrigation, and other adaptations comes at a cost. And wealth is not equally divided. Flood defences for sea level rise this century may cost developed countries just 0.1% of GDP but will cost developing countries 1%.[51]

Hurricane Katrina in 2005 may have been the costliest climate related disaster on record at $125 billion but thankfully the death toll was limited to 1,700 people. Compare this to Cyclone Nargis in Myanmar in 2008, where at least 140,000 people perished, yet the losses were just $4 billion. Wealthy countries have the infrastructure, early warning systems, and organisation required to deal with climate stresses. Poor countries don't.

Making the matter more unjust is that the wealth of developed countries has accumulated over generations, driven by the industrial revolution, and burning of fossil fuels. The more carbon each region has burnt, the richer its citizens have become. North America and Western Europe have 7% of the global population yet emit 25% of annual carbon emissions today and are collectively responsible for 43% of all human emissions over the last 200 years. As Lord Stern put it, this is "double inequity, those who suffer most contributed least".

The 10% of the global population living in the wealthiest countries of the world emit 30% of all greenhouse gas emissions, take home 50% of all global income, and have accumulated over 80% of the wealth in the world. The 10% of the population living in the poorest countries of the world emit just 2% of all CO_2, earn 1% of global income, and have no accumulated wealth – they live hand to mouth.

Novelist Amitav Ghosh, writes, "Although different groups of people have contributed to it in vastly different measure, global warming is ultimately the product of the totality of human actions over time. Every human being who has ever lived has played a part in making us the dominant species on this planet, and in this sense every human being, past and present, has contributed to the present cycle of climate change".[52] But how do we assign responsibility today for human actions over all of time?

Justice is about getting what you deserve.[53] If rich countries got rich from emissions in the past, and stay rich with emissions in the present, then shouldn't those countries bear a larger share of the cost for cleaning up the mess or pay compensation to the poorest countries worst hit by climate change? This is the idea of climate justice. Who bears the cost of upfront investment to create a better system for all? Here are the options:

Grandfathering: Use each nations' emissions today as a benchmark and set percentage reduction targets in the future. This seems unfair given that it doesn't account for the accumulated emissions of the past, the inequality of emissions today, and the wealth those emissions have created. Why should the US hold just 4% of the world's population, 7% of dryland, and yet be allocated 15% of the atmosphere to dump CO_2?

Egalitarianism: Take the remaining carbon budget and divide it up for each person on the planet whilst letting each individual trade their allocation if they have too much or too little based on their lifestyle. This would create a large transfer of wealth from rich to poor but doesn't account for past emissions, ability to pay, ability to actually reduce emissions, or how to account for embedded emissions.

Polluter pays: We could count back to the start of the industrial revolution or maybe to 1990 when the world really grasped the potential damage of emitting greenhouse gases and share the clean-up cost based on past emissions. This seems fairer but what about those countries which emitted lots of CO_2 but their ancestors squandered the wealth and left no benefit to the inhabitants today?

Beneficiary pays: Maybe instead we share the costs based who derived the greatest wealth from emissions in the past. This accounts for historic emissions, inequality, and ability to pay. But what about countries like China which account for nearly one third of emissions today yet just 10% of emissions in the past: how do we get to net-zero if the largest polluters in the present aren't pushing as hard as possible?

The transition to a zero-carbon world is one that requires the cooperation and compliance of all global governments, corporates, and individuals. There is no global planner capable of serving up justice, only global negotiation, and each country will tend towards arguing for the system most beneficial for themselves.

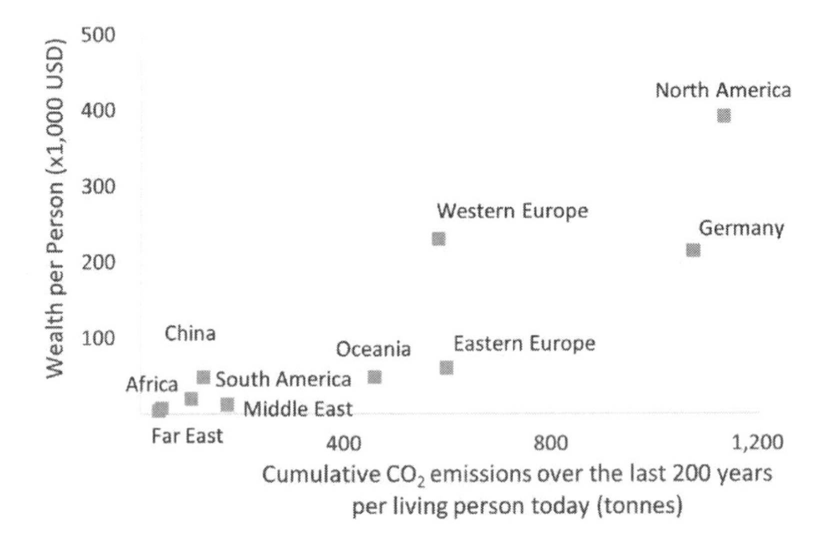

Cumulative CO_2 emissions by region for the last 200 years divided by the current population versus the average wealth per person in the region (assets minus debt). Cumulative CO_2 emissions from Tom Boden et al. CDIAC.[54] *Wealth per person from Credit Suisse Wealth Report 2018.*[38]

Hurdles to Action – Bad Actors: The global energy industry, like many others, is rife with individuals, corporations, and governments which have pushed ethical, moral, and legal boundaries to further their own agenda. This goes beyond simple self-interest; these are bad actors who either deliberately or inadvertently create hurdles to change.[55]

In their book *Merchants of Doubt,* Naomi Oreskes and Erik M. Conway meticulously piece together threads of historical evidence to show "how a handful of scientists obscured the truth on issues from tobacco smoke to global warming".[56] The book follows the actions of a small group of physicists who rose to prominence working on the atomic bomb during the start of the Cold War, and who came to occupy very senior scientific positions in the US agencies through the latter half of the twentieth century. Perhaps disillusioned by public concern over nuclear research, warped by the fear of communist infiltration through growing regulation, or through pure self-interest, this handful of individuals fought a decades-long campaign of disinformation, funded by the tobacco and fossil fuels industry, which first slowed action on smoking and then turned to impairing progress on acid rain, ozone, and climate change. These individuals exploited their scientific credentials, political connections, and public profiles, to manipulate the scientific process and spread misleading information with the intention of seeding enough doubt in the public and political conscious to slow or even halt change on major threats to the public good.

At the heart of their disinformation campaign was their exploitation of the scientific process. Contrary to popular belief, science doesn't deal in absolute truth: science is a moving consensus based on the balance of evidence. And that consensus clearly showed the harmful effects of smoking, the destruction of the ozone layer, and warming of the climate. Yet these bad actors knew exactly how to "cherry pick" any small uncertainties to delay action. They pushed their minority theories as though it were a two-sided debate, despite the fact that nearly all scientists who actually worked in these fields were thinking differently. They shouted about their half-baked ideas in the media with attention grabbing headlines whilst the credible scientists denounced these theories in scientific journals which the public never read. They downplayed the dangers by using the range of uncertainties to highlight the best possible outcome, or claim, that economic analysis showed solutions were too costly. The scientific community then had to expend valuable resources trying to dismiss or debunk these theories for confused politicians, so-called paralysis by over-analysis. These bad actors weren't just turning the dials of boom and boom, they were actually sabotaging the machine.

Sincere scepticism is good and is a fundamental part of the scientific process, but this was insincere scepticism, twisting the scientific consensus to suit their own agenda. Many of the activities of these physicists (and other individuals) were sponsored and/or supported by think tanks like the Heartland Institute, the Marshall Institute, the American Legislative Exchange Council, and the CATO Institute.[57,58] The trail of money and influence often led back to the tobacco industry and major fossil fuel companies.

The Center for Media and Democracy and *The Guardian*[59] found documentation that suggests the United States' largest coal producer, Peabody Energy, had financial ties with a large network of climate-denial-linked think tanks including the American Legislative Exchange Council (ALEC).[60,61] ExxonMobil's 'Worldwide Giving' report (2006)[62] and Exxon's 'Dimensions' report (1998)[63] show financial contributions to the CATO Institute and the Heartland Institute amongst many others.[64,65] The CATO Institute itself was co-founded by Charles G. Koch (of oil and industrial conglomerate Koch Industries) in 1977 and senior management of Koch have (on and off) served on the board ever since.[66,67]

In 2012, the Heartland Institute ran a billboard campaign with a mugshot of mass murderer Ted Kaczinsky, the Unabomber, with the tagline, "I still believe in global warming. Do you?"[68] It was an attempt to associate climate change advocacy with dangerous radicalism.[69] In 2017, the Institute mailed out 300,000 unsolicited copies of their textbook and DVD, "Why scientists disagree about global warming", to K-12 and college science teachers across America. The book even used the acronym of the self-created non-governmental international panel on climate change (NIPCC) to add to the confusion.[70]

The actions of bad actors have long added (and are still adding) to the divide and tensions over climate change.[71]

Lining Up to Clear the Bar – Overcoming the Hurdles to Action

The hurdles to action and reaching net-zero are many, but they are certainly not insurmountable. The science behind the attribution of specific events to climate change continues to improve. The last IPCC report already asserted with high confidence that everything from deadly heat waves in France, the increase of category 5 hurricanes hitting the US, to the failure of Russian grain harvests, are all associated with growing climate change hazards. The tragedy of the commons and misaligned economic interests may show why markets alone are unequipped to find the optimal road to net-zero, but in the next chapter we will examine how the economic and political tools have already been developed to expedite the process.

We are 30 years on from the start of global climate negotiations and the world is realising that climate action isn't a one-time game of prisoners' dilemma, but that it's an iterative process played over and over again. This has re-written the rules and climate negotiations are now following

Robert Axelrod's 'Evolution of Co-operation'.[72] Players are learning that long-term thinking and reciprocity bring the best outcome, that peer pressure can effectively penalise free-riders, and that forgiveness creates lasting accord but past actions will not be forgotten.[73] Or as Michael Liebreich, the founder of New Energy Finance (now BloombergNEF), wrote in 2012, "Be Nice, Retaliatory, Forgiving and Clear . . . instead of shooting for an unachievable top-down climate deal, we should focus on accelerating the inevitable emergence of domestic action on climate change". Following the failings of the Copenhagen Climate Summit in 2009, the world learnt that a one time, top-down deal wasn't going to work. Christiana Figueres was appointed head of the UN Framework Convention on Climate Change (UNFCCC) in July 2010. She rebuilt the negotiation process using iterative, bottom-up commitments by country, designed to ratchet towards a 1.5-2ºC aligned emissions reduction trajectory. Her ground-breaking work culminated in the non-legally binding Paris agreement which allowed President Obama to sign up the US without congressional approval, but it also meant that President Trump could leave. Yet, in the absence of the US, pledges from other nations have only grown stronger, and the world has gladly welcomed President Biden back to the negotiating table. It seems we are learning that collective action is in our best interests.

We are now lining up ready to clear the bar, and by recognising that a net-zero system will be cheaper, cleaner, safer, and more reliable, we leapfrog most of the remaining hurdles. Those regions with less to lose from global warming will still benefit from cleaner air and better health. Those wealthier countries better equipped against climate damage will still reap the rewards of moving to a cheaper, more efficient, and less volatile energy system. Attribution of damages to global warming, or the competition between present needs and future needs, become less relevant because moving to net-zero is both better now and better in the future. Participation and cooperation, though still essential, should become organic. Once the world is one quarter of the way through, the transition net-zero becomes the cheapest option. It becomes uneconomic, unethical, and unimaginable to remain bound to fossil fuels. Freeriding no longer makes sense, the common good and self-interests align. Concentrated action for those that can afford the upfront spending (for bigger savings down the line) brings benefits to all. The top five biggest emitting countries are responsible for 60% of all emissions and those countries with the top 10% wealthiest of the world's population emit 30% of all emissions. If just one third of the world can act, the rest will naturally follow, and the odds are looking better.

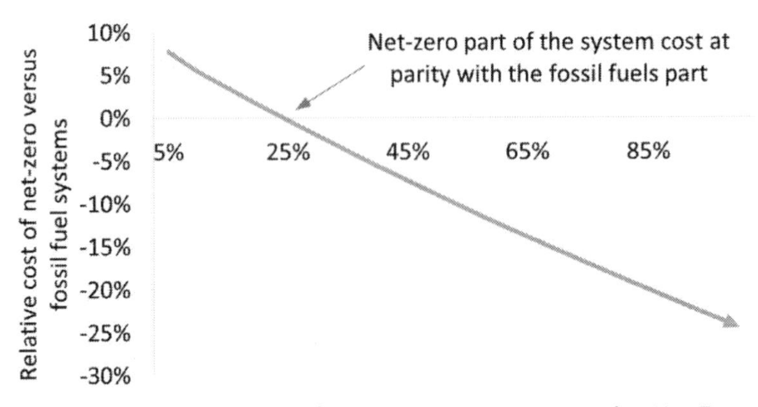

Net-Zero Energy System cost (buying and running cost of both supply and demand) divided by Fossil Fuel Energy System cost. Net-zero gets cheaper with scale as supply costs and capital costs decline, and efficiencies improve. Author's calculations based on modelling in chapters 7 to 12.

And yes, there may be bad actors who want to stand in the way of progress, driven by self-interest, marginalisation, or fear, but it's important not to get distracted from action by throwing blame. It's easy for all sides to get caught up in their own vision of the world as we reverberate around the echo chambers of social media, tailored news, and partisan politics.[74] Were the luddites who smashed Arkwright's spinning wheels to halt the beginnings of the industrial revolution bad actors? Apportioning blame is a fuzzy business and, as Winston Churchill once said, "history is written by the victors". Without cooperation towards net-zero, however, we all lose. Exxon, once the most profitable company ever, has dropped out of the main US stock index. Peabody coal filed for chapter 11 bankruptcy in 2016[75], and the CATO Institute has disbanded its climate denial arm[76]. Climate justice gets served, but we mustn't allow the misdeeds of executive management or politicians to disenfranchise hardworking employees or citizens: a just transition is a fair transition for all.

Also remember the most disenfranchised of all are the poorest 10% of the world's population who have had no impact on historical emissions and account for just 2% today. Climate Justice means prioritising access to energy, clean cooking, safe heating, sanitation, and connection to the modern world as fast as possible. The best option may be switching dirty wood or dung fuel for natural gas heating and cooking or connecting to a nearby fossil fuel powered grid. Alternatively, it may mean deploying cheap solar plus batteries and electric heating, cooking, and light in rural areas far from grid infrastructure. Either way, the additional carbon emissions will be small and once wealth becomes significant, net-zero will be the cheapest, fastest, and easiest way to run the revamped economy.

Recognising the benefits of net-zero reshapes the debate, pre-bunks (rather than debunks) the doubt, and turns climate change solutions from an intergenerational zero-sum game to a call for action with a win-win outcome.[77]

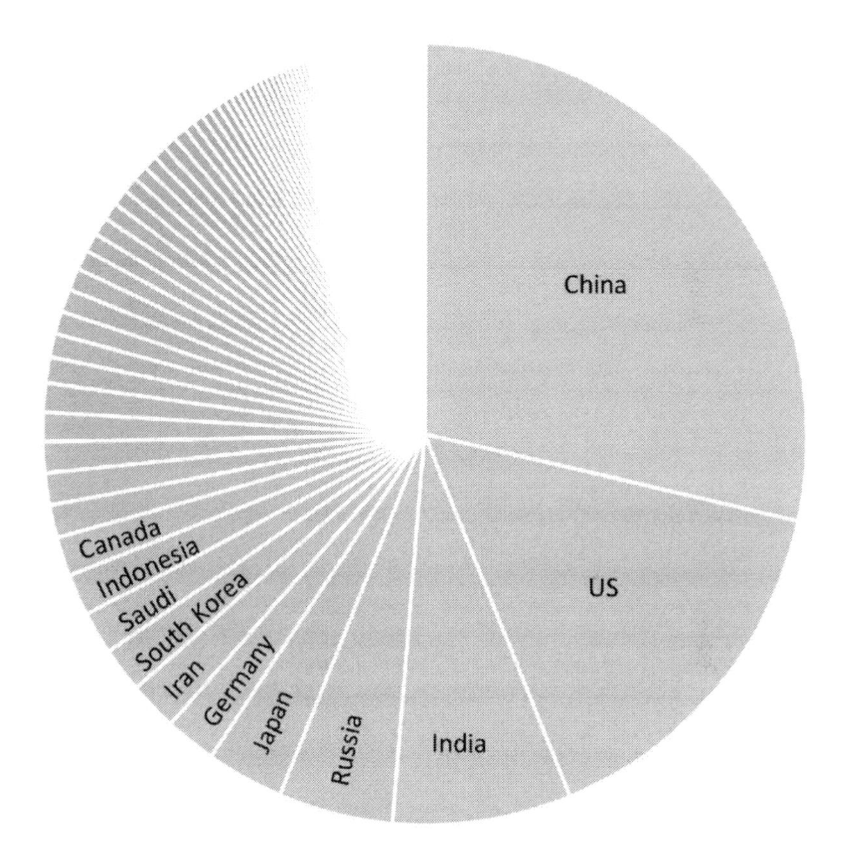

Share of CO_2 emissions by country today [78]

Chapter 15: Plotting a Good Course

The Politics of Change

In this chapter we will review the economic, political, and cultural tools of change which can overcome the hurdles to action and accelerate the transition to net-zero.

Translating Goodwill and Promises into Action

Awareness and willingness to act is certainly on the rise and it feels as though we may be on the verge of accelerated change. The scientific community has found a unified voice, stating in no uncertain terms that the world is warming with dire consequences unless stopped. The United Nations, World Economic Forum, and 26 countries representing nearly 1 billion people, have declared a climate crisis.[1,2,3] Greta Thunberg, the Sunrise Movement, and Extinction Rebellion are mobilising young and old alike with protests all over the world.[4] Major media outlets such as the BBC, *The Guardian*, and Bloomberg are ramping coverage of climate change.[5] The investment community is increasingly leaning on companies to comply with Paris Accord targets or to divest from fossil fuel companies completely driven by movements such as the Climate Action 100+ or Bill McKibben's 350.org.[6,7,8,9] The environment is now a top priority issue for voters and politicians across the world. More than 120 nations, covering well over half of global GDP and emissions, have pledged to reach net-zero by the middle of this century.[10,11] Sweden, the UK, France, Denmark, New Zealand, and Hungary have made their pledges legally binding.

But translating awareness, protests, company, and political pledges into action will be the challenge of the decades to come and it's the role of all global governments, corporates, and individuals to overcome their fears and enact the necessary change.

Government: The organisation of human socio-economic systems once relied upon a collective belief or trust in gods to create common purpose and cooperation. The idea of a divine ruler unified large numbers of humans, allowing the priests or clerics to go about the day-to-day business, dividing the finite resources as they saw fit.[12] Today, we place our trust in national governments to unify us, and in elected officials to manage both the distribution and growth of wealth.[13] Politicians must walk a tightrope, balancing the will of the people, expert advice and moral judgement alongside party and personal interest. They must plan for both the short term and the long term needs and desires of a nation whilst fighting a five-year re-election cycle. Plotting a pathway to please the party, constituents, community, employees, business, non-profits, and donors whilst remaining true to their moral compass.

Corporate: The British East India Company was the first commercial corporation owned by partners and not an individual, granted Royal Charter by Queen Elizabeth I in 1600. At its peak, the East India Company ruled more than one fifth of the World's population with a private army of a quarter of a million soldiers. By the late nineteenth century, the UK had begun to grant companies limited liability[i] and the US recognised corporations as legal entities in their own right. Thatcher and Reagan expanded the corporate agenda through the 1980s, promoting global trade, turbocharged by developments in global telecommunications and global cargo shipping, and laying the foundations for the modern trans-national companies we know today. Many global corporates hold greater power than the governments of many nation states. However, the CEOs and senior management of these modern behemoths are beholden to partners, shareholders, government, customers, community, and employees. They must define their corporate purpose, balancing benefits to society through the products and services they offer, the impact they have on the environment, global resources and suppliers, returns for shareholders, and the quality of employment. They must abide by laws set by national governments, global trading agreements and, in many respects, also govern themselves to a self-imposed moral standard. Corporates must balance short term performance with longer term strategy from daily targets, to planning and investing in equipment and people, which will endure for decades.

Individuals: Just as governments have replaced God's law, religious order has been replaced by humanism and free will. Liberalism emphasises the freedom of the individual and trust in capital markets, and socialism emphasises equality and the good of the people and trust in the union. Nation states, governments, and corporations cannot exist without these

---

i Limited liability means shareholder's personal assets are separate or protected from company problems. If you own shares in a company and it goes bust, the banks who loaned the company money cannot come knocking at your door.

kinds of belief, nor without trust and commitment from their people. In return for our allegiance, we are entitled to protection, pay, and influence. Individuals have responsibility for collectively guiding governments, steering corporations, and shaping the human and natural world through their individual actions. Family, social, news, political, and business networks are responsible for integrating and redistributing information, creating culture. We are simultaneously citizens, employees, consumers, owners, and shareholders, each with power to shape the world through our information networks and through our voting power, work output, and our purchases. Individuals must strike a balance between self-interest and the common good, between immediate needs and future needs, and between the present generation and that of future generations.

The Tools of Change: Policies and Actions to Help Overcome Market Failures[14]

For most of this book we have treated the world as a unified entity, yet humans are still tribal in nature. We may have replaced clan warfare with the competition of capital markets, yet we still identify ourselves as part of a particular clan be it political, technological, corporate, national, cultural, or religious. The competing desires of each collective create conflicting views on how to navigate climate change. But we are also coming to better understand the benefits of action, the complexities of cooperation, and the necessary tools to overcome market failures and expedite change. No matter your views on personal, political, or economic freedoms, there are solutions spanning the spectrum which can optimise the transition.[15,16]

Command and Control Regulations: A Tool to Legally Force Change

These are a set of legal restrictions or contractual obligations enforced by international bodies, governments, or corporates. Regulations are designed to change behaviour where the free market is failing to provide a solution. Environmental standards have been used since the 1960s to protect land, air, and water from destruction, contamination, and pollution, yet little has been done to control the emissions of carbon dioxide. Rules-led regulation can be used to set absolute emissions levels, limit consumption, ban fossil fuels technologies, and enforce efficiency standards. Potential policies range from energy efficiency standards for consumer products or buildings, tailpipe emissions limits on new vehicles, or clean energy standards governing the penetration of zero carbon electricity. Command and control regulation creates forced change and is able to overcome most market failures, but it also circumvents the market mechanisms which create cost-effective, competitive solutions so may provide a less than optimal outcome if done badly.

The Montreal Protocol was an international command and control regulation, described by the late UN secretary Kofi Annan as "perhaps the single most successful international agreement". First ratified in 1987, the Montreal Protocol set a timetable to reduce the production of ozone depleting substances, such as CFCs, 13 years after scientists first figured out the ozone layer might be thinning and two years after a hole was discovered over Antarctica.[17] Now ratified by 197 parties, Montreal was the first universal treaty in the United Nations history. 30 years on and the hole in the ozone layer is starting to shrink, with the UN expecting full recovery by 2060. The Montreal Protocol has prevented hundreds of millions of cases of skin cancer and cataracts, CFCs have been replaced with ozone friendly HFCs, and the Kigali amendment is the next phase designed to eliminate HFCs production by 2050 due to their action as potent greenhouse gases.

A more mis-guided government policy was US bioethanol. George Bush implemented the renewable fuel standards nearly 15 years ago. These regulations require a minimum amount of bioethanol blended in to transportation fuel to reduce US dependence on foreign imports and to reduce CO_2 emissions. However, as we saw in chapter seven, first generation ethanol emits nearly as much CO_2 as gasoline plus it takes up valuable agricultural land. US bioethanol now uses around 40% of all US corn production and more than 10% of all US crop land to provide less than 10% of transport energy needs. Development of second-generation production has stalled, and quotas have been abandoned as the economics of the technology struggles. Building out an ethanol industry only makes sense if there is a clear path to transition towards second and third generation biofuels, but command and control regulation has failed thus far in moving towards this goal.

Market Structure: A Tool to Accommodate and Support Change

The provision of electricity will become ever more important through a net-zero transition. Energy demand will halve but electricity demand will grow two or three times over. The way electricity markets are structured will become an important decision for governments, operators, and utilities companies.

Since the first grids were completed in Europe and the US over a century ago, the electricity industry has proved a natural monopoly with the same companies owning the generation and distribution of power offering little opportunities or room for new competition to enter the market. At first this seemed okay because power prices were getting cheaper through economies of scale, and the companies were still growing, competing, and innovating for new business. But by the second half of the 20th century, electrical demand was starting to become saturated and the industry had become more focussed on saving money than on innovation. Energy shortages started to push power prices higher and government regulation eventually intervened and came to regulate the monopolies by setting electricity prices or capping revenues and profits. The downside was that utilities have been left with little reason to develop new improved technologies.[18] Electricity is a commodity with no premium pricing; it takes decades to bring a new technology to market. Fossil fuel technologies are still subsidised, and operators are naturally risk adverse because they make no greater profit from taking risk and yet take all the blame if the grid fails. Research and development spending in energy accounts for just 0.5% of sales, with two thirds coming from government.[19] This compares to 3% of sales in most industries.

However, the structure and distribution of electricity generation is starting to change. Renewables like wind and solar have added more distributed generation and independent projects which compete with the incumbent utilities to sell power. However, electricity is still the only major commodity in the world with no significant storage and this has some unusual impacts on the structure of the market. Most electricity generation gets paid on long term contracts set at a pre-agreed price, but the more volatile short-term needs are usually allocated through an auction or bidding process one hour or one day before the power is needed. If it's sunny or windy in the time period being auctioned, wind and solar will always bid the lowest price and win the supply because they have no extra cost for generation.[ii] Renewables are increasingly pushing coal and gas generation off the grid, reducing fossil fuel capacity factors, and making the investment in new coal or gas facilities untenable. A great result for emissions but also a potential problem if too many fossil fuel plants are shut down because there is no back up generation when it is cloudy, still, and electricity demand is high.[20]

Some governments have turned to making large capacity payments to keep these fossil fuel plants available to produce power if needed, whilst

---

ii Renewables are said to be zero marginal cost because they use no fuel and the sun, wind, and water are free; there is no cost difference if the energy is used or not.

others like the Electricity Reliability Council of Texas (ERCOT) have simply opened up access for all generation and let the market prices plot a course.[iii]

Adding electricity storage to the grid will ultimately solve this dilemma. But the economics of adding storage will only work if capacity payments are rolled back and, for the first time in history, we manage to create truly competitive electricity markets with unhindered connection and access, competitive bidding, and a return to innovation. Alternatively, markets could move towards a more planned system, using capacity payments based on the dispatchability of power to deploy renewables plus storage.

Creating the right market structure will be incredibly important to ensure safe, reliable, and low-cost energy in the future.

Information Reporting and Labelling to Empower Change

Enforced or voluntary reporting requirements force countries, regions, or companies to publish agreed information. Third party agencies and non-profits such as the Carbon Disclosure Project (CDP), Task Force on Climate-related Financial Disclosures (TCFD), or Sustainability Accounting Standards Board (SASB) are pushing to encourage investors, companies, and regions to disclose more information on their environmental impact, to build a climate change strategy around transition and physical risks and opportunities, and to embed the environment into company reporting right alongside their financial accounting.[21] Better information and informed strategy helps track progress, create awareness, increase transparency, and leads to better investment decisions and policy.

Similarly, product labelling is another important tool that governments can enforce, or companies can adopt, to help inform consumers on quality, energy efficiency, or carbon emissions associated with their products. In turn, this helps to inform buying decisions and overcome

---

iii You may associate Texas with the oil industry, but it also has three times more wind power than any other US state. The development of wind power in Texas was rather serendipitous, as Leah Stokes points out in her book Short Circuiting Policy the state passed a bill designed to allow better grid access so small natural gas generators could sell into the market, but a small amendment requiring a few percentage points of zero carbon electricity was slipped in. Easy access coupled with zero carbon requirements created new cheap renewable electricity taking advantage of Texas's great wind resources and lots of jobs in rural areas. Wind energy proved a hit, the renewables allocation got upped, permitting and transmission lines were built and fast tracked and Texas, the home of fossil fuels, now generates 20% of its electricity from wind and the only new electricity generation capacity in the que for the next few years is wind and solar.

aversion to change. Reporting and labelling seeks to overcome apathy and status quo bias through filling knowledge gaps in the market. Individuals, corporates, and governments become better informed about the consequences of their actions.

Subsidies to Support the Economics of Change

The public promotion of net-zero solutions through tax breaks, capital allowances, loans, or direct payments. Subsidies can be used to help develop a new market by making novel technologies cheaper and more accessible for consumers. Subsidies are designed to encourage demand, scale production, and ultimately drive costs down the experience curve, allowing capital markets to take over. Subsidies are essentially a tax in reverse. They encourage consumption of a particular good or service whilst transferring money from governments to corporates or individuals. Subsidies can be very effective, but they can also go very wrong if poorly thought out.

Subsidies have been successful in China where installed solar has grown from next to nothing to over one third of installed global solar power in the last ten years. The government started out by providing balanced solar subsidies of between $5-15c per kWh, tax incentives, cheap loans, and longer-term installation targets.[22] State-backed and private Chinese companies acquired equipment and expertise from all over the world and dramatically ramped production firstly for local markets, and then extending their reach and exporting to Germany, Southern Europe, and the US. China is now the world's solar factory, making around two thirds of all panels.[23] The 80-90% cost reductions over the last decade were kick started by subsidies in the EU but have been turbo charged by the rapid scale up of the Chinese market, which is now moving to become subsidy free, as solar electricity reaches cost parity with coal.

Equally, subsidy programs can stimulate overconsumption and create financial problems for governments. The cash for ash scandal was a prime example. Northern Ireland implemented the renewable heat incentive in 2012 which provided subsidies for businesses switching to wood pellet boilers, heat pumps, or solar thermal. However, the payments were so generous, with no caps, they more than covered the cost of the fuel. Registered businesses could not only recoup the equipment costs but could then profit from heating. The more they heated, the more money they made. This led to reports of farmers installing heat pumps or pellet burners in chicken coops or empty sheds and running them 24/7 no matter the weather. The government ended the scheme in 2016, but not before enough contracts had been signed for 20-year payments which will cost Northern Ireland hundreds of millions of euros more than budgeted.[24] In the meantime, Arlene Foster, the minister accountable for the scheme, had become Northern Ireland's First Minister. After the scandal broke, she refused to step down, which (amongst other issues) led to Martin McGuinness, the power sharing Deputy First Minister, to resign with the consequence that the Executive Office in Northern Ireland collapsed for three years.[25,26] What should have been a simple piece of subsidy legislation went very wrong due to a lack of impartial information gathering and a lack of enforcement and monitoring.

Research and Development to Support Technological Change

Research and development money given by governments, corporates, or individuals is another way to support the development of new technology. Early-stage scientific research, technical development, demonstration projects, and public procurement can all help early-stage innovation. These measures support technologies through the valley of death as they scale up towards the commercial experience curve and they further help to expedite technological potential.[iv]

The 20% of net-zero technologies which require further fundamental research and development include next generation nuclear, carbon capture, hydrogen, and biofuels. Well balanced participation from government, academia, big business, and individual entrepreneurs is essential. We can view the research into these technologies as an insurance policy towards full decarbonisation, should we reach a limit on deploying wind and solar electricity.

Many of the most significant innovations through history have sprung from government labs or universities. Recent examples include the internet, Google's search engine, smartphones, GPS, super-computers, sequencing the human genome, MRI scanners, and civil aviation.[27,28] These industries were enabled by pioneering research in both government-funded and University labs. In fact, if you ask the CEOs of most multi-billion-dollar companies where most of their money goes, it's usually the D (development) not the R (fundamental research). Businesses tend to be risk averse and so they struggle with the high failure rate of early research.

An example of a highly successful energy research and development collaboration is lithium-ion batteries. Early prototypes were first

iv The US government has spent $230 billion over the last 70 years developing energy technology with half on nuclear, one quarter on fossil fuels, and one quarter on renewables and energy efficiency.

developed by Michael Whittingham at Exxon in the 70s, but safety concerns led the company to abandon the project. Professor John Goodenough at Oxford University picked up the work in the late 1970s, improving safety and energy density before Akira Yoshino at Sony built on Goodenough's work in the late 80s using the company's expertise in film coating cassette tapes. Sony launched the first commercial lithium-ion battery in 1991, now the mainstay technology powering most portable devices, power tools, electric cars, and a promising solution for grid electricity storage. Whittingham, Goodenough, and Yoshino won the Nobel Prize in Chemistry in 2019.[29]

A rather more notorious example of early-stage funding involved a company called Solyndra. Set up in 2005 and based in Freemont, California, Solyndra was a tech company with a disruptive product for the solar PV market. They wanted to produce solar panels from thin films of CIGS (copper, indium, gallium, selenide) coated on tubes, rather than flat, silicon-based panels. The idea was to save on expensive silicon metal and to boost efficiency by capturing more light using the curved design. By 2009, the American government, convinced the company was onto something, fast tracked a loan by the administration, despite analyst concerns around cash flow. By September, then Vice President Biden and Energy Secretary Steven Chu presided over the ground-breaking official ceremony for the new factory. However, by 1 September 2011, Solyndra had filed for bankruptcy, 1,000 workers had lost their jobs, the US government had lost $500 million, and Silicon Valley investors were down $1.3 billion.[30] What went wrong?

Well, Solyndra had solved a problem which didn't really exist. Yes, at the time silicon was expensive, but production capacity only really existed to supply silicon for computer chips. The rapid acceleration in solar demand through the late 2000s forced mass shortages. Prices skyrocketed from $50 to $400 per kg as new production wasn't built fast enough to keep pace with demand. But this was an economics problem and not a technology problem. In response to the high prices came large swathes of new Chinese capacity and by 2010 prices were back below $50 per kg and solar silicon wafer production was now firmly on the experience curve. Solyndra's more complex tubular panels couldn't compete with cheap silicon and demand for their product dried up. Chinese silicon producers continued to scale up, reduce waste, improve efficiency, and quality of manufacture. Silicon now sells for less than $7 per kg and is one of the main reasons that today's solar panels are so cheap and getting cheaper still. Solyndra is a cautionary tale of trying to pick a winning technology.[v,31] Yet failures are inevitable and it's often overlooked that the same Department of Energy loan program also lent $500 million to Tesla which was fully repaid nine years early[vi] and helped create the most valuable car company in the world.[32,33]

Structured Financing as a Tool to Address Market Failures

Sourcing the money is step one but structuring the way in which the funds are provided to individuals, corporates, or governments, will help accelerate a rapid deployment of net-zero solutions. Creating tailored financing structures can address market inefficiencies such as lack of access to cash, parabolic discounting, and renter-tenant problems by spreading the upfront cost over a suitable number of repayments and linking the repayments to the beneficiary.

The US has a program called Property Assessed Clean Energy (PACE) which provides loans to homeowners to finance the upfront cost of energy efficiency improvements or renewable energy installations. The loan is tied to the property rather than the individual and paid back though property tax or utility bills: the costs and benefits are tied together and it doesn't matter if you sell or rent the property.[14]

In a similar fashion, many corporates want to move towards net-zero but would rather use the funds on their balance sheets to grow their core business rather than paying for solar panels or new energy efficient equipment. No money, upfront, or off-balance-sheet financing arrangements allow the company to repay the upgrades through ongoing expenses rather than paying upfront. In many cases the energy savings are more than enough to cover the extra ongoing cost.

Founded in the United States in 2003 by Jigar Shah, SunEdison was a pioneer of no money upfront solar. SunEdison developed a business model which connected corporates who wanted solar energy but didn't like the upfront cost with banks and institutional investors who had money to deploy. The company pioneered power purchase agreements which standardised repayments over 25 years and they installed the equipment. This innovative financing model proved a hit, and SunEdison soon signed up household names like Wholefoods, IKEA, and Staples and built relationships with major banks such as Goldman Sachs and Wells Fargo.

---

v Ironically, the next big development in silicon solar is bifacial panels which absorb sunlight on both sides.

vi The program issued 22 loans in total of which 20 were repaid in full.

By 2008, SunEdison was one of the largest solar companies in the US and a year later was bought out by a silicon cell manufacturer called MEMC who, perhaps wary of growing Chinese competition in solar silicon, wanted to diversify into solar development. The company took the SunEdison business model and pushed out an ambitious development pipeline all across the world. Conscious they needed to resell more and more of these solar developments to investors, MEMC (which was now renamed SunEdison) developed two yieldco companies and listed them on the stock exchange. Terraform Global and Terraform Power were companies formed from a collection of SunEdison solar developments which were already up and running and that any investor could now buy or sell. Yieldcos had been around a while in Europe and had proved a successful way to separate the low risk running of solar farms or wind projects from the higher risk development stage. Low risk yieldcos can raise cheap money from the market to buy newly finished projects and pay steady dividends to investors.

Unfortunately, the US market got overexcited and wanted more projects and more growth, pushing the yieldcos to expand and take advantage of tax incentives. More cautious investors, who owned these yieldcos for the steady dividend, became wary that the Terraforms were overpaying for SunEdison projects to help plug a hole in SunEdison cash flows and they dumped the stock. SunEdison's money dried up, the company filed for bankruptcy, the shares dropped over 90%, and all three companies were ripped apart and sold off.[20] What started as a hugely successful business innovation under Jigar Shah had become a cautionary tale of market hubris or over exuberance.

Cap and Trade Brings Climate Damage into the Economic System

A way to bring carbon dioxide emissions and the cost of damages into the economic system by pricing carbon and addressing the tragedy of the commons and the problem of externalities. Cap and Trade systems set an upper limit on the emissions of a particular company, industry, or country: the cap. If the company or country exceeds the emissions cap, they can buy credits from an entity that has emitted less than their cap: the trade. If companies don't comply, they are fined. The initial allowances can be given freely or sold and the tradeable credits are priced based upon the supply and demand balance. Cap and Trade is a mix of command and capital markets. The government sets the overall limit but the ability to trade credits means the market can efficiently allocate where emissions are best used. The benefits of Cap and Trade systems are a defined overall limit and the use of efficient allocation by market trading, with the changes largely hidden from citizens and therefore more palatable to politicians. The downside is finding a suitable limit, and this can be difficult. Large swings in the price of credits can make financial planning hard for companies and the system is tough to monitor and enforce for all but the largest businesses.

If only larger companies are included or emissions are given away freely (grandfathered), the system also becomes less economically efficient and allocations are vulnerable to corruption. Think for a moment about budget and premium airlines. Budget airlines are low cost because they pack lots of passengers on to every plane which creates a much lower carbon footprint per passenger when compared to a premium airline whose business class clients can stretch out on a nice comfortable bed. If allowances are grandfathered based on current emissions, the government is not penalising the premium airline for a higher carbon footprint and locks in the carbon inefficiency. The same principle holds for any industry.

Cap and Trade systems have been used very successfully in the past. In the 1980s, US president Ronald Reagan and the Environmental Protection Agency (EPA) put in place the first cap and trade system to phase out leaded gasoline. At the time, lead chemicals were being added to fuel to prevent inefficient firing of the engine cylinders, thus improving power and fuel efficiency in cars. However, lead was fouling the newly added catalytic convertors and lead emissions were found to be capable of crossing the blood-brain barrier and causing poisoning. The saying 'mad as a hatter' comes from the induced madness of top hat makers who would line the inside of their products with lead and routinely suffer from delusions, memory loss, hallucinations, brain swelling, and death. Reagan's Cap and Trade initiative was launched in 1982 and just five years later the full reductions were complete.[34] The program was considered phenomenally successful, moving faster than expected and costing 20% less than alternative solutions.[vii]

The European Union's Emissions Trading System for CO_2 is the largest carbon pricing regime to date. Started in 2005, the scheme initially covered about half of European CO_2 emissions from 11,500 installations across electricity generation and industry. It later included aviation and shipping. The program has had mixed results. Carbon credits have traded anywhere between zero and €30 per tonne as a lack of data, energy

vii Cap and Trade went on to be adopted by George HW Bush to cut acid rain causing sulphur emissions from coal plants. The regulation cut SO_2 emissions by 36% between 1990 and 2004 despite 25% increases in coal fired power plants.

price volatility, free allocations, and volatile economic growth has made setting the limits difficult.[35] CO_2 emissions since 2005 have declined by nearly one third for the half of emissions covered by the Cap and Trade scheme.[36] The other half of European emissions, covered under the voluntary effort sharing scheme, have declined by 10%. Overall emissions are down 20% in the last 15 years and even adjusting for the additional $150 billion of net annual imports from China throughout the period (with embedded emissions) reductions are still 15-18%. Certainly, lessons have been learnt around the need for good data, the ability to roll credits into the next year, possible price floors, consistent allocation policy, and wider participation. Uncertainty around the carbon credit price has ultimately proved the biggest shortcoming of the European scheme, leading to short term behavioural changes but less influence on long term investments.

Carbon Taxation Brings Climate Damage into the Economic System

This is another form of carbon pricing which applies a direct tax to CO_2: the polluter pays principle. A tax is applied to upstream fossil fuel producers or carbon emitting industries based on the amount of CO_2 which their products release. The cost of the tax gets incorporated into the price of the coal, oil, or gas and is passed through the supply chain to the consumer. Goods and services with higher CO_2 emissions become more expensive and consumers are incentivised towards buying alternative lower emissions products. It makes net-zero technology more competitive and more attractive.

Cap and Trade has been favoured by politicians, but carbon taxation is generally favoured by economists. Carbon taxes should be easier to implement because they are enforced on a smaller number of upstream producers and don't require complicated emissions monitoring. Most countries already levy fuel duties, so the government systems are already in place. The tax is simple and efficient, captures all polluters, and is transparent, with less scope therefore for corruption. It is also informationally and economically efficient: the price increases reflect the level of emissions, so no decisions need be made on where to allocate credits or where to grant exceptions. The main drawback of a carbon tax is the lack of a defined limit, so the level of the tax must be set correctly to encourage decarbonisation and the consumer must respond to the higher prices. Some argue that national carbon taxes create disadvantages for domestic industry and the polluting activities will just relocate to different parts of the world, but research has shown that so-called carbon leakage is at most 20% and easily overcome using an import carbon tax on countries with no carbon pricing. An import tax also encourages the exporting country to adopt a carbon price, so they receive the tax payments rather than the government of the importing country. But there are limits to the carbon tax. It is only effective on the 80% of technologies on the commercial experience curve, and it suffers from many of the same market failures we have already discussed, so other policy tools are still required.

A carbon tax could start low with defined increases through time, and some level of flexibility – this allows businesses to forward plan but also ensures ongoing decarbonisation. The other issue is the emotive potency of the word "tax", which is why politicians are hesitant. The fact that tax is easier to grasp in simple terms and more visible to the consumer has proven a political downfall: no politician wants to be seen to be raising taxes. However, the money raised through the tax could be spent by government, directly redistributed to the people in a carbon cheque, or used to reduce other distortive taxes such as income tax (surely it is better to discourage carbon emissions rather than discourage hard work). If reimbursed to the people in a direct payment, the only cost is the administration of the scheme and the relative cost of adopting zero emissions technology versus fossil fuels technology. Although we have already shown this to be slightly more expensive today, it soon becomes the cheaper option.

The level of carbon tax can be set in two basic ways. The first, used by most economists, calculates the social cost of carbon. This is the future economic damage caused by one extra tonne of CO_2 emissions today when compared to the baseline scenario (in today's money). You pay compensation[viii] for dumping pollution in the atmosphere. Which sounds sensible, but the problem is that no economist can agree on how big those damages are. Chapter five showed why forecasting climate damage is uncertain and subjective, and economists have calculated a suitable social cost of carbon at anything from \$0 to \$1,000 per tonne of CO_2. The Obama administration set the price of carbon at \$45 compared to the Trump administration which set the number as little as \$1 per tonne

viii A Pigovian tax is a tax on any activity that generates negative externalities and is set at the marginal cost (incremental or extra cost) to correct for the damage done.

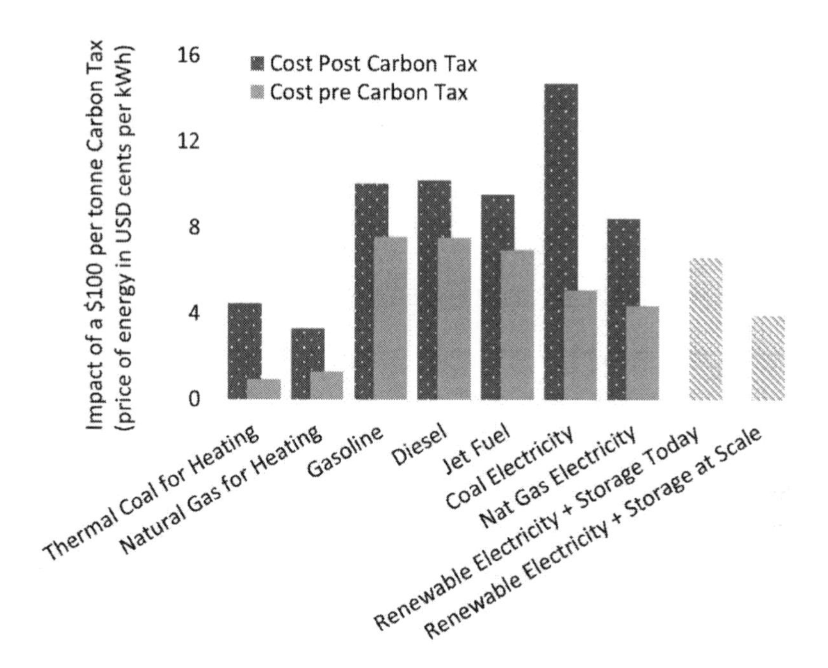

Wholesale cost of energy with and without a $100 per tonne carbon tax. Coal is most impacted. $100 per tonne makes renewable electricity with storage the cheapest electricity option today and once renewables scale up, driving the price down, they will be competitive with coal and gas for even the most inefficient resistive heating uses. Author's calculations based on emissions values and cost data given in chapters 7 and 8.

Using the damages model from chapter five suggests a $75 per tonne CO_2 social cost of carbon which includes no monetary value for lives lost, personal suffering, and the destruction of the natural world. Monetising death is a hard value judgement to make, but our model can still tell us that, in addition to $75 of market damages per tonne, every two high carbon lifestyles will force one person into suffering, and for every 40 high carbon lifestyles, we sentence one person to death.[ix] A sobering thought.

The second way to price carbon is simply to estimate a high enough level to encourage the switch to net-zero technologies. The price starts low, addressing the easiest parts of the transition first, and increases over time to eventually force all areas of the economy to switch with

ix High carbon lifestyle is 70 years of emitting 10 tonnes of CO_2 per year or 700 tonnes of carbon. These assumptions use our damage model which assumes we do nothing about climate change and make minimal adaptations but also assumes no dangerous tipping point is breached.

enough flexibility to ensure a good pace of change. Instead of trying to compensate for damages we are using the tax to expedite the transition to a better system.

Based on our analysis over chapters seven to twelve, we can calculate the required carbon tax for net-zero technology to compete with fossil fuels at a systems levels (buying and running costs). Based on today's technology, a carbon price of between $20 and $50 per tonne of CO_2 should decarbonise around half of all emissions. Once the technologies are fully scaled, $50 per tonne should ensure that 90% of all net-zero technologies are more economically favourable. The remaining 10% require $100 per tonne or more in order to try and fully transition the entire economy.

The last IPCC report (2014) estimated a carbon price starting at $25 per tonne and ramping to around $140 per tonne is required to keep CO_2e below 600ppm by 2100 (~50% chance of warming less than 2ºC).[37] The model ensemble from the IPCC AR5 report had solar and offshore wind electricity at $11-12c per kWh (it is now less than half that price) and only some of the model pathways electrified transport and even then, only did it in the second half of the century.[38,39] Our scenario cuts carbon emissions significantly faster with a lower carbon price because of the huge price declines in wind, solar and batteries over the last ten years and further gains as the technologies scale. The IPCC notes "most of the scenarios developed since the 1970s for energy and climate change make exogenous assumptions about the rate of technological change". In other words, the cost of net-zero technology is input by hand each year rather than letting costs decline with scale. The costs are based on expert analysis or energy forecasting bodies which have always underestimated the rate of progress in renewables.

On the supply side, coal electricity falls first, being the dirtiest fuel. Oil next and gas remains competitive for the longest. On the demand side, once the technology scales, all types of transport, lighting, heating, cooling, agriculture, and many industrial processes will be more attractive without any carbon tax. Water heating, textiles, chemicals, fertiliser, and cooking are some of the more economically challenging areas for net-zero and will continue to require carbon taxation or other policy support.

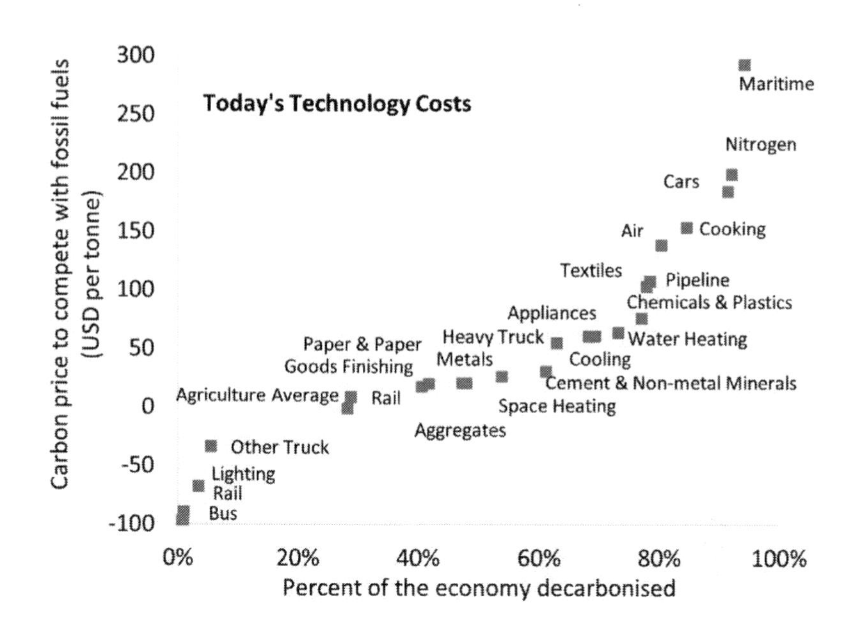

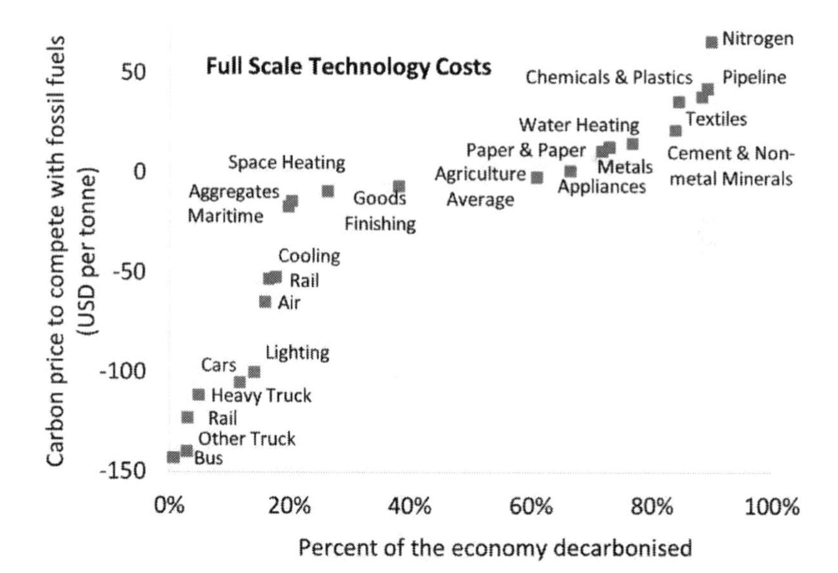

Carbon Price ($ per tonne CO_2) required for net-zero technology options to cost the same as the Fossil Fuel equivalent. Negative price means net-zero is the cheaper option before carbon pricing. X-axes show cumulative percentage of the economy moved towards zero carbon by adding all the technology areas from the left to right. Top Chart uses today's net-zero technology costs. Bottom chart uses net-zero technology costs once scaled. Author's calculations based on modelling from chapters 7 to 12.

Fuel levies and taxes already add economic disincentive to certain fossil fuels. The OECD 'Taxing Energy' report shows only gasoline and diesel in OECD countries pay anywhere close to the social cost of carbon with around $80 per tonne effective carbon tax on liquid fuels. Coal and gas are taxed next to nothing.[40]

But to complicate matters further, fossil fuels subsidies (mostly in developing countries) give that money back and more. Total subsidies of $500 billion per year paid to the fossil fuel industry (compared to renewables subsidies of $200 billion) represents the equivalent of an $8 per tonne or $0.5c per kWh financial incentive to burn carbon and must be eliminated.[41]

Sweden has enforced a carbon tax since 1991 and this is the cornerstone of its environmental policy. The country set the initial price at €24 and has slowly ramped up to €114 per tonne.[42] Throughout the last 30 years, the Swedish economy has grown by over 80% and greenhouse gas emissions have dropped by 26%. It is the second best[x] performing country in the block. Sweden has the strictest emissions target[xi] of all EU countries and yet is one of only three member states on track to meet their 2030 goal.

Not all schemes have been so successful. Australia's Labour minority government implemented a hybrid carbon tax / Cap and Trade scheme in 2012 starting at AUS$23 per tonne for certain emitters. However, the scheme was opposed by Tony Abbott, the opposition leader at the time, who came to power and repealed the program in 2014. The scheme never had cross party support, did not have broad participation across all industries, and the money was redistributed through lower income tax. These issues meant the carbon tax became a political tool with too much government interference and as it was not equally enforced, it was therefore economically inefficient: the higher energy costs were more visible than the lower income tax benefits. One of the economic benefits of a carbon tax is the visible impact it has on prices in the economy, but this is also the main political drawback. As Danny Cullenward and David G. Victor write in 'Making Climate Policy Work': "Policymakers rely on regulation and other forms of industrial policy because they know that the costs of regulations are less visible to voters and that extensive regulation makes it easier to shift costs and benefits as needed to address political opposition."[43]

So finally, using a simple example, let's think about how a carbon tax could be implemented with less political backlash by using a direct

---

x Greece dropped carbon emissions since 1990 by nearly 30% due to economic decline.

xi Sweden has a target of 40% emissions cuts vs 2005 levels by 2030.

reimbursement to the people. The average individual in the UK is responsible for ten tonnes of CO_2 emissions per year[xii] and therefore a carbon tax of £75 per tonne will cost the average person £750 per year.[44,45] But the distribution of emissions is not equal.[46] The poorest 10% of households emit just 5 tonnes of CO_2 per year – their bills would increase by £400 or 5% of income. The richest 10% emit 16 tonnes and their bills would increase by £1,200 or 2% of income.[47] If the monies collected are not given back to the people, the tax is regressive, disproportionately hurting poorer individuals because they lose a greater percentage of income. But if the collected tax were to be reimbursed equally in an annual payment, each person would be repaid £750. The poorest 10% become £350 better off (5% of income), the richest 10% are £450 worse off (1% of income – a figure they would barely notice). A carbon tax could be engineered not only as a highly effective environmental policy but also a politically attractive progressive wealth redistribution tool.[48]

Corporate and Shareholder Purpose to Align Business with Net-Zero

Moving away from government intervention, capital markets need to play a role in addressing issues such as global warming. The idea is that risks posed by climate change become incorporated into business planning. Products and services naturally move towards a solution driven by the optimal economic outcome. For this to become an effective solution, companies must correctly report and integrate climate risks and realign their corporate purpose – the set of goals towards which the company operates. As management consultant, Peter Drucker, put it, "what gets measured, gets managed".[49]

Corporate purpose should bring positive change for all stakeholders, profits for the business owners, returns for shareholders, good quality employment, fair treatment of suppliers, and create a positive impact on society from consumption of the goods and services sold. They must also balance these economic and social benefits with the impact of their operations on the environment, the natural world, and resources of the planet.[50]

In the 1980s corporate purpose became heavily skewed towards maximising shareholder value, driven by the ideas of economist Milton Friedman and the Chicago School of Economics. In the view

xii The average UK individual emits just over 10 tonnes of carbon per year, 5 tonnes is domestic emissions from transport, industry, agriculture, and households, the other 5 tonnes is imported emissions from goods and international aviation.

of Friedman, shareholders owned the company and therefore employed management, so it should be the shareholders who ultimately choose how and where to allocate profits towards social causes. As Friedman put it, "Few trends could so thoroughly undermine the very foundations of our free society as the acceptance by corporate officials of a social responsibility other than to make as much money for their stockholders as possible".[51]

Friedman believed businesses should make money, governments should decide how to tax it and spend it. But when Freidman wrote his book, the world was an entirely different place. Today's shareholders have more diversification but less vested interest in the companies they own[xiii], and globalisation has taken many multinational corporations out of the reach of national governments. Whilst we must strive for a greater level of global cooperation and strengthen international laws, we must also push corporates to strengthen their social responsibilities.

Maximising shareholder value and profit growth remain the primary objectives of most companies today. But an increasing number of corporate scandals, the banking crisis, increasing wealth inequality, and environmental concerns, are creating a growing consensus that corporates need to rebalance priorities and take a more active role in society. As Tim Cook, CEO of Apple, puts it, "People should have values, so by extension, a company should. And one of the things you do is give back… We give back through our work in the environment, in running the company on renewable energy. We give back in job creation."

Equally, shareholders themselves still have the responsibility to take an active role in social-environmental solutions. Investment managers at pension funds allocate the money we, as individuals, save for retirement into shares, bonds, or other assets of companies and governments. They have a fiduciary duty (financial responsibility) to provide the best possible outcome for their members. This means growing the value of the invested money whilst minimising risk, traditionally done by analysing and forecasting financial performance, but increasingly environmental, social, and governance factors (ESG) are being included in the analysis. Those companies that have the least environmental impact and do the most social good, as well as being most ethically run compared to their peers, have shown to outperform all others.[52]

Over 80% of studies find that sustainable business practices improve operational performance, company cash flows, and ultimately investor

xiii The average holding in the shares of a public company is estimated to be four months today, down from eight years in the 1960s. [From Sustainable Investing, A Path to a new Horizon, Routledge, 2021, p192] returns.[53] ESG helps to identify the most efficient operations and to avoid consumer backlash or corporate scandal. Incorporating ESG factors supports the broader umbrella of sustainable investing which includes many different strategies.[xiv]

Public companies, private companies, shareholders, and financial institutions need to collectively realign corporate purpose to better reward ESG values and re-allocate funding to companies deploying net-zero and doing social good. If done well this not only benefits the environment and society but should also prove economically beneficial. Badly run companies whose products and services have a negative impact should find it increasingly hard or increasingly expensive (higher interest) to borrow money or sell shares. Well run, ethical companies with a positive environmental and social impact should find it easier to access capital. Good companies are able to grow faster and do more good, bad companies shrink and do less harm.

From a climate change perspective, those companies aligning themselves with net-zero solutions are enabling an energy system transformation which will be cheaper, safer, and less volatile and will ultimately be rewarded. Those companies that continue to invest in a long-term future for fossil fuels are wasting money, destroying shareholder value, and will cease to exist unless they realign their corporate purpose and re-allocate their investments.

Investors, financial intermediaries, and corporates are showing encouraging early signs towards realigning corporate purpose through sustainable practices and ESG standards. The Principle of Responsible Investment initiative (PRI), launched by the UN in 2006, has over $90 trillion of assets (nearly half of financial capital) and over 3,000 investors signed up and following a framework which strives to engage ESG issues aligned with the Paris Accord and with the United Nations Sustainable Development Goals (SDGs).[54] The Climate 100+ investor initiative is pushing the top 100+ fossil fuel emitting companies and 60 other strategically important businesses to curb emissions, improve

---

xiv Strategies include: Negative Screening, excluding sectors or companies below a certain ESG benchmark; Positive Screening, only investing in sectors, companies, or projects above a specific ESG benchmark; Sustainability Themed Investing, investing in companies or assets related to sustainable themes such as low carbon energy; Impact/Community Investing, providing investment and financing to non-publicly traded businesses, communities, or individuals who would otherwise struggle to find the money and will create clear social or sustainable benefits; Corporate Engagement and Shareholder Action, or the use of unified shareholder power to influence corporate behaviour through voting, board seats, proposals, and pressure on management; Philanthropy Funding, projects with social and environmental benefits, without financial return.

governance, and strengthen climate related disclosure. These investors represent over $40 trillion of assets under management (20% of financial capital). The UN net-zero asset owners' alliance has total assets under management of $4 trillion solely invested in companies on track to become net-zero by 2050 or earlier. Some of the biggest corporate emitters have signed up to net-zero by 2050, including emissions from their operations and the emissions of their products. They include oil companies Repsol and BP, Miners Vale and BHP Billiton, Quantas Airlines, Electric utilities companies Xcel, Southern, Enel, RWE, DTE and Duke Energy, Steelmaker TyssenKrupp and Heidelberg Cement. Microsoft have vowed to become carbon negative by 2030 and offset all the emissions the company has ever produced.

Whilst more corporates are engaging with climate, a 2019 study using the Arabesque S-ray sustainability big data tool highlighted that only 18% of the world's largest public companies have long term plans aligned with the 2050 Paris climate targets, and over one third still don't report greenhouse gas emissions.[55] There is even less visibility with state-owned and private companies which represent over 70% of global emissions and hold over 90% of remaining fossil fuel reserves.

Consumer Choice and Individual Action to Drive Change from the Bottom Up

There are 100 companies operating today which are responsible for half of all human CO_2e emissions over the last 200 years and account for three quarters of all current emissions. These are the coal, oil, and gas companies who find, extract, and sell fossil fuels which are burnt to provide the consumer with transport, consumer products, basic amenities, and food. 60% are government owned companies, 30% are owned by shareholders, and ten per cent are private. These statistics make it very easy of course to place singular blame for global warming on government inaction or corporate greed, but we need to remember that we ourselves are the citizens and the consumers who elect the politicians and buy the goods and services. Governments or companies cannot exist without the trust, commitment, and spending power of the people.

Individuals have power over all of these companies through consumer choice. We have sufficient options to choose low carbon and influence corporate purpose; when companies see revenues in decline, they will very quickly change their business model. We can exercise our influence on government by voting for politicians who prioritise net-zero policies; when governments start to lose voters, they will very quickly change policy. We can invest our pensions in sustainable funds and divest polluting companies; when investment managers see money outflows, they will very quickly reposition their portfolios.

Recognising that we as individuals are both responsible for climate change and have the power to affect change is vital to action. Every single individual, all over the world, has at least some control over the global energy system. From lifestyle changes, political voting, or corporate influence, there are many routes towards pushing for a cleaner energy future.

Some changes are simple and effective whilst others are more difficult. Every individual leads a different life and has different opportunities and limitations for change that are defined by wealth, ownership, public and personal infrastructure, governmental democracy, and living requirements. I don't want to guess at which specific actions are best for you. Hopefully, having read this far, you understand the necessary actions for making a difference and the impact they will have. If I had to summarise four broad-based routes of action, in order of importance, they would be:

Electrify Everything: The fastest and cheapest way to net-zero is through electrifying as much energy demand as possible – choose electric cars over gasoline, choose heat pumps over gas boilers, and choose electric cookers over gas or solid fuels. Power with low carbon electricity tariffs or home solar generation if possible. Force energy demand from fossil fuels towards electricity and force that electricity to come from net-zero sources.

Push for Net-Zero: Vote for credible net-zero aligned politicians, direct savings into sustainable funds, and choose low carbon consumer products – think sustainable wood, local sourcing, low carbon certification, and long-lasting durable goods. Signal to government that climate change and air pollution must be a top priority and to business that environmental damage is no longer an economic externality but damaging to their bottom line.

Reduce Meat and Dairy: Big meat eaters (Americas, Australia, Europe) need to halve their consumption of animal-reared meat and dairy to buy us more time and for agriculture to have a chance of reaching net-zero. Halve consumption and choose more poultry or fish over beef and lamb. Make it unnecessary and unprofitable to clear any more of the world's forests and improve your diet and health at the same time.

Reduce Unnecessary Consumption: Help to slow change. I won't tell you to consume less and change your lifestyle, I leave that to personal choice, but think about unnecessary consumption which brings you no benefit: idling cars, pointless journeys, equipment running for no good reason, poor insulation, waste, or inefficient products. Cutting these saves on emissions and saves money. There is little reason not to make these changes.

And remember, we are all climate change hypocrites because we live in a system powered by fossil fuels with a global population too large to take off-grid. But we have all the solutions necessary to replace fossil fuels with zero carbon supply and demand. Individuals can push the demand side as governments and industry switch the supply side. Neither should wait for the other to act, but both must push on together.

No Size Fits All – Selecting the Best Tools of Change

The world is a spectrum of personal, economic, and political freedoms which range from the authoritarian rule of dictators to the liberal expression of anarchy and everything in between. Most people reading this book will be doing so in what they consider to be a liberal capitalist democratic system where we as individuals make most economic decisions (rather than the state) and where free markets, private ownership, voting, individual rights, and freedom of speech are held to be of upmost importance. Most of the 195 countries in the world are considered liberal today but roll back to 50 years ago and that number was just 30. Most countries use a hybrid socio-economic system with a degree of both social and private ownership combined with elements of free markets and a centrally planned economy. Stock markets sit alongside state pensions, private wealth is backed up by welfare payments, and markets are partially regulated by governments.

	social ownership	private ownership
planned	socialist planned economy	command capitalism
market	market socialism	market capitalism

Regardless of the balance between liberalism and socialism our three agents of change remain the same: governments, corporates, and individuals. As Professors of Political Science and Energy, Michael Aklin and Johannes Urpelainen, write: "Environmentalists, clean technology entrepreneurs, and green parties can only succeed if political institutions give them access, public opinion is favourable, and the alliance of fossil fuel producers and heavy industry is vulnerable to political change."[56]

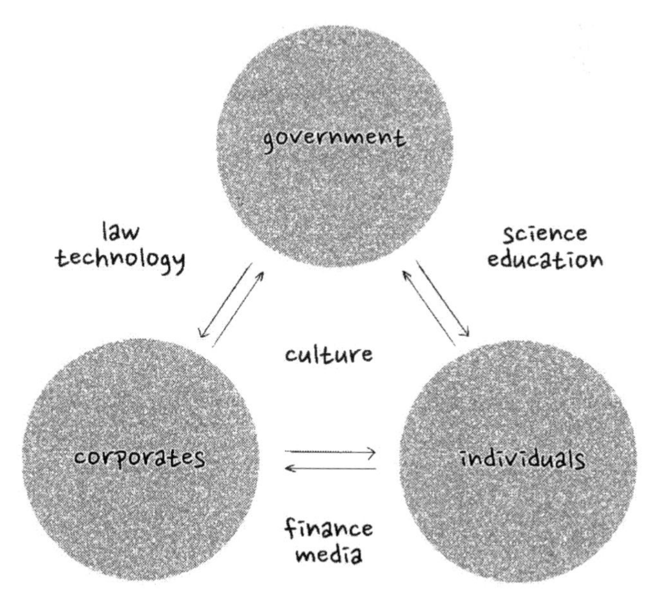

Through science, education, finance, media, law, and technology we can change culture and transform our energy system for the better. The tools of change span from *centrally planned* command and control regulation to *market based* consumer choice. Each tribe must choose the most effective tools on their own and, just as I couldn't dictate the best individual actions, so I can't dictate the best choice of socio-economic policy.

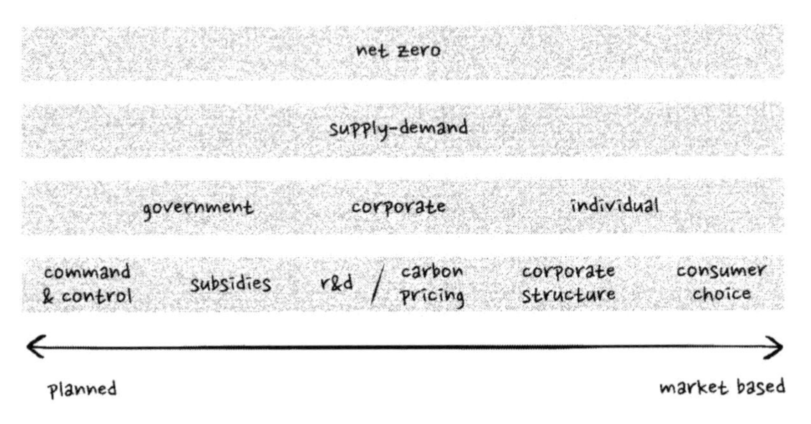

The net-zero transition we have outlined through this book is a rough cut, is far from perfect, and is just one of many options, but thinking through the necessary changes gives us a good idea of how we can use our tools of change to good effect. There is no silver bullet. Government, corporates, and individuals alone cannot transition to net-zero. It requires a combination of both systems change and cultural change. The policies, actions, and stories over the last sections illustrate that while we already have the necessary tools to expedite change, we must use them carefully.

Non-partisan cooperation towards net-zero ensures an effective transition even through political, social, or economic change. Individuals must recognise their role in changing energy demand; not just blaming government inaction or big business profiteering. Corporates should strive for net-zero as the best sustainable economic outcome; greenwashing should not be used as a marketing tool. Politics should be used to transition the energy system; climate change should not be used as an excuse to change politics. As Noam Chomsky states in his book *Climate Crisis and the Global Green New Deal*, "Dismantling capitalism is impossible within the time frame necessary for taking urgent [climate change] action, which requires a major... mobilization, if severe crisis is to be averted."[57]

Policy and action must target the required outcome of reaching zero emissions rather than getting lost in the detail: why the Montreal

Protocol worked, and corn ethanol did not. Changes should be made upstream, all-inclusive, and as efficient as possible, like a carbon tax. Targets must be clear, progressive, and long term, so businesses can plan through the economic cycle and we do not repeat the boom and bust of solar subsidies creating dysfunctional supply chains. Information gathering, monitoring, and enforcement, is essential to avoid another cash for ash scandal and, where possible, policies should have a degree of flexibility to adapt to market changes. Governments should try to avoid picking winners or setting prices and should focus on oncomes, leaving the detail to the efficiencies of markets so that we avoid another Solyndra but we don't miss the next Tesla. Policy should instead be used to create a level playing field for new technologies through research and development, reducing soft costs, and breaking monopolies, so that disruptive innovations such as lithium-ion batteries can find their way to the grid. Finally, all policy and actions must be well integrated. In the early part of the last decade Germany found out that a combination of heavily subsidised solar, rolling back nuclear, and an overly generous carbon allowance opened the door for new dirty coal power plants: well-intentioned measures can go unexpectedly wrong. Systems thinking is vital.

Transitioning to Net-Zero Through the Economic Cycle

The transition to net-zero must progress unhindered by economic change and supported by the tools at our disposal. Whether the economy is strong or weak, whether markets are bullish or bearish, the transition must continue.

Economic growth is driven by two fundamental forces. One is the steady increase of productivity with improving technology, organisational structures, and knowledge. The other is the boom and bust of debt cycles that we touched on in chapter five.[58] The tools deployed through the transition to net-zero should be structured to accommodate these elements.

Just like productivity improvements, carbon pricing and research and development should remain true to course through the transition. Steadily increasing carbon prices provide direction and rate of change. Research and development supports the full productivity improvements of net-zero. Setting a true course for each of these levers through the economic cycle ensures maximum benefits at optimum speed.

Bull markets accommodate demand side change. Consumer confidence is high, businesses are expanding, and credit is freely available. Individuals and industry can fund the higher upfront costs of electric vehicles, heat pumps, and industrial refits, and will benefit from savings in the future. These are productive investments which can help extend and manage the cycle. Regulation, subsidies, finance, and tax incentives should be used to bolster demand side change during the boom.

Bear markets accommodate supply side change. Inevitably, the productive use of capital across the economy becomes harder to find as a bear market approaches. Debt repayments start to outweigh additional income, growth slows, confidence drops, and credit dries up. At this point in the economic cycle, central banks lower interest rates on short term borrowing (by other banks), passing lower interest rates to customers. As interest rates approach zero, central banks will print money (Quantitative Easing) and use it to buy bonds or other financial assets. This pushes more money into the economy which reduces longer term interest rates (because there is more money available for those that want to borrow) and drives price inflation (because there is more money for the same quantity of goods and services). Lower interest rates and expectation of future price inflation should drive more borrowing and more spending. At this point, governments can raise cheap debt to bolster supply side change, directly investing in infrastructure and incentivising industry to follow suit. The rate at which grid networks, EV charging equipment, and wind and solar installations are built can be given a boost. Jobs can be created, restoring confidence, and bolstering the economy. Lower system costs for citizens in the future will further accelerate economic output and drive higher tax revenues for governments to repay their borrowing.

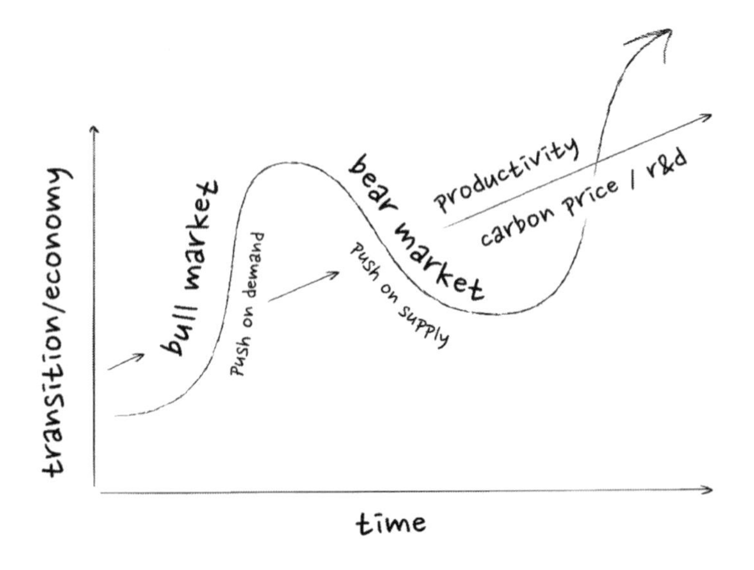

As I write this, we are deep into the second wave of the COVID-19 global pandemic and we are very much in the depths of a bear market. Stock markets declined 30% at their lows, and unemployment figures are approaching levels not seen since the great recession in the 1930s.[59] Oil futures plummeted to negative $42 per barrel in April 2020 as traders were paying to try and give away oil as storage space ran out.[60] Global air travel is not expected to recover for years (if ever) and many industries are scrambling for bail-out money.

Central banks around the world have cut interest rates to zero and pledged to print up over $5 trillion new money as governments embark on once-in-a-generation stimulus plans. So far, a total of $12 trillion has been pledged following the economic devastation of the coronavirus but as yet only $0.5 trillion is earmarked for net-zero projects.[61,62,63]

As Dr Fatih Birol, Executive Director of the International Energy Agency, puts it: "Governments can achieve both short-term economic gains and long-term benefits by making clean energy part of their stimulus plans".[64] He notes the support that clean energy gave to the recovery after the Credit Crunch in 2009, which transformed wind and solar economics, propped up the construction industry through efficiency stimulus, and enhanced the resilience of energy networks. The key to successful stimulus for a net-zero transition is to target opportunities where access to money or the upfront economics present a bottleneck, but the solutions are not technically demanding, have existing legal frameworks, and are socially accepted. Supporting rapid deployment, job stimulus, and economic benefits.

Quick supply side stimulus for rapidly deployable modular technology. Post 2009 solar, onshore wind, and smart grid were the success stories with existing subsidy systems and technology ready to slide down the experience curve. This time round, offshore wind, lithium-ion, and hydrogen can be added to the list. Governments should shore up the balance sheets of the relevant energy companies, roll out incentive structures, and mandate demand. Subsidies should follow the experience curve to avoid boom-bust economics this time around.

Lower demand side hurdles so individuals can electrify once the economy improves. Quick, easy, and scalable projects such as home insulation, EV charge points, and smart metering lower the hurdles to the adoption of heat pumps, electric cars, and demand side response once the economy and consumer spending improves. Employment might be directed into tree planting in order to boost the negative emissions offset for decades to come.

Selective investments into longer term, more complex engineering projects. Carbon capture projects post-2009 have made limited progress over the last ten years. Embarking on big, bespoke engineering projects is higher risk and slow. Investments should be carefully selected to focus on long term research and development goals and infrastructure: carbon capture in cement and fertilisers, third generation biofuels, and international grid connectivity.

Net-zero conditions for bailouts. Strategically important industries may need further bailing out. Financial support should come with net-zero conditions. Airlines, oil and gas and food services can be pushed to lower carbon intensity.

Once stimulus has boosted the economy and recovery gradually takes hold, central banks will need to sell back bonds and reduce their balance sheets, while governments will need to pay back debt. There will likely be growing need to address wealth inequality across the economy: quantitative easing may prove an effective monetary tool, but it tends disproportionally to inflate financial asset prices[xv] which are disproportionally held by the wealthy.[65]

Carbon pricing acts as a suitable tool for generating revenue for the government to support debt repayments or a carbon dividend to support redistribution of wealth. Implementing a carbon tax through an economic recovery should provide greater acceptance and provides a flexible income stream for governments to help manage the budget and the economy.

The net-zero transition opens a new and productive investment opportunity which can help to regenerate the economy. Governments can borrow to sustain growth in the near term with the knowledge that lower energy system costs will pay back the debts in the future. The required upfront investment creates jobs and stimulates economic activity and can be financed using historically low interest rates, and will payback through a cheaper energy system, greater economic output, and more tax revenues in the future.

---

xv The top 10% in the US have increased their share of total US wealth from 70% to 75% in the last 10 years following large QE since the credit crunch.

Expediting Change: Optimising the Speed of the Transition

Recognising the benefits of net-zero is key to enabling governments, corporates, and individuals to enact change. The hurdles to action are blown away. You don't believe in climate change or don't care about other countries? Well net-zero will be cheaper, and it reduces air pollution in your own. Think net-zero technology is too expensive? Well, total buying and running costs will end up 25% cheaper than existing systems. Don't want to risk change? Once the transition is underway, remaining bound to fossil fuels becomes uneconomic, unethical, and unimaginable.

The faster the world transitions to a net-zero economy the better – technology gets cheaper with scale, not time. Push hard on renewables, electrify everything, and support the development of storage and next generation technologies, using our tools of change. There is no one-size-fits-all, but simply recognising the benefits of net-zero creates non-partisan agreement on action.

Concerted action for those that can afford higher upfront cost today (for savings down the line) makes the transition affordable for all. Countries with average income levels in the highest bracket emit one third of carbon emissions. Scale net-zero in these regions and the experience curve makes net-zero the cheapest option for all. The rest of the world amplifies change by adopting cost competitive technologies along the way.

As an individual, the best way to support the transition is to electrify your energy use, vote or invest in net-zero, reduce meat and dairy consumption, and spread the message on the benefits. Change the debate. You are as big a part of the problem and solution as national governments and global corporates.

The time to act is now. Renewables are reaching parity with fossil fuel electricity already; consumer choice allows for far greater electrification and we are embarking upon once-in-a-generation fiscal stimulus and money printing operation. Net-zero will stimulate jobs, boost the economy, and will pay back down the line with a cheaper, safer, and more reliable energy systems eliminating nearly eight million premature deaths from air pollution and avoiding the worst impacts of climate change.

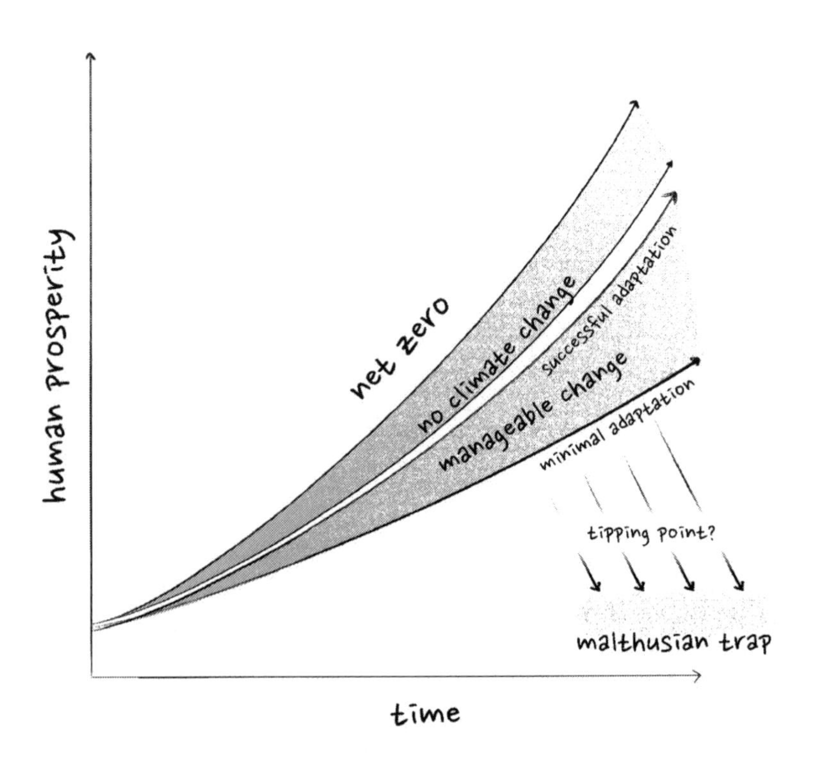
human prosperity
net zero
no climate change
successful adaptation
manageable change
minimal adaptation
tipping point?
malthusian trap
time

Chapter 16: Betting the Planet

The Future of Humanity

Paul Ehrlich and Julian Simon agreed upon their wager over the future of planet Earth in 1980, staking $1,000 on the change in a basket of metals prices over the decade. Ehrlich was convinced exponential population increases and Earth's finite resources would push prices higher; Simon was convinced that improving technology and market efficiencies would lower the value of most commodities. During the 1980s, the global population swelled by a record 800 million people and the world was thrown into not just one, but two energy crises, yet by 1990, the price of all five metals were substantially lower. Simon had won the bet and Ehrlich reluctantly mailed him a cheque.

Yet the arguments over the future of planet Earth rage on between neo-Malthusian views and cornucopian ideals. We have learnt how by turning the dials of doom or boom, we can paint two vastly different visions of the future using exactly the same science, economics, and technology. But as we continue to accumulate knowledge, the range of these dials is beginning to narrow: the likely warming, the discounted damages, and the rate of technological change, are becoming clear. I would argue we are now at a juncture where the possible outcomes are sufficiently narrow to show that a rapid transition to net-zero will create a win-win outcome. The information is not perfect, the details are not final, but the road is illuminated. Net-zero accelerates the cornucopian ideals of human progress whilst creating a sustainable society living within the planet's boundaries and fulfilling our moral obligations to nature.

Humans, acting as part of the evolutionary processes of life on Earth, have already irreversibly changed the planet. We cannot restore the ancient forests, return plundered resources, or un-breed species and why would we want to? On a planet that has gone through 4.5 billion years of change how would we decide where to reset? We must instead look forward. The cornucopian ideals of growth can continue, and just as coal and machinery replaced the burning of forests and slave labour in the nineteenth century, so net-zero solutions and artificial intelligence will replace fossil fuels and undesirable labour through the twenty-first. We can build a more inclusive world where each person on the planet can have the highest quality of life and, once the population stabilises and our physical needs are met, growth moves to the non-resource economy. Ecological economist and the founder of the steady state economics movement Herman Daly wrote in 1976, "Only two things are held constant [in a steady state economy] – the stock of human bodies, and the total stock or inventory of artifacts. Technology, information, wisdom, goodness, genetic characteristics, distribution of wealth and income, product mix, etc. are *not.*"[1]

The developments of language, mathematics, and record keeping have already moved humanity beyond the evolutionary limits of compiling information in DNA alone.[2] As the value of civilisation moves beyond simply satisfying our biological and physical needs, the exchange of ideas becomes our predominant income and the accumulation of knowledge our capital. The infinite nature of understanding may ultimately render the concepts of money and growth meaningless relics of the twentieth century. As futurist Ramez Naam puts it: "A wheel can only be used in one place and one time, but the design for a wheel can be shared with an infinite number of people, all of whom can benefit from it. Ideas aren't zero-sum. That means the world isn't zero-sum. One person or nation's gain doesn't have to be another's loss. By creating new ideas, we can enrich all of us on the planet, while impoverishing none."[3]

Net-zero creates a sustainable way to power civilisation and to provide sufficient resources to successfully thrive within our planetary limits. Net-zero lets us walk a softer path over the coming centuries whilst we master the iterative prisoners' dilemma and learn to co-operate as a single species and hopefully navigate the dangers of known-unknowns, those existential threats such as nuclear war, pandemics, or climate change.

It took 4.5 billion years for humans to evolve, but only another three million years to harness the energy of combustion. Beyond the next century, as we start to think in millennia, we must carefully evaluate our tumultuous relationship with energy. Can we reach a level of global cooperation to safely unlock vast quantities of atomic energy? And to what end do we use that power? If we choose to follow a path of compounding energy growth in the long-term then expanding our reach through the galaxy may become a necessity. Einstein warned, "the most powerful force in the Universe is compound interest" and with an annual growth rate of just 1% it would take less than 4,500 years to consume the Earth and less than 9,000 years to consume every atom in our galaxy to satisfy our thirst for energy.[i]

I started this book by asking you to think of our road to net-zero as a back-of-the-envelope solution to a Fermi problem and I want to finish with a famous Fermi paradox.

There are an estimated 400 billion stars and six billion Earth-like planets in our galaxy, of which many have already existed at least one billion years longer than Earth.[4,5] The Fermi paradox asks why have we found no signs of intelligent life?[ii] Or as Fermi put it, "Where is everybody?"[6] Why has no advanced alien civilisation already colonised the Earth and exploited our sun for their expanding energy needs?

Economist Robin Hansen questioned whether there is some 'Great Filter' which makes the evolution of intelligent life highly improbable or that has destroyed every technologically advanced civilisation before interstellar contact?[7] Has our natural world already overcome near impossible odds to blossom with life or are we headed for the same extinction event, some unknown-unknown, that all past civilisations have fallen foul of already? Or is the galaxy *already* teeming with life, an abundance of technologically advanced beings perfectly satisfied living within their celestial limits?

Net-zero may offer a solution for the coming centuries but, in the long term, do we come full circle, back to our visions of doom or boom?

---

i Final energy use today is 100 billion kWh which is the equivalent of 4,000 kg of mass if converted at 100% efficiency to useful energy. Compounding at 1% for 4,500 years this increases to an annual energy consumption of $1*10^{23}$ kg (2% of the mass in the Earth) and a cumulative $1*10^{25}$ kg used (nearly double the mass of Earth). Though it is hard to imagine what we would do with all this energy, as an example take space travel, it would require 700 years' worth of today's annual energy use to accelerate the international space station to 99% the speed of light yet this still only consumes 3 million kg of pure mass/energy ($3*10^{6}$ kg). But then I imagine a farmer living 10 thousand years ago at the beginnings of the agricultural revolution could not imagine what humanity would do with 100,000 times more energy today. [Relativistic mass calculation from: Lumen, "Relativistic Energy", [online], accessed 2020.]

ii Earth-like planets must be rocky and orbit a sun like star at the correct distance for water and atmosphere to form. The Milky Way is 13.5 billion years old and 100,000 light years across (it would take a radio signal 100,000 years to traverse).

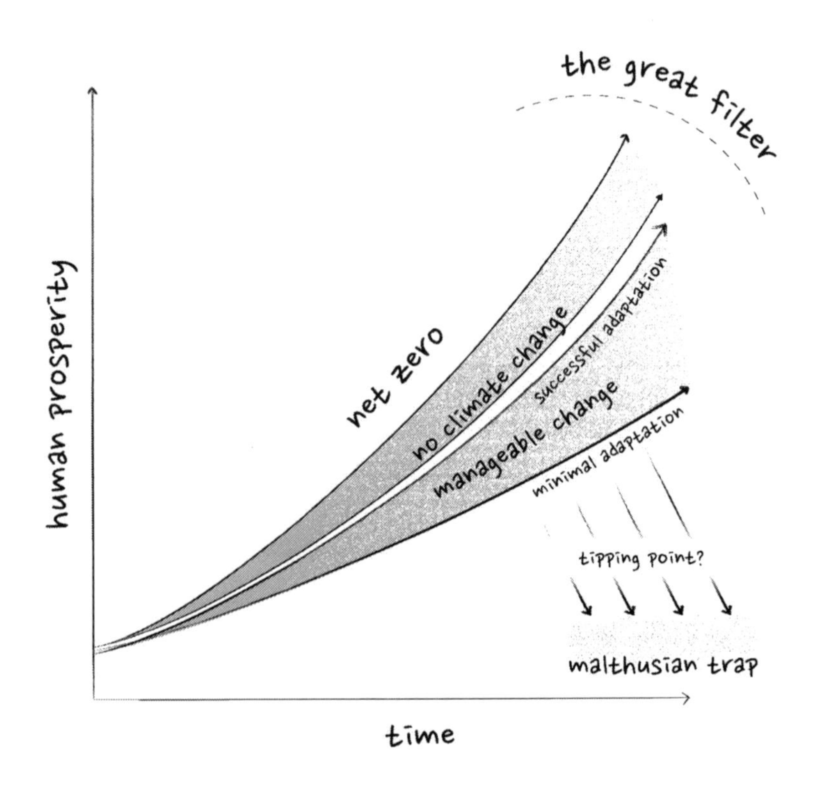
the great filter
human prosperity
net zero
no climate change
successful adaptation
manageable change
minimal adaptation
tipping point?
malthusian trap
time

Afterword: Hedgehogs and Foxes

Understanding the Climate Divide

Through the pages of this book, we have come across many characters with many different views on climate change. From Ehrlich to Simon, doomsters to boomsters, and a whole spectrum of views in between. Today, a quick scroll through the latest news, academic journals, or climate Twitter quickly tells you the divide on climate change is still very real.

To portray this divide as between climate denial and climate science is an oversimplification and one which no longer really rings true. The weight of objective scientific consensus is now too strong to construct a credible denial argument. Simply put, human emissions of CO_2 are warming the planet and creating climate change, but the subtleties and the magnitude of the Earth's response, the value we should place on the welfare of future generations, and the extent to which technologies can offset change, are still fertile battle grounds.

Maybe we could paint the division as a struggle between left and right and between liberty and authority. Aligning on one side are William Godwin's anarchic vision, Julian Simon's neo-liberal economics, big oil and the gaggle of Cold War physicists spreading doubt to halt the perceived threat of Communism. On the other side are Malthusian population controls, the Club of Rome's limit to growth, and the environmental movement calling for more regulation. But to paint the division in such simple terms is unhelpful and, I believe, untrue. The issue is more complex, with battles waged across the political spectrum over anything from nuclear power, carbon capture, and climate engineering, to the genetic modification of crops. For me, as somewhat of an outsider looking in, the division and inaction on climate change seems to run deeper than just politics and seems to stem from how we construct our innate view of reality and how society validates that version of the world.

In 1951, written in an obscure essay, Jewish philosopher Isaiah Berlin borrowed a line from Greek Poet Archilochus stating, "The fox knows many things, but the hedgehog knows one big thing".[1] Berlin used this as a metaphor for describing the two fundamental positions of the human psyche. Asserting that people are either foxes who draw on a wide variety of experiences, "pursue many ends" and can hold multiple, even contradictory, world views at the same time; or people are hedgehogs who view the world through a "single, universal, organising principle", one big idea.

Thirty years later, in the mid-1980s, Philip E. Tetlock, a social scientist at the University of Pennsylvania, began a decades long experiment to try and understand what makes individuals better or worse at forecasting the future. He first recruited hundreds of serious professionals from academia, think tanks, government, the International Monetary Fund, and the media. They were routinely given questions on just about any world topic, anything from where the stock market will trade in 12 months to the outcome of conflict in the Middle East. After 20 years of meticulously keeping score on the accuracy of those forecasts, Tetlock concluded that "the average expert was roughly as accurate as a dart-throwing chimpanzee".[2]

But within that average existed two sub-groups. One group whose forecasts were worse than random, they would *actually* lose to the dart-throwing chimp, and a second group who did marginally better than a random guess. Impressed by the research, and following failings during the Iraq war, the intelligence agencies in the US became interested in using the project to improve their forecasting ability. The Good Judgement Project was born, pitting experts in US intelligence agencies against thousands of ordinary volunteers from all walks of life. This time round the volunteers were ranked on the accuracy of their forecasts and those with a top 2% score elevated to 'superforecaster' status. The experts saved face with scores slightly better than merely a random guess, but it turned out the super-forecasters, made up of retired teachers, former ballerinas, or computer programmers, were able consistently to beat the US intelligence agencies at their own game. Year after year the same superforcasters generated better predictions despite having no formal training or access to privileged information and technical capabilities.

The difference between the good forecasters and the bad forecasters was not education, experience, expertise, PhDs, seniority, or international acclaim, but it was the way in which those individuals fundamentally thought about the world. It turned out the foxes nearly always beat the hedgehogs in predicting the future.

Super-forecasters can understand all sides of an argument, apply the wisdom of crowds, and aggregate lots of information, think in terms of probabilities, and avoid internal or external biases. They will always consider counter narratives, learn from mistakes, and adapt their predictions, as necessary. They are super foxes.

Hedgehogs, on the other hand, show a poor ability to forecast the future. They conceive only their view of the world, one big idea, and they shape the information or trends to fit their construct, dismissing contradictory lines of evidence.

Yet whilst foxes sit back and predict the future, it is hedgehogs that tend to try and shape the world. From dictators to aspirational politicians, tycoons, business magnates, stock picking experts, celebrities, life-coaches, and media pundits. Those who rise to fame and yield the most influence on the world around us tend to be hedgehogs with one simple idea or world vision which we can easily latch onto.

The debate on the future of humanity, sustainability of the planet, and climate change has been dominated by the singular views of hedgehogs. Malthus's one big idea of population control, William Godwin's idea of anarchy and utopia. As David Epstein points out in his book *Range*, Ehrlich and Simon may have started out as foxes, but they pushed each other to become hedgehogs with polar opposite views of doom and boom and the accompanying fame and influence.[3]

Isiah Berlin didn't consider either the fox or the hedgehog to be better or worse, just different. He attributed genius in the world to many hedgehogs but also warned of the dangers of a single unifying idea which he asserted could lead "to disastrous consequences in social, personal and political life – this is the price that may be paid for forms of genius which may well be profounder than any other. I may have more personal sympathy with foxes: I may think that they are more politically enlightened, tolerant and humane: but this does not imply they are otherwise more valuable."

Perhaps Freidrich Hayek put it more bluntly in his 1944 free-market, liberalist doctrine *The Road to Serfdom*, writing: "From the saintly and single-minded idealist to the fanatic is often but a step". He was voicing his distrust of socialist central planning, but one could argue the same single-minded, hedgehog vision of liberalism could be just as dangerous.[4]

The fox is ultimately better at forecasting how the world will change but accepts we cannot exactly predict the future. The hedgehog may be worse at getting the forecast right, but believes the future is knowable and wants to drive change. Should we try and act as global planner and drive towards the best possible future whilst running the risk of getting it badly wrong? Or should we play our cards as they come but perhaps settle for a lesser outcome? These competing views strike at the heart of the debate on climate change.

For me, personally, a decade spent as a scientist taught me how to better understand the world and the value of being a fox, but another ten years at an investment bank taught me how to solicit action and the benefits of being a hedgehog. I wrote this book to illuminate all sides of the debate, but I also presented a singular solution. British statistician George Box once stated: "essentially, all models are wrong, but some are useful" and at best my vision of net-zero is a rough-cut guide to many possible solutions. I hope this book conveys the importance of objectively evaluating evidence, remaining open to all ideas, and holding onto a flexible vision of the future. But I also hope, as the idea of net-zero becomes ubiquitous, it can act as an all-encompassing vision for a better world. Net-zero holds the complexity, subtleties, and flexibility demanded by a fox, yet wrapped in the simple messaging of a hedgehog.

May the foxes reading this become a little more hedgehog, and the hedgehogs a little more fox.

– THE END –

Recommended Resources

These are my recommendations of the most useful books, resources, blogs, and social media profiles on climate change and the road to net-zero.

Synopsis: Net-Zero – Empowering Action Towards a Sustainable Future

In the synopsis I ran through climate change to net-zero, start to finish, yet this is just one of many possible futures. I ask you to keep an open mind and empathise with other world views. A great book to get you thinking like a fox is *Superforecasting* by Philip E. Tetlock and Dan Gardner. I would also recommend *The Bet* by Paul Sabin who tells the story of Ehrlich and Simon and gives the reader a great insight into the psychology of the climate debate. And #climatetwitter is as good a place as any to start exploring the many views on climate science and solutions.

Chapter 1: Third Rock from the Sun – Human Prosperity and Planetary Pressures

The Human Planet by Mark Maslin and Simon L. Lewis provides a fantastic account of planetary history and the beginnings of the Anthropocene. Angus Maddison's *Contours of the World Economy* is about the best record of the history of wealth creation over the last 3,000 years you will find. NASA and NOAA websites host much of the climate data if you want to explore temperature, CO_2, and sea levels for yourself. *Factfulness* by Hans Rosling is great book for framing the world using data rather than preconceptions and bias, and the amazing website 'Our World in Data' lets you explore this theme to your heart's content. I would also recommend you follow the website creators Max Roser and Hannah Ritchie on Twitter.

Chapters 2 and 3: Carbon Dioxide and Climate – Earth's Geological Past to the Epoch of Humans / Forecasting Change – The Science of Warming

Spencer Weart's book *The Discovery of Global Warming* is an excellent history of climate change science and much of it can be found for free on his website. The two best books I have read on the science of global warming are Professor Andrew Dessler's *Introduction to Modern Climate Change* and *Global Warming Understanding the Forecast* by Ocean Chemist David Archer. The best book I have read on the modelling of climate is by the father of climate modelling, Syukuro Manabe, called *Beyond Global Warming*. The IPCC reports can be downloaded from the IPCC website; there are both synthesis reports (a few hundred pages of politically negotiated summary points) and full reports from all three working groups (thousands of pages on science, adaptation, and mitigation). I would wait until you learn the core concepts before you dive into the full reports as they are complicated and detailed. The Carbon Brief website hosted by Simon Evans has excellent resources and explainers on all aspects of climate change and sign up to his daily newsletter to keep informed on the latest happenings across the world. Vox also has fantastic 'explainers' on many technicalities of climate science. I would also recommend following climate scientists Stefan Rahmstorf, Kate Marvel, and Zeke Hausfather on Twitter for considered opinion on the latest talking points. You can also follow climatologist Glen Peters on Twitter for the latest thinking on possible emissions trajectories.

Chapters 4 and 5: Fearing Change – The Physical Impacts from a Warming Planet / Damage Assessment – Quantifying the Social and Economic Costs of Climate Change

I found the book *The Thinking Person's Guide to Climate Change* by the editor at the National Centre for Atmospheric Research, Robert Henson, provides a nicely balanced take on the physical effects of climate change and Tim Smedley's *Clearing the Air* provides a global take on air pollution. The 'Stern Report' can be downloaded from the UK government website, it is over 15 years old but is still one of the most comprehensive reports on the science, economics, and politics of climate change and written in (mostly) accessible language. *Climate Economics* by Richard Tol is a maths heavy but good resource which explains the economics of climate change and challenges many of Lord Stern's views.

Chapter 6: Opening for Planetary Caretaker – Building a Climate Change Strategy

The iea.org website hosts information such as the IEA Sankey Diagram which provides a great overview of final energy supply and demand. The BP statistical review report and excel file is free to download and provides detailed energy supply information. The Royal Society 'Geoengineering the climate' gives a good overview of climate engineering.

Chapters 7 and 8: Sustainable Energy – Supply Options with Net-Zero Emissions / Energy Storage and Distribution – Building a Safe, Reliable, and Cost Competitive Network

The late David Mackay (Regus Professor of Engineering at the University of Cambridge) wrote a book called *Sustainable Energy – Without the Hot Air* in 2009, a no-nonsense guide to the fundamental engineering, thermodynamics, and physical limits of decarbonising the economy. His book was the inspiration for much of the technology analysis in chapters 8 to 11. The IRENA (International Renewable Energy Agency) website hosts especially useful cost data and net-zero energy system reports. Jesse Jenkins is an assistant professor of engineering at Princeton and works on energy modelling; you can watch his succinct lectures on reaching net-zero on YouTube. Bloomberg New Energy Finance is a great resource for the cutting-edge clean technology developments and follow the original founder of NEF, Michael Liebreich, on Twitter for interesting and often controversial opinions on the latest energy goings on. The Energy Gang podcast is an excellent (though US centric) resource to keep up with net-zero news and one of the former hosts, Jigar Shah, founder of SunEdison and now Director of Loans at the DOE, provides great content and commentary on LinkedIn. David Roberts runs a newsletter called Volts with great insight into technical energy policy and technology. If you really want to dig into the levelised cost of energy sources and how to evaluate projects, then Ed Bodmer's website (edbodmer.com) is a fantastic resource.

Chapters 9 to 12: Re-engineering the Economy – Running on Zero Carbon Energy

The Ren 21 global status reports provide a detailed account of how the net-zero transition is progressing. Mark Z. Jacobson's *100% Clean, Renewable Energy and Storage for Everything* textbook is a great summary of the engineering and technology required to get to a fully renewable energy system. #energytwitter is a good follow for the latest papers and up to date information.

Chapter 13: The Net-Zero Economy – Socio-Economic Benefits

Citi bank's 'Energy Darwinism 2' report from 2014 is a little dated on the costs now but serves as a good template for thinking about the financial implications of a global energy transition, it's free to download and worth a read. Mark Z. Jacobson's '100% wind, water, and solar over 143 countries' scientific paper is another excellent, if still controversial to many, take on the costs and benefits of going zero carbon. Follow Mark Z. Jacobson on social media and you will get an insight into the battle between 100% renewables and electrify everything versus nuclear, biofuels, hydrogen, and carbon capture!

Chapter 14: A World Trying to Transition – Progress and Hurdles to Action

The UNFCC NDC staging website hosts the latest agreed emissions goals for each country in the world. Naomi Oreskes and Erik M. Conway's *Merchants of Doubt* is a must read to understand the actions of bad actors trying to delay climate action – though a few names are included who, in my opinion, acted in good faith, but have a genuinely different world view and it would be unfair to label them bad actors – but I will let you decide for yourself. The book *Climate Justice* by Christian Seidel and Dominic Roser gives a good account of the practical and moral issues of dealing with climate.

Chapter 15: Plotting a Good Course – The Politics of Change

The emissions gap report by the UN Environmental Programme tells you the latest thinking on likely temperature outcomes for the planet. Designing Climate Solutions: A Policy Guide for Low-Carbon *Energy* by Hal Harvey, Jeffrey Rissman, and Robbie Orvis provides a succinct set of policy options to expedite change.

About the Author

Dr Mathew Hampshire-Waugh has spent the last ten years working as an equity analyst at global investment bank, Credit Suisse. He resigned his role as director in 2019 to commit to writing on climate change and new energy technology full time.

During a decade as an investment banker, the author worked with the top executives of many multi-billion-dollar companies and built relationships with many of the world's largest investment managers. Mathew's work centred on forecasting technology trends, financial performance, and the intrinsic value of companies involved in markets including renewable energy, electric cars, battery technology, and biofuels. His role was to publish and pitch share price recommendations to the world's largest institutional investors, hedge funds, and private wealth managers.

Prior to his career in the banking industry, Hampshire-Waugh gained his doctorate in materials chemistry from University College London, where he worked on novel coatings and nanomaterials for use in energy saving glazing and solar panel design. During his doctorate Mathew registered a patent for an efficiency enhancing coating for solar modules, published numerous scientific papers, and engaged in public speaking, consultancy, and media outreach for the BBC, Teachers TV, and other outlets.

The Author's Road to Net-Zero

My personal decarbonisation journey started when I realised that after moving out of a small flat in London into a large hundred-year-old house in the suburbs, and having recently married and had two children, my carbon footprint had ballooned. The average annual CO_2 emissions per person in the UK is just over ten tonnes per year, if you include domestic emissions plus imported goods, international air travel, and shipping. My family of four were emitting well over 45 tonnes of CO_2 per year; something had to change.

Over the last few years, we have cut our emissions by more than half and we have done it without sacrificing quality of life. The biggest cuts came from replacing our old diesel car with an all-electric vehicle, adding 16 kW or 52 solar panels and battery storage to the house, replacing the old gas boiler for a heat pump in the swimming pool and adding smart zoned heating to the radiators. We eat vegetarian or vegan for half the week and have added simple switches or timers to lighting, fans, and pumps to reduce unnecessary running time. We have better insulated the house and we try to choose low carbon products and avoid unnecessary consumption. We now emit five tonnes per person per year, half the UK average emissions with no compromise on lifestyle.

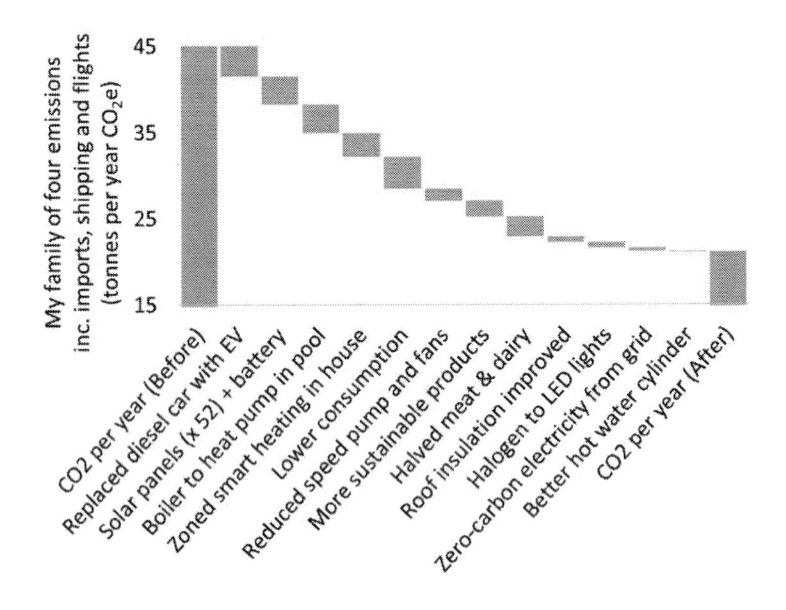

Annual CO_2 emissions (tonnes CO_2e) for my family of four including imports, shipping, and international flights. Breakdown: Before was 15t from the house, 13t transport, 10t food and 9t goods and services. After: 4.5t house, 4.5t transport, 8t food and 6t goods and services. Author's calculations based on modelling in chapter 7 to 12 and utilities information etc.

The changes to diet, insulation, timers, consumption, energy supplier, and lighting cost less than £250 in total and save £1,000 per year and 11 tonnes of annual CO_2 emissions.

The car, solar panels, battery, heat pump, and smart heating required an upfront investment of £40,000 above the fossil fuel alternative (or doing nothing), but we save £5,000 per year in gas, electricity, diesel, and water costs. The investment will pay for itself in eight years and will continue saving us money and 15 tonnes of annual emissions for a further 20 years.

The £24,000 investment for the solar panels and battery works out at

13-14 pence per kWh (without subsidy) compared to retail electricity in the UK at 14-15 pence per kWh. The solar panels provide enough energy to power the house, heat the pool and hot water, and charge the car for seven months of the year, even with an east/west facing roof and the gloomy British weather.[i] The 300-mile range battery in our electric car added £10k to the price but saves over £2,000 per year in fuel, a five-year payback.

Yes, we still emit more CO_2 than one third of the global population. Yes, we are still climate change hypocrites, but our journey continues. Of the remaining 21 tonnes of CO_2, we can shave off another five by replacing the boiler in the house with a heat pump. Beyond that, we will continue to vote for net-zero policies, invest in sustainable funds, and work towards and source zero carbon options from agriculture, air travel, and industrial goods.

I hope after reading this book you will join me on the road to net-zero.

Mathew Hampshire-Waugh

For more learning resources on *Climate Change and the road to Net-Zero* go to **net-zero.blog**

For our guides on 'decarbonising your life' and to explore profiles of companies and organisations engaged in climate change and the road net-zero go to **net-zero.life**

To enquire about workhops, speaking engagements or working with the author visit **net-zero.consulting**

---

i The UK receives almost the lowest amount of solar radiation on Earth. Using a solar tracking system, you can capture more solar energy in the arctic than in the UK. And yet home solar panels can still deliver electricity cheaper than the grid.

Index

A

B

C

D

K

L

M

N

O

P

Q

R

S

T

U

W

References

Preface: A Story Unwritten – Planet Earth and Humans

1 Johan Rockstrom et al., "A safe operating space for humanity", Nature, 461, 2009, 472-475.

Synopsis: Net-Zero – Empowering Action Towards a Sustainable Future

1 Thomas Robert Malthus, *An Essay on the Principle of Population*, J. Johnson (London), 1798.

2 Angus Maddisson, *Contours of the World Economy 1-2030AD*, Oxford University Press, 2007. ($1000 = $500 in 1990 international USD).

3 Max Roser, Hannah Ritchie and Esteban Ortiz-Ospina, Our World in Data, [online], accessed 2020.

4 United Nations, [online], un.org, accessed 2020.

5 Richard E. Zeebe et al. "Anthropogenic carbon release rate unprecedented during the past 66 million years", Nature Geoscience, 9, 2016, 325-329.

6 Munich Re NatCatSERVICE, "Data on natural disasters since 1980", [online], munichre.com, 2020, accessed 2020.

7 WHO, "Air Pollution", World Health Organisation, [online], who.int, 2020, accessed 2020.

8 The Royal Statistical Society, "George Box: a model statistician", Sept 2010.

9 BP, Statistical Review of World Energy 68th edition, 2019.

10 Climate Emergency Declaration, [online], climateemergencydeclaration.org, 7 Feb 2021, accessed 2021.

11 Shuja Asrar, 'Can you guess the age of a leader based on average age of country's population', The Times of India, Mar 5, 2021.

12 Nick Winder, "Integrative Research as Appreciative System", Systems Research and Behavioural Science, 22, 2005, 299-309.

13 C. S. Holling, "Understanding the Complexity of Economic, Ecological, and Social Systems", Ecosystems, 4, 2001, 390-405.

14 Kate Raworth, *Doughnut Economics*, Random House Business, 2017.

15 Mariana Mazzucato, *The Value of Everything*, Penguin, 2018, p76.

Chapter 1: Third Rock from the Sun – Human Prosperity and Planetary Pressures

1 Becky Oskin, "How the Earth formed", livescience.com, [online], June 2014, accessed 2020.

2 Simon L. Lewis and Mark A. Maslin, *The Human Planet*, Pelican Books, 2018.

3 Michael Marshall, "Timeline: The Evolution of Life" in New Scientist, 2009.

4 Rick Gore, "Rise of Mammals" in National Geographic, [online], accessed 2020.

5 Ross J. Salawitch et al., "Paris Climate Agreement: Beacon of Hope", Springer Climate, (2017), 3.

6 Angus Maddisson, *Contours of the World Economy 1-2030AD*, Oxford University Press, 2007. ($1000 = $500 in 1990 international USD)

7 Max Roser, "Economic Growth" published online at OurWorldInData.org., [Online Resource], published 2020, accessed 2020.

8 BBC, "Do we really live longer than our ancestors?", October 3rd, 2018, [online], accessed 2020.

9 Caleb E. Finch, "Evolution of the human lifespan and diseases of aging", PNAS, 107 (2010).

10 Thomas Robert Malthus, *An Essay on the Principle of Population*, J. Johnson (London), 1798.

11 Max Roser, Hannah Ritchie and Esteban Ortiz-Ospina, "World Population Growth", Our World in Data, [online], 2013, accessed 2020.

12 Lawrence Davis, "Chapter 83: Efficiency of the Human Body" in" Body Physics: Motion to Metabolism", Press Books, [online], accessed June 2020.

13 Angus Maddisson, *Contours of the World Economy 1-2030AD*, Oxford University Press, 2007, 13-60.

14 Nicolaus Copernicus, On the Revolutions of the Heavenly Spheres, Great Books of the Western World, 16, translated by Charles Glenn Wallis, 1952, 497–838.

15 Vaclav Smil, *Energy*, One World Press, 2017.

16 History.com Editors, Industrial Revolution, History, [online], A&E Television Networks, Sept 2019, accessed 2020.

17 The Editors of Encyclopaedia Britannica, "Edwin Drake", britannica.com, [online], Nov 2020, accessed 2020.

18 The Editors of Encyclopaedia Britannica, "Sir Charles Algernon Parsons", britannica.com, [online], Jun 2020, accessed 2020.

19 Sir Humphry Davy, *Elements of Agricultural Chemistry*, Richard Griffin and Company, 1807.

20 Gregory T. Cushman, "Guano and the opening of the Pacific world: a global ecological history", Cambridge University Press, 2013.

21 Erisman, Jan Willem; Sutton, Mark A.; Galloway, James; Klimont, Zbigniew; Winiwarter, Wilfried (28 September 2008). "How a century of ammonia synthesis changed the world", Nature Geoscience, 1 (10): 636–639.

22 Michael Aaron Dennis, "Jack Kilby", britannica.com, [online], Nov 2020, accessed 2020.

23 Eugene A. Avallone et. al, (ed), *Marks' Standard Handbook for Mechanical Engineers 11th Edition*, Mc-Graw Hill, New York (2007), 9-4.

24 Angus Maddisson, *Contours of the World Economy 1-2030AD*, Oxford University Press, 2007, 13-60.

25 Bank of England, "Gross Domestic Product Real-Time Database" in Statistics, [online], accessed 2020.

26 Hannah Ritchie and Max Roser, "Energy", [online], OurWorldInData.org, 2014, accessed 2020.

27 E. Cook, "The Flow of Energy in an Industrial Society" in Scientific American, 1971, 135.

28 Vaclav Smil, *Energy Transitions: Global and National Perspectives 2nd edition*, ABC-CLIO, (2017).

29 Sally Shuttleworth and Berris Charlney, "Science periodicals in the nineteenth and twenty-first centuries", Notes and records of the Royal Society of London, 70, 2016, 297-304.

30 Max Roser, "Employment in Agriculture", OurWorldInData.org, [online], 2013, accessed 2020.

31 Max Roser, Esteban Ortiz-Ospina, and Hannah Ritchie, "Life Expectancy", OurWorldInData.org, [online], 2020, accessed 2020.

32 Max Roser, Hannah Ritchie, and Bernadeta Dadonaite, "Child and Infant Mortality", OurWorldInData.org, [online], 2020, accessed 2020.

33 Max Roser and Esteban Ortiz-Ospina, "Income Inequality", OurWorldInData.org, [online], 2013, accessed 2020.

34 Max Roser, "Employment in Agriculture", OurWorldInData.org, [online], 2013, accessed 2020.

35 Max Roser and Esteban Ortiz-Ospina, "Global Education", OurWorldInData.org, [online], 2016, accessed 2020.

36 Max Roser, Hannah Ritchie, and Bernadeta Dadonaite, "Child and Infant Mortality", OurWorldInData.org, [online], 2020, accessed 2020.

37 Hans Rosling, Ola Rosling and Anna Rosling Ronnulund, *Factfulness*, Sceptre, 2018.

38 Gapminder.org, "Dollar Street", [online], accessed June 2020.

39 Tomas Hellebrant and Paolo Mauro, "The Future of Worldwide Income Distribution", Working Paper 15-7, PIIE, April 2015.

40 Hannah Ritchie and Max Roser, "Land Use", OurWorldInData.org, [online], 2020, accessed 2020.

41 Fridolin Krausmann et al., "Global HANPP trends", PNAS, 110 (2013), 10324-10329.

42 M.Shahbandeh, "Global number of pigs 2012-2020", Statistica.com, [online], 2020, accessed April 2020.

43 M.Shahbandeh, "Cattle population worldwide 2012-2020", Statistica.com, [online], 2020, accessed April 2020.

44 M.Shahbandeh, "Global number of chickens 1990-2018", Statistica.com, [online], 2020, accessed April 2020.

45 Karl Marx, *Das Kapital*, 1867.

46 Paul R Ehrlich, *The Population Bomb*, Ballantine Books, 1971.

47 Paul R. Ehrlich and Anne H. Ehrlich, *The Population Explosion*, Pocket Books, 1991.

48 Donella H. Meadows, *The Limits to Growth*, Protomac Associates, 1972.

49 Carey W. King, *The Economic Superorganism*, Springer, 2021.

50 Donella H. Meadows, *Limits to Growth The 30-Year Update 3rd Edition*, Chelsea Green Publishing, 2004.

51 US Government, The Global 2000 Report, Ninety-Sixth Congress Second Session, September 4, 1980.

52 Julian Simon and Herman Khan, *The Resourceful Earth: A Response to Global 2000 First Edition*, Wiley-Blackwell, 1984.

53 Paul Sabin, *The Bet*, Yale University Press, 2013.

54 John Tierney, "Betting on the Planet" in The New York Times, 1990.

55 Tim Jackson and Peter A. Victor, "Unraveling the claims for (and against) green growth", Science, 6468 (2019), 950-951.

56 Herman Kahn, Leon Martel, and William Brown, *The next 200 years: A scenario for America and the World*, Morrow, 1976.

57 Max Roser and Hannah Ritchie, "Food Supply", OurWorldInData.org, [online], 2013, accessed 2020.

58 Stephen Broadberry et al., *British Economic Growth 1270-1870*, Cambridge University Press, 2015, 80-129.

59 J.M.K.C. Donev et al., "Hubbert's peak", energyeducation.ca, [online], 2015, accessed 2020.

60 BP, Statistical Review of World Energy 68th edition, 2019.

61 World Energy Council, World Energy Resources, 2016.

62 NOAA, "ESRL Data", Scripps Institution of Oceanography, 2020.

63 IPCC AR5, "Figure SPM.2", Climate Change 2014: Synthesis Report, IPCC, 2014, 5.

64 Jonathan Safran Foer, *We are the Weather*, Penguin, 2020.

65 Gerald Mayr et al., "A Paleocene pengion from New Zealand substantiates multiple origins of gigantism in fossil Sphenisciformes", Nature Communications 8, 2017.

Chapter 2: Carbon Dioxide and Climate – Earth's Geological Past to the Epoch of Humans

1 NOAA, "Happy 200th birthday to Eunice Foote, hidden climate science pioneer", climate.gov, [online], 2019, accessed 2020.

2 Eunice Newton Foote, "On the Heat in the Sun's Rays", American Association for the Advancement of Science, Aug 23, 1856.

3 Raymond P. Sorenson, "Eunice Foote's Pioneering Research on CO_2 and Climate Warming", AAPG, 2011.

4 Thomas R. Anderson, "CO_2, the greenhouse effect and global warming: from the pioneering work of Arrhenius and Callendar to today's Earth System Models", Endevour, 3 (2016), 178-187.

5 Svante Arrhenius, "On the Influence of Carbonic Acid in the Air upon the Temperature of the Ground", Philosophical Magazine and Journal of Science, 1896.

6 E. O. Hulburt, "The temperature of the Lower Atmosphere of the Earth", Phys. Rev., 38, 1931, p1876.

7 Guy S. Callendar, "The Artificial Production of Carbon Dioxide and its Influence on Temperature", Royal Meteorological Society, 275 (1938).

8 Spencer Weart, *The Discovery of Global Warming Second Edition*, Harvard University Press, 2008.

9 Jay Madigan, "The NASA Earth's Energy Budget Poster", nasa.gov, [online], 2014, accessed 2020.

10 Andrew Dessler, ***Introduction to Modern Climate Change***, Cambridge University Press, 2016.

11 UK Government, "Background on the Greenhouse Effect", National Archives, [online], accessed 2020.

12 Andrew Dessler, "How the greenhouse effect works", [online], YouTube.com, 2013, accessed 2020.

13 American Chemical Society, "ACS Climate Science Toolkit", [online], acs.org, accessed 2020.

14 NASA Earth Observatory, "The Carbon Cycle", [online], nasa.gov, 2011, accessed 2020.

15 John Mason, "Understanding the long term carbon cycle: weathering of rocks", [online], skepticalscience.com, 2013, accessed 2020.

16 Douwe et al.,"Plate Tectonic controls on atmospheric CO_2 levels since the Triassic", PNAS, 2014, 4380-4385.

17 Andrew Dessler, ***Introduction to Modern Climate Change***, Cambridge University Press, 2016, 72.

18 P. Ciais et al., "Carbon and Other Biogeochemical Cycles", Climate Change 2013: The Physical Science Basis. Working Group I to the Fifth Assessment Report of the IPCC, 2013, 498.

19 James Lovelock, *The Revenge of Gaia*, Penguin, 2007.

20 Robert Henson, *The Thinking Person's Guide to Climate Change*, 2nd edition, AMS books, 2019.

21 NASA Earth Observatory, "El Nino", [online], nasa.gov, accessed 2020.

22 Simon L. Lewis and Mark A. Maslin, *The Human Planet*, Pelican Books, 2018.

23 Michael Greshko, "What are mass extinctions, and what causes them?", [online], nationalgeographic.com, 2019, accessed 2020.

24 J. D. Hays, John Imbrie, N. J. Shackleton, "Variations in the Earth's Orbit: Pacemaker of the Ice Ages", Science, 4270 (1976), 1121-1132.

25 J. Jouzel et al., "EPICA Dome C Ice Core 800KYr Deuterium Data and Temperature Estimates", IGBP PAGES/World Data Center for Paleoclimatology, NOAA, 2007.

26 Jonathan Safran Foer, *We are the Weather*, Penguin, 2020.

27 Paul Crutzen, "Geology of Mankind", *Nature*, 415, 23(2002).

28 Tom Boden and Bob Andres, "Global CO_2 emissions from fossil fuel burning, cement manufacture, and gas flaring 1751-2014", Carbon Dioxide Information Analysis Center, Oak Ridge Laboratory, 2017.

29 Holli Riebeek, "The Carbon Cycle", [online], NASA Earth Observatory, nasa.gov, 2011, accessed 2020.

30 NOAA, Climate Education Resources, Carbon Cycle, [online], accessed 2020.

31 J. Jouzel et al. "EPICA Dome C Ice Core 800KYr Deuterium Data and Temperature Estimates", IGBP PAGES/World Data Center for Paleoclimatology, NOAA, 2007.

32 D. M. Etheridge et al., "Law Dome Atmospheric CO_2 Data", IGBP PAGES/World Data Center for Paleoclimatology, NOAA, 2001.

33 NASA Goddard Institute for Space Studies, "Land-Ocean Temperature Index", [online], nasa.gov, 2020, accessed 2020.

34 NOAA, "ESRL Data", Scripps Institution of Oceanography, 2020.

35 Richard E. Zeebe et al. "Anthropogenic carbon release rate unprecedented during the past 66 million years", Nature Geoscience, 9, 2016, 325-329.

36 Philip Kokic etal., "A probabilistic analysis of human influence on recent record global mean temperature changes", Climate Risk Management, 3, 2014, 1-12.

37 YaleEnvironment360, "1 Billion Animals Estimated Dead in Australian Wildfires", [online], e360.yale.edu 2020, accessed 2020.

38 Jeff Masters, "Venice Has its worst flood in 53 years", [online], scientificamerican.com, 2019, accessed 2020.

39 IPCC AR5, Climate Change 2014: Synthesis Report, IPCC, 2014.

40 Munich Re NatCatSERVICE, "Data on natural disasters since 1980", [online], munichre.com, 2020, accessed 2020.

41 Joseph Romm, *Climate Change What Everyone Needs to Know*, Second Edition, Oxford University Press, 2018.

42 Simon L. Lewis and Mark A. Maslin, *The Human Planet*, Pelican, 2018, p8.

Chapter 3: Forecasting Change – The Science of Warming

1 Svante Arrhenius, "On the Influence of Carbonic Acid in the Air upon the Temperature of the Ground", Philosophical Magazine and Journal of Science, 1896.

2 Spencer Weart, *The Discovery of Global Warming Second Edition*, Harvard University Press, 2008.

3 Guy S. Callendar, "The Artificial Production of Carbon Dioxide and Its Influence on Temperature", Royal Meteorological Society, 275 (1938).

4 IPCC AR5, Climate Change 2014: Synthesis Report, IPCC, 2014, 151.

5 Syukuro Manabe and Anthony J. Broccoli, *Beyond Global Warming*, Princeton University Press, 2020, p138.

6 David Archer, "Global Warming: Understanding the forecasts", Wiley-Blackwell, 2009.

7 Roz Pidcock, "The most influencial climate change papers of all time", Carbon Brief Poll, [online], carbonbreif.org, 2015, accessed 2020.

8 Syukuro Manabe and Richard T Wetherald, "Thermal Equilibrium of the Atmosphere with a Given Distribution of Relative Humidity", Journal of the Atmospheric Sciences, 1967.

9 Andrew Gettelman and Richard B. Rood, *Demystifying Climate Models*, Springer, Berlin, 2016.

10 Robert Henson, *The Thinking Person's Guide to Climate Change*, 2nd edition 2019, AMS books

11 Andrew Dessler, *Introduction to Modern Climate Change*, Cambridge University Press, 2016.

12 IPCC, "Climate Change 2013: The Physical Science Basis. Contribution of Working Group I to the Fifth Assessment Report of the IPCC, Cambridge University Press, 2013, 683.

13 Stefan Rahmstorf, "Anthropogenic climate change: revisiting the facts", chapter three in *Global Warming: Looking Beyond Kyoto*, Brookings Institution Press, 2008, 34-54.

14 Kristina Pistone et al, "Observational determination of Albedo decreased caused by vanishing arctic sea", PNAS, 2014, 3322-3326.

15 M. R. Raupach et al., "The declining uptake rate of atmospheric CO_2 by land and ocean sinks", BioGeosciences, 13 (2013), 3453-3475.

16 Jia Yang et al., "Spatial and temporal patterns of global burned area in response to anthropogenic and environmental factors", Journal of Geophysical Research, 3 (2014).

17 Jolly, W., Cochrane, M., Freeborn, P. et al., "Climate-induced variations in global wildfire danger from 1979 to 2013", Nat Commun 6, 7537 (2015).

18 Holli Riebeek, "The Carbon Cycle", [online], NASA Earth Observatory, nasa.gov, 2011, accessed 2020.

19 Takeshi Lse et al., "High Sensitivity of peat decomposition to climate change through water table feedback", Nature Geoscience, 2008.

20 Edward Schuur and Benjamin Abbott, "High risk of permafrost thaw", Nature, 480 (2011), 32-33.

21 Kevin Schaefer et al., "Amount and timing of permafrost carbon release in response to climate warming", Tellus B, 2 (2011).

22 Robert McSweeney, "Nine tipping points that could be triggered by Climate Change", [online], carbonbrief.org, 2020, accessed 2020.

23 Gregory Cooper et al. "Regime shifts occur disproportionately faster in larger ecosystems", Nature Communications, 11 (2020).

24 Nature editorial, "Ocean circulations is changing, and we need to know why", nature 556, 149 (2018).

25 IPCC, "Climate Change 2013: The Physical Science Basis. Contribution of Working Group I to the Fifth Assessment Report of the IPCC", Cambridge University Press, 2013, 683.

26 Gunnar Myhre et al., "New estimates of radiative forcing due to well mixed greenhouse gases", Advanced Earth and Space Science, 14 (1998).

27 IPCC AR5, Climate Change 2014: Synthesis Report, IPCC, 2014, 155.

28 S. C. Sherwood et al., "An assessment of Earth's climate sensitivity using multiple lines of evidence", Reviews of Geophysics, 4 (2020).

29 IPCC 3rd Assesment Report, "Climate Change 2001: The Scientific Basis. Contribution of Working Group I to the Third Assessment Report of the IPCC", Cambridge University Press, 881.

30 American Chemical Society, "The Keeling Curve: Carbon Dioxide Measurements at Mauna Loa", National Historic Chemical Landmarks, [online], acs.org, 2015, accessed 2020.

31 United Nations, "World Population Prospects", [online], popuilation.un.org, 2019, accessed 2020.

32 OECD, "The economic consequences of Climate Change", OECD Publishing, Paris, 2015.

33 Keywan Riahi et al., "The Shared Socioeconomic Pathways and their energy, land use, and greenhouse gas emissions implications: An overview", Global Environmental Change, 42, 2017, 153-168.

34 Glen Peters, "A critical look at baseline climate scenarios", CICERO, Feb 2021.

35 Bill Mckibben, *Eaarth*, Macmillan USA, 2011.

36 James H. Butler and Stephen A. Montzka, "Annual Greenhouse Gas Index (AGGI)", noaa.gov, [online], 2020, accessed 2020.

37 Met Office, "What is 'climate sensitivity'?", [online], metoffice.gov, accessed 2020.

38 Masakazu Yoshimori et al., "A review of progress towards understanding the transient global mean surface temperature response to radiative peturbation", Progress in Earth and Planetary Science, 3, 21 (2016).

39 NASA Goddard Institute for Space Studies, "Land-Ocean Temperature Index GISTEMP", [online], nasa.gov, 2020, accessed 2020.

40 Church and White, "Sea-level rise from the late 19th to the early 21st Century", Surveys in Geophysics, 2011.

41 Etheridge et al., "Law Dome Atmospheric CO_2 Data", IGBP PAGES/World Data Center for Paleoclimatology NOAA, 2001.

42 NOAA, "ESRL Data", Scripps Institution of Oceanography, 2020.

43 Keywan Riahi et al., "RCP 8.5-A scenario of comparatively high greenhouse gas emissions", Climatic Change, 109(33), 2011.

44 Iain Colin Prentice et al., "Biosphere Feedbacks and Climate Change", Grantham Institute Briefing Paper 12, 2015.

45 David A. McKay, "Fact-Check: Will 2C of Global Warming trigger rapid runaway feedbacks?", [online], climatetippingpoints.info, 2019, accessed 2020.

46 Will Steffan et al., "Trajectories of the Earth System in the Anthropocene", PNAS, 115 (2018), 8252-8259.

47 J. S. Sawyer, "Man-made Carbon Dioxide and the 'greenhouse' Effect", Nature, Vol 239 (1972).

48 Wallace S. Broecker, "Climatic change: are we on the brink of a pronounced global warming?", Science 189 4201 (1975): 460-463.

49 James Hansen et al., "Climate Impact of Increasing Atmospheric Carbon Dioxide", Science, 213 (1981), 957-966.

50 BP, Statistical Review of World Energy 68th edition, 2019.

51 World Energy Council, World Energy Resources, 2016.

Chapter 4: Fearing Change – The Physical Impacts from a Warming Planet

1 Alan Buis, "A Degree of Concern: Why Global Temperatures Matter", [online], climate.nasa.gov, 2019, accessed 2020.

2 Daniel Kahneman and Amos Tversky, "Prospect Theory: An Analysis of Decision under Risk", Econometrica, Vol. 47, No. 2 (1979), 263-291.

3 Anders Levermann et al., "The multimillennial sea-level commitment of global warming", PNAS, 110 (34), 2013, 13745-13750.

4 IPCC, "Climate Change 2013: The Physical Science Basis. Contribution of Working Group I to the Fifth Assessment Report of the IPCC", Chapter 13: Sea Level Change, Cambridge University Press, 2013.

5 World Bank, "Population living in areas where elevation is below 5 meters", [online], data.worldbank.org, 2013, accessed 2020.

6 World Ocean Review, "A report on the World's Oceans, A Battle for the Coast", Chapter 3, [online], world oceanreview.com, 2010, accessed 2020.

7 Gordon McGranahan et al., "The rising tide: assessing the risks of climate change and human settlements in low elevation coastal zones", Environment and Urbanisation, 19 (1), 2007, 17-37.

8 Joel E. Cohen and Christopher Small, "Hypsographic demography: the distribution of human population by altitude", PNAS, 95(24), 1998, 14009–14014.

9 International Displacement Monitoring Centre, "Assessing the impacts of climate change on flood displacement risk", IDMC, 2019.

10 Joseph Romm, *Climate Change What Everyone Needs to Know*, Second Edition, Oxford University Press, 2018.

11 Met Office, "Tropical Cyclone Facts", [online], metoffice.gov, accessed 2020.

12 WMO, "Statement on Tropical Cylones and Climate Change", WMO International Workshop on Tropical Cyclones, 2006.

13 K. Emanuel, "Increasing destructiveness of tropical cyclones over the past 30 years", Nature 436 (2005), 686–688.

14 NOAA, "Temperature and Precipitation Maps", [online], ncdc.noaa.gov, 2020, accessed 2020.

15 Munich Re NatCatSERVICE, "Data on natural disasters since 1980", [online], munichre.com, 2020, accessed 2020.

16 Wikipedia contributors, "List of natural disasters by death toll", Wikipedia, The Free Encyclopaedia, 2020, accessed 2020.

17 Cyclone Nargis, Storm Development and Background, Coolgeography.co.uk, [online], accessed 2020.

18 Moving Ideas, "City Information: Death Valley CA City Data", [online], movingideas.org, accessed 2020.

19 BBC, "Highest Temperature on Earth as Death Valley US hits 54.4C", [online], bbc.co.uk, 17 Aug 2020, accessed 2020.

20 World Bank, "Climate Change Knowledge Portal: Historical Data", Country Temperature CRU, [online], datacatalog.worldbank.org, 2011, accessed 2020.

21 International Organisation of Migration, "Extreme Heat and Migration", The UN Migration Agency, 2017.

22 James Hansen, Makiko Sato, and Reto Ruedy, "Perception of climate change", PNAS, 109 (2012), 24125-2423.

23 World Bank, "Confronting the New Climate Normal", Turn down the Heat Series, Written by the Potsdam Institute for Climate Impact Research and Climate Analytics, 2012.

24 N. Andela, "A human-driven decline in global burned area", Science, 356, 6345 (2017), 1356-1362.

25 Daisy Dunne, "How climate change is affecting wildfires around the world", [online], carbonbrief.org, 2020, accessed 2020.

26 C. Price, "Thunderstorms, Lightning and Climate Change", *Lightning: Principles, Instruments and Applications*, Springer, 2009, 521-535.

27 W. Matt Jolly et al., "Climate-induced variations in global wildfire danger from 1979 to 2013", Nature Communications 6 (2015), 7537.

28 Munich Re NatCatSERVICE, "Data on natural disasters since 1980", [online], munichre.com, 2020, accessed 2020.

29 University of Cambridge, "New Approaches to help business tackle climate change", [online], cam.ac.uk, 2020, accessed 2020.

30 Matthew E. Falagas et al. "Seasonality of mortality: the September phenomenon in Mediterranean countries", CMAJ, 181 (8), 2019, 484-486.

31 P. G. Dixon, et al., "Heat Mortality Versus Cold Mortality: A Study of Conflicting Databases in the United States", Bull. Amer. Meteor. Soc., 86, 2005, 937–944.

32 Simo Näyhä, "Environmental temperature and mortality", International Journal of Circumpolar Health, 64 (5), 2005, 451-458.

33 IPCC, "Climate Change 2014: Impacts, Adaptation, and Vulnerability. Part A: Global and Sectoral Aspects", Contribution of Working Group II to the Fifth Assessment Report of the IPCC, Cambridge University Press, 2014, 721.

34 Tim Smedley, *Clearing the Air: The Beginning and End of Air Pollution*, Bloomsbury, 2019.

35 WHO, "Air Pollution", World Health Organisation, [online], who.int, 2020, accessed 2020.

36 Anil Markandya and Paul Wilkington, Electricity Generation and Health, The Lancet, 2007, Vol 370, Issue 9591

37 Aaron J. Cohen et al., "Estimates of 25 year trends of the global burden of disease attributable to ambient air pollution", The Lancet, 389 (10082), 2017, 1907-1918.

38 Karn Vohra et al., "Global mortality from outdoor fine particle pollution generated by fossil fuel combustion: Results from GEOS-Chem", Environmental Research, 2021, 110754.

39 Hannah Ritchie and Max Roser, "Air Pollution", [online], OurWorldInData.org, 2020, accessed 2020.

40 WHO, "More than 90% of the World's children breathe toxic air every day", [online], World Health Organisation, 29 October 2018, accessed 2020.

41 Michael Greenstone and Claire Qing Fan, "Introducing the Air Quality Life Index", Energy Policy Institute at the University of Chicago, 2018.

42 Jos Lelieveld et al., "Loss of life expectancy from air pollution compared to other risk factors: a worldwide perspective", Cardiovascular Research, 116 11), 2020, 1910-1917.

43 ERG Kings College London, "London Air", [online], londonair.org, accessed 2020.

44 WAQI, "World's Air Pollution: Real-time Air Quality Index", [online], waqi.info, accessed 2020.

45 OECD, "The economic consequences of outdoor air pollution", 2016.

46 De Yun Wang et al., "Quality of Life and Economic Burden of Respiratory Disease in Asia-Pacific-Asia-Pacific Burden of Respiratory Diseases Study", Value Health Reg Issues, 9, 2016, 72-77.

47 Zaichun Zhu et al., "Greening of the Earth and its drivers", Nature Climate Change, 6, 2016, 791–795.

48 L. Hunter Lovins, *A Finer Future*, New Society Publishers, 2018.

49 IPCC 4th Assesment Report. "Climate Change 2007: Synthesis Report. Contribution of Working Groups I, II and III to the Fourth Assessment Report of the IPCC", Geneva, Switzerland, 104.

50 Frances Moore et al, "Economic impacts of climate change on agriculture: a comparison of process-based and statistical yield models", Environmental Research Letters, 12 (6), 2017.

51 Francesco Tubiello et al., "Climate Change Response Strategies for Agriculture: Challenges and Opportunities for the 21st Century", Agriculture and Rural Development Discussion Paper No. 42, The International Bank for Reconstruction and Development / The World Bank, 2008.

52 United Nations, "SDG 6 Synthesis Report 2018 on Water and Sanitation", UN Water, 2015.

53 United Nations, "Water & Climate Change", [online], unwater.org, accessed 2020.

54 Peter H. Gleick, "The World's Water", Pacific Institute, vol 8, 2014.

55 Donella H. Meadows, ***Limits to Growth The 30-Year Update 3rd Edition***, Chelsea Green Publishing, 2004, 68.

56 P. Torcellini et al., "Consumptive Water Use for US Power Production", NREL/TP 550 33905, 2003.

57 VOSS Foundation, "Global Water Resources", UNEP, [online], vossfoundation.org, 2006.

58 Jacob Schewe, "Multimodel assessment of water scarcity under climate change", PNAS, 111 (9), 2014, 3245-3250.

59 Hafsa Ahmed Munia et al., "Future transboundary water stress and its drivers under climate change: a global study", Earth's Future, 8 (7), 2020.

60 FAO, "Water and Food Security", [online], fao.org, accessed 2020.

61 IPCC 5th Assesment Report."Climate Change 2014: Synthesis Report. Contribution of Working Groups I, II and III to the Fifth Assessment Report of the IPCC", Geneva, Switzerland, 151 pp.

62 F. T. Portmann et al., "Impact of climate change on renewable groundwater resources: assessing the benefits of avoided greenhouse gas emissions using selected CMIP5 climate projections". Environmental Research Letters, 8(2), 024023.

63 Jacob Silverman et al., "Coral reefs may start dissolving when atmospheric CO_2 doubles", Geophysical Research Letters, 36, 2009.

64 Secretariat of the Convention on Biological Diversity, "An Updated Synthesis of the Impacts of Ocean Acidification on Marine Biodiversity", Montreal, Technical Series No. 75, 2014.

65 Mckinsey Global Institute, "Climate risk and response: Physical hazards and socioeconomic impacts", Mckinsey and Company, 2020.

66 IPCC Global Warming of 1.5 C special report, Oct 2018. IPCC, 2019: IPCC Special Report on the Ocean and Cryosphere in a Changing Climate, 2018.

67 Luke M. Brander et al., "The economic impact of ocean acidification on coral reefs", Climate Change Economics, 03 (01), 2012.

68 Christopher Wanjek, "Food at work: Workplace solutions for malnutrition, obesity and chronic diseases", International Labour Office, 2005.

69 Joseph Romm, *Climate Change: What Everyone Needs to Know*, Second Edition, Oxford University Press, 2018.

70 Yann Chavaillaz et al., "Exposure to excessive heat and impacts on labour productivity linked to cumulative CO_2 emissions", Scientific Reports, 9, 13711, 2019.

71 Mckinsey Global Institute, "Climate risk and response: Physical hazards and socioeconomic impacts", Mckinsey and Company, 2020.

72 Mckinsey Global Institute, "Climate risk and response: Physical hazards and socioeconomic impacts", Mckinsey and Company, 2020.

73 OECD, "The economic consequences of outdoor air pollution", 2016.

74 Dominic Kailash Nath Waughray (WEF) and Marco Lambertini (WWF), "Human Activity is eroding the World's ecological foundations", World Economic Forum, 2020.

Chapter 5: Damage Assessment – Quantifying the Social and Economic Costs of Climate Change

1 World Bank, "GDP", [online], data.worldbank.org, 2020, accessed 2020.

2 Credit Suise Research Institute, "Global wealth report 2020", 2020.

3 Glenn-Marie Lange, "The Changing wealth of nations 2018", World Bank Group, 2018.

4 Thomas Piketty, *Capital in the 21st Century*, Harvard UP, Reprint Edition, 2017.

5 Mark Lynas, *Six Degrees*, Harper Perennial, 2007, p159.

6 OECD, "The economic consequences of Climate Change", OECD Publishing, Paris, 2015.

7 Philip Schofield, "The Legal and Political Legacy of Jeremy Bentham", Annual Review of Law and Social Science, 9, 2013, 51-70.

8 UCL, "Auto-Icon", Bentham Project, [online], ucl.ac.uk, accessed 2020.

9 The Committee for the Prize in Economic Sciences in Memory of Alfred Nobel, "Economic Growth, Technological Change, and Climate Change", 2018.

10 William D. Nordhaus, "Social cost of carbon in DICE model", PNAS, 114 (7), 2017, 1518-1523.

11 Moritz Drupp et al., "Discounting Disentangled", London School of Economics Grantham Research Institute, working paper 195, 2015.
12 Christian Gollier, *Pricing the Planet's Future*, Princeton University Press, 2013, 11.
13 Nicholas Stern, "The Economics of Climate Change", The Stern Review, Cambridge, 2007.
14 Elroy Dimson, Paul Marsh and Mike Staunton, "Credit Suisse Global Investment Returns Yearbook 2018", Credit Suisse Research Institute, 2018.
15 William D. Nordhaus, "A Review of the Stern Review on the Economics of Climate Change", Journal of Economic Literature, 45 (3), 2007, 686-702.
16 Richard Tol, *Climate Economics*, Edward Elgar Publishing, 2014.
17 Roberto Roson and Martina Sartori, "Estimation of Climate Change Damage Functions for 140 Regions in the GTAP 9 Data Base", Journal of Global Economic Analysis, 1(2), 2016, 78-115.
18 Angus Maddison, *Contours of the World Economy*, Oxford University Press, 2007.
19 Credit Suisse, "Global Wealth Report 2020", Credit Suisse Research Institute, 2020.
20 David Anthoff et al. "Equity Weighting and the Marginal Damage Costs of Climate Change", Ecological Economics, 86(3), 2009, 836-849.
21 Yves Balasko, "Pareto optima, welfare weights, and smooth equilibrium analysis", Journal of Economic Dynamics and Control, 21 (2-3), 1997, 473-503.
22 Vincent Biausque for the OECD, "The value of statistical life a meta-analysis", Working Party on National Environmental Policies, 2010.
23 World Health Organisation, "Mortality and Global Health Estimates", The Global Health Observatory, [online], who.int, accessed 2020.
24 World Health Organisation, "Fact Sheet Cancer", [online], who.int, 2018, accessed 2020.
25 World Health Organisation, "The Atlas of Hearth Disease and Stroke", [online], who.int, accessed 2020.
26 WHO, "Air Pollution", World Health Organisation, [online], who.int, 2020, accessed 2020.
27 Hannah Ritchie and Max Roser, "Causes of Death", [online], OurWorldInData.org, 2020, accessed 2020.
28 Angela Mawle, "Climate change, human health, and unsustainable development", Public Health Policy, 31, 2010, 272–277.
29 Joint Monitoring Programme Progress Report for Water Supply, Sanitation and Hygiene (JMP), "2.3 Billion People Lack Basic Sanitation Undermining Health Progress", WHO and UNICEF, 2013.
30 BBC News, "Which country pays the most bribes", [online], bbc.co.uk, 2013, accessed 2020.
31 The Global Nutrition Report, "The Burden of Malnutrition", Chapter 2, 2018.
32 John Elfein, "Prevelance of Infectious Disease by Country in 2019", [online], stastistica.com, 2020, accessed 2020.
33 Max Roser and Esteban Ortiz-Ospina, "Global Extreme Poverty", [online], OurWorldInData.org, 2020, accessed 2020.
34 Gregory A. Roth et al., "Global, Regional, and National Burden of Cardiovascular Diseases for 10 Causes, 1990 to 2015", Journal of the American College of Cardiology, 70(1), 2017, 1-25.
35 Joan B Soriano et al., "Global, regional, and national deaths, prevalence, disability-adjusted life years, and years lived with disability for chronic obstructive pulmonary disease and asthma, 1990–2015. A systematic analysis for the Global Burden of Disease Study 2015", The Lancet Respiratory Medicine, 5(9), 2017, Pages 691-706.
36 S. Harrendorf et al., "International Statistics on Crime and Justice", European Institute for Crime Prevention and Control, 2010.
37 The UN Refugee Agency, "Figures at a Glance", [online], unhcr.org, 2020, accessed 2020.
38 Avert, "Global HIV and AIDS Statistics", [online], avert.org, 2018, accessed 2020.
39 United Nations Office on Drugs and Crime, "World Drug Report", 2017.
40 Cleusa P. Ferri et al., "Global Prevalence of Dementia: a Delphi consensus study", Lancet, 366(9502), 2005, 2112-2117.
41 Ke Xu et al, "Global Spending on Health: A World in Transition", World Health Organisation, 2019.
42 Jim Gee and Professor Mark Button, "The financial cost of fraud 2019", Crowe and University of Portsmouth, 2019.
43 R. Murphy, "The cost of tax abuse. A briefing paper on the cost of tax evasion worldwide", Tax Justice Network, 2011.
44 SIPRI Military Expenditure Database, "World military expenditure grows to $1.8 trillion in 2018", [online], 2019, accessed 2020.
45 Graham Farrell and Ken Clark, "What does the world spend on Criminal Justice", The European Institute for Crime Prevention and Control (HEUNI), 2004.
46 UN office on Drugs and Crime, "Transnational organized crime: the globalized illegal economy", [online], accessed 2020.
47 Patrick J. Michaels and Paul C. Knappenberger, Lukewarming, Cato Institute, 2016.
48 Bjorn Lomborg, *Cool It*, Knopf Publishing Group, 2007.
49 Thomas C Shelling, "Dynamic Models of Segregation", Journal of Mathematical Sociology, 1, 1971, 143-186.
50 James Hansen, Makiko Sato and Reto Ruedy, "Perception of climate change", PNAS, 109 (37), 2012, 2415-2423.

51 Mckinsey Global Institute, "Climate risk and response: Physical hazards and socioeconomic impacts", Mckinsey and Company, 2020.

52 James Hansen, Makiko Sato, and Reto Ruedy, "The New Climate Dice: Public Perception of climate change", NASA, August 2012.

53 Hoegh-Guldberg et al., "Impacts of 1.5°C Global Warming on Natural and Human Systems", Global Warming of 1.5°C, IPCC special report, 2018.

54 David Wallace-Wells, *The Uninhabitable Earth*, Penguin Random House, 2019.

55 Naomi Oreskes and Erik M. Conway, *Merchants of Doubt*, Bloomsbury, 2010.

56 Kay Giesecke and Baeho kim, "Risk Analysis of Collateralized Debt Obligations", Operations Research, 59(1), 2011, 32-49.

57 Roman Tomasic and Folarin Akinbami, "The Role of Trust in Maintaining the Resilience of Financial Markets", Journal of Corporate Law Studies, 11:2, 2015, 369-394.

58 Berkshire Hathaway INC, Shareholder Letter, 2002.

59 Neal Deckant, "Criticisms of CDO in the wake of the goldman sachs scandal", Review of Banking and Financial Law; 30, 2011, 407.

60 Paul J. Davies, "Half of all CDO of ABS failed", Financial Times, February 2009.

61 Christophe McGlade and Paul Ekins, "The geographical distribution of fossil fuels unused when limiting global warming to 2 °C", Nature, 517, 2015, 187–190.

62 Citi Global Perspectives and Solutions, "Energy Darwinism 2, Why a Low Carbon Future Doesn't Have to Cost the Earth", August 2015.

63 Carbon Tracker Initiative, "Unburnable Carbon - Are the world's financial markets carrying a carbon bubble?", 2011.

64 Mark Carney Channel 4 News, "Climate Change Interview", July 31, 2019.

65 Richard S. J. Tol, "Economic impacts of climate change," Department of Economics, University of Sussex Business School, Working Paper Series 7515, 2015.

66 Matthew E. Kahn et al., "Long-Term Macroeconomic Effects of Climate Change: A Cross Country Analysis", International Monetary Fund, 2019.

67 Weitzman M. L., "Fat-tailed uncertainty in the economics of catastrophic climate change", Review of Environmental Economics and Policy, 5(2), 2011, 275 – 292.

Chapter 6: Opening for Planetary Caretaker – Building a Climate Change Strategy

1 Pallav Purohit and Lena Hoglund-Isaksson, "Global emissions of fluorinated greenhouse gases 2005-2050 with abatement potentials and costs", Atmospheric Chemistry and Physics 17(4), 2017, 2795-2816.

2 IPCC AR4, P. Forster et al., "Changes in Atmospheric Constituents and in Radiative Forcing", Contribution of Working Group I to the Fourth Assessment Report of the IPCC, Cambridge University Press, Chapter 2, 2007, 212.

3 P. Smith et al., "Agriculture, Forestry and Other Land Use (AFOLU)", Contribution of Working Group III to the Fifth Assessment Report of the Intergovernmental Panel on Climate ChangeIPCC AR 5, Chapter 11, Cambridge University Press, 2014.

4 BP, Statistical Review of World Energy, 2019, 68th edition.

5 IPCC AR5, "Figure SPM.2", Climate Change 2014: Synthesis Report, IPCC, 2014, 5.

6 Woodbank Communications Ltd, "The Energy Supply Chain", [online], mpoweruk.com, accessed 2020.

7 IEA, "World Final Consumption 2018", [online], iea.org, accessed 2020.

8 Andrew Robbie, "Global CO_2 emissions from cement production", Earth Systems Science Data, 10, 2018, 195-217.

9 Geoffrey A. Ozin and Mireille F. Ghoussoub, The Story of CO_2, Aevo UTP, 2020, p7.

10 World Bank, "GDP per Capita (current US$)", 2018, [online], data.worldbank.org, accessed 2020.

11 G. Janssens-Maenhout etal., "Fossil CO_2 and GHG emissions of all world countries", Publications Office of the European Union, Luxembourg, 2017.

12 IPCC, "Climate Change 2014: Impacts, Adaptation, and Vulnerability. Part A: Global and Sectoral Aspects", Contribution of Working Group II to the Fifth Assessment Report of the IPCC, Cambridge University Press, 2014, 977.

13 James Smith et al., "Forest Volume-to-Biomass Models and Estimates of Mass for Live and Standing Dead Trees of US Forests", General Technical Report US Department of Agriculture, 2003.

14 Food and Agriculture Organization of the United Nations, "Land Use", [online], fao.org/faostat, accessed 2020.

15 Thomas W. Crowther et al., "The global tree restoration potential", Science, 365(6448), 2019, 76-79.

16 R Parajuli et al, Woodland Owner Notes, "Is Reforestation a Profitable Investment? An Economic Analysis", [online], content.ces.ncsu.edu, 2019, accessed 2020.

17 David M. Summers et al., "The costs of reforestation: A spatial model of the costs of establishing environmental and carbon plantings", Land Use Policy, 44, 2015, 110-121.

18 Suzanna Dayne, "'Grain is for green': How China is swapping farmland for forest", [online], forestnews.cifor.org, 2017.

19 Filip Meysman, "Negative CO_2 Emissions via Enhanced Silicate Weathering in Coastal Environments", Biology Letters, 2017.

20 L. Taylor et al. "Enhanced weathering strategies for stabilizing climate and averting ocean acidification", Nature Climate Change 6, 2016, 402-406.

21 National Minerals Information Center, "Crushed Stone Statistics and Information", [online], usgs.gov, 2020, accessed 2020.

22 Ocean Nutrition Corporation, [online], oceannourishment.com, accessed 2020.

23 Kelsey Piper, "The climate renegade", [online], vox.com, Jun 4 2019, accessed 2020.

24 Jeff Tollefson, "Iron-Dumping Ocean Experiment Sparks Controversy", Nature, 2017.

25 Martin Lukacs, 'World's biggest geoengineering experiment 'violates' UN rules', The Guardian, Oct, 2012.

26 Ian S.F. Jones, "The cost of carbon management using ocean nourishment", International Journal of Climate Change Strategies and Management, 6 (4), 2014, 391-400.

27 David MacKay, *Sustainable Energy without the Hot Air*, UIT Cambridge, 2009, 245.

28 Kurt Zenz House et al., "Economic and energetic analysis of capturing CO_2 from ambient air", PNAS, 108 (51), 2011, 20428-20433.

29 David W. Keith et al., "A Process for Capturing CO_2 from the Atmosphere", 2(8), 2018, 1573-1594.

30 Carbon Engineering, "Direct Air Capture Presentation 2017", [online], carbonengineering.com, accessed 2020.

31 Bill Gates, How to Avoid a Climate Disaster, Allen Lane, 2021, p95 and p63.

32 Global Thermostat, "The GT Solution", [online], globalthermostat.com, accessed 2020.

33 Ben Soltoff, "Inside ExxonMobil's hookup with carbon removal venture Global Thermostat", [online], greenbiz.com, Aug 2019, accessed 2020.

34 DRAX, "Bioenergy and Carbon Capture", [online], drax.com, 2018, accessed 2020.

35 Martin Blunt, "Briefing paper No4. Carbon Dioxide Storage", Grantham Institude for Climate Change Imperial College London, 2010.

36 The Royal Society, "Geoengineering the climate, science, governance and uncertainty", 2009.

37 S. Self et al., "The atmospheric impact of the 1991 Mount Pinatubo Eruption", 1993.

38 Kate Marvel, Essay, *All We Can Save*, One World, 2020.

39 IPCC AR5, "2014: Drivers, Trends and Mitigation. In: Climate Change 2014: Mitigation of Climate Change", Contribution of Working Group III to the Fifth Assessment Report of the IPCC, Cambridge University Press, Chapter 5, 2014.

40 IPCC AR5, "Figure SPM.2", Climate Change 2014: Synthesis Report, IPCC, 2014, 5.

41 World Bank, "GDP per Capita (current US$)", 2018, [online], data.worldbank.org, accessed 2020.

42 BP, Statistical Review of World Energy, 2019, 68th edition.

43 United Nations, "World Population Prospects", [online], popuilation.un.org, 2019, accessed 2020.

44 Max Roser, "Fertility Rate", [online], OurWorldInData.org, 2020, accessed 2020.

45 Paul Sabin, *The Bet*, Yale University Press, 2015.

46 Amartya Sen, *Development as Freedom*, Oxford University Press, 1999, p226.

47 Dominic Roser and Christian Seidel, *Climate Justice: An Introduction*, Routledge, 2016.

Chapter 7: Sustainable Energy – Supply Options with Net-Zero Emissions

1 Albert Einstein, "Does the inertia of a body depend upon its energy-content", 1905.

2 Mark Tiele Westra, "Fusion in the Universe: the power of the Sun", The European Journal for Science Teachers, 3, 2006.

3 Andrew Dessler, *Introduction to Modern Climate Change*, Second Edition, Cambridge University Press, 2015.

4 University of Michigan, "Nuclear Energy Factsheet", Centre for Sustainable Systems, [online], ccs.umich.edu, 2020, accessed 2020.

5 Radioactivity.eu.com, "Nuclear Energy, 200 million electronvolts", Editions De Physique, [online], radioactivity.eu.com, accessed 2020.

6 National Academy of Sciences, "Nuclear Technologies Timeline", [online], greatachievements.org, accessed 2020.

7 Luke Richardson, "The History of Solar Energy", [online], energysage.com, 2018.

8 Alan Chodos, "Bell Labs Demonstrates the First Practical Silicon Solar Cell", This month in Physics History, APS News, 18 (4), 2009.

9 Ed Bodmer, "Levelised Cost of Energy and Carrying Charges", [online], edbodmer.com, accessed 2020.

10 BCG, "The experience curve", [online], bcg.com, accessed 2020.

11 David MacKay, *Sustainable Energy without the Hot Air*, UIT Cambridge, 2009.

12 Vaclav Smil, *Energy*, One World Press, 2017.

13 European Technology and Innovation Platform Bioenergy, "Transesterification to biodiesel", [online], etipbioenergy.eu, accessed 2020.

14 Tim Simmons, "CO_2 Emissions from Stationary Combustion of Fossil Fuels", IPCC Good Practice Guidance and Uncertainty Management in National Greenhouse Gas Inventories, 2006.

15 A. K. Van Harmelen and W. W. R. Koch, "CO_2 emission factors for fuels in the Netherlands", TNO Environment, Energy and Process Innovation, 2002.

16 Ethan Warner and Garvin A. Heath, "Life Cycle Greenhouse Gas Emissions of Nuclear Electricity Generation", Systematic Review and Harmonization, Journal of Industrial Ecology, 16(81), 2012.

17 Devon Holst, "Embodied Energy and Solar Cells", Green Chemistry Initiative Blog, [online], 2017, accessed 2020.

18 I. Nawaz, G. N. Tiwari, "Embodied energy analysis of photovoltaic (PV) system based on macro- and micro-level", Energy Policy, 34 (17), 2006, 3144-3152.

19 Wacker Chemie, "Before There is Solar Energy there is Wacker", [online], wacker.com, 2017, accessed 2020.

20 Vasilis M. Fthenakis et al., "Emissions from Photovoltaic Life Cycles", Environmental Science Technology, 42(6), 2008, 2168-2174.

21 B Monsen et al., "CO_2 Emissions from the Production of Ferrosilicon and Silicon metal in Norway", 1998.

22 Industrial Efficiency Technology Database, "Iron and Steel", [online], iipinetwork.org, accessed 2020.

23 Industrial Efficiency Technology Database, "Cement", [online], iipinetwork.org, accessed 2020.

24 Mike Ashby, "Material property data for engineering materials", University of Cambridge Engineering Department, 4th edition, 2016.

25 IHA, "2018 Hydropower status report", International Hydropower Association, 2018.

26 Paul Hollis, "Fertilizer recommendations proving valid even with increasing corn yields", [online], farmprogress.com, 2013, accessed 2020.

27 USDA Economic Research Service, "Fertiliser Use and Price", [online], ers.usda.gov, 2019, accessed 2020.

28 Ryan Davis et al., "Process Design and Economics for the Production of Algal Biomass", NREL technical paper, TP-5100-64772, 2016.

29 NREL, "Life Cycle Assessment Harmonisation", [online], nrel.gov, accessed 2020.

30 Bert Metz et al., "Carbon Dioxide Capture and Storage", IPCC report, Cambridge University Press, 2005.

31 N Florin and P Fennell, "Carbon Capture Technology: future fossil fuel use and mitigating climate change", Imperial College London Grantham Institute, Briefing Paper No 3, 2010.

32 M.Blunt, "Carbon Dioxide Storage", Imperial College London Grantham Institute, Briefing Paper No 4, 2010.

33 British Geological Survey, "Mineral Planning Factsheet, Underground Storage", [online], 2.bgs.ac.uk, Feb 2008, accessed 2020.

34 Sylvie Cornot-Gandolphe, "Underground Gas Storage in the World - 2018 Status", Cedigaz Insights, 2018.

35 EDF, "About Hinkley Point C", [online], edfenergy.com, accessed 2020.

36 John Zactruba, "How to Calculate the Coal Quantity Used in a Power Plant", Energy and Power Plants, Bright Hub Engineering, [online], brighthubengineering.com, acessed 2020.

37 International Atomic Energy Agency, "Estimation of Global Inventories of Radioactive Waste and other Radioactive Materials", 2008.

38 Elizabeth K. Ervin, "Nuclear Energy: Statistics", University of Mississippi School of Engineering, 2009.

39 EIA, "Drilling productivity report", [online], eia.gov, accessed 2020.

40 Andrew A. Pericak et al., "Mapping the yearly extent of surface coal mining in Central Appalachia using Landsat and Google Earth Engine", PLOS, July 2018.

41 S. Sanchez-Carbajal, "Optimum Array Spacing in Grid-Connected Photovoltaic Systems Considered in Technical and Economic Factors", International Journal of Photoenergy, 1486749, 2019.

42 Power Technology, "Alta Wind Energy Center (AWEC), California", [online], power-technology.com, acessed 2020.

43 Wikipedia contributors, "Alta Wind Energy Center", Wikipedia, The Free Encyclopedia, accessed 2020.

44 Wikipedia contributors, "Walney Wind Farm", Wikipedia, The Free Encyclopedia, [online], accessed 2020.

45 Orsted, "Our Wind Farms, Walney, Fact Sheet", [online], orsted.co.uk, acessed 2020.

46 IRENA, "Geothermal Power: Technology Brief", 2017.

47 James Smith et al., "Forest Volume-to-Biomass Models and Estimates of Mass for Live and Standing Dead Trees of US Forests", General Technical Report US Department of Agriculture, 2003.

48 The Editors of Encyclopaedia Britannica, "Three Gorges Dam", Encyclopedia Britannica, 2020.

49 Michael E. Webber, *Thirst for Power*, Yale University Press, 2016.

50 Wikipedia contributors, "Three Gorges Dam", Wikipedia, The Free Encyclopedia, accessed 2020.

51 John van Zalk and Paul Behrens, "The spatial extent of renewable and non-renewable power generation: A review and meta-analysis of power densities and their application in the U.S.", Energy Policy, 123, 2018, 83-91.

52 Food and Agriculture Organization of the United Nations, "Land Use", [online], fao.org/faostat, accessed 2020.

53 AZ Quotes, [online], azquotes.com, accessed 2020.

54 Rachel Maddow, *Blowout*, Penguin Random House, 2019.

55 United States Energy Information Administration, "Liquid Fuels and Natural Gas in the Americas", January 2014.

56 USGS, 'Trends in Hydraulic Fracturing Distributions and Treatment Fluids, Additives, Proppants, and Water volumes', report 5131, 2014.

57 Ryan Schultz et al. "Hydraulic Fracturing- Induced Seismicity", Reviews of Geophysics, 58(3), 2020.

58 CBS News, "Can you drink fracking fluid? One gas exec did", August 22, 2011.

59 BP, Statistical Review of World Energy, 2019, 68th edition.

60 World Energy Council, "World Energy Resources: 2016", WEC, 2016.

61 Nuclear Energy Agency and International Atomic Energy Agency, "Uranium 2016: Resouces Production and Demand", OECD, 2016.

62 International Atomic Energy Agency, "Analysis of Uranium Supply to 2050", IAEA, 2001.

63 James Conca, "Uranium Seawater Extraction Makes Nuclear Power Completely Renewable", [online], forbes.com, 2016, accessed 2020.

64 James D. Newton, "Uncommon Friends: Life with Thomas Edison, Henry Ford, Harvey Firestone, Alexis Carrel, and Charles Lindbergh", Harcourt Brace Jovanovich, San Diego, California, 1987, 31.

65 Fiona Macleod, "Reflections on Banqiao", The Chemical Engineer, [online], 8th August 2019, accessed 2020.

66 Richard Pallardy, "Deepwater Horizon oil spill", Encyclopedia Bitannica, 2020.

67 The Two-Way, "Oil Execs Grilled on Copycat Emergency Plans", June 15th, 2010.

68 Dorothy L Robinson, "Air pollution in UK: the public health problem that won't go away", British Medical Journal, 350, 2015.

69 United States Nuclear Regulatory Commission, "Backgrounder on the Three Mile Island Accident", NRC Library, [online], nrc.gov, accessed 2020.

70 United Nations Scientific Committee on the Effects of Atomic Radiation, "Sources and Effects of Ionizing Radiation", Annex D Health Effects due to Radiation from the Chernobyl Accident, 2008, 45-221.

71 World Health Organisation, "Health effects of the Chernobyl accident and special health care programmes", 2006.

72 World Nuclear Association, "Fukushima Daiichi Accident", [online], world-nuclear.org, accessed 2020.

73 David Kushner, "The real story of Stuxnet", IEEE Spectrum, [online], spectrum.ieee.org, 26 Feb, 2013.

74 Anil Markandya et al., "Electricity Generation and Health", The Lancet, 370(9591), 2007, 979-990.

75 Brian Wang, "Deaths by Energy Source in Forbes", [online], nextbigfuture.com, June 2012, accessed 2020.

76 Jos Lelieveld et al., "Loss of life expectancy from air pollution compared to other risk factors: a worldwide perspective", Cardiovascular Research, 116(11), 2020, 1910-1917.

77 World Health Organisation, "Air Pollution", [online], who.int, accessed 2020.

78 David Howell, Energy Empires in Collision, Gilgamesh publishing, 2017.

79 The Editors of Encyclopaedia Britannica, "John D. Rockefeller", Encyclopaedia Britannica, 2021.

80 Javier Blas and Jack Farchy, *The World for Sale*, Random House Business, 2021, p44.

81 Anna-Alexandra Marhold, "WTO law and economics and restrictive practices in energy trade: The case of the OPEC cartel", *The Journal of World Energy Law and Business*, 9(6), 2016, 475–494.

82 IEA, "Key World Energy Statistics 2019", IEA, 2019.

83 Index Mundi, "Commodity prices, energy", [online], indexmundi.com, accessed 2020.

84 Markets Insider, "Uranium", [online], markets.businessinsider.com, accessed 2020.

85 Ingeborg Graabak and Magnus Korpas, "Variability Characteristics of European Wind and Solar Power Resources - a Review", Energies, 9(6), 2016, 449.

86 E Ela et al., "Impacts of Variability and Uncertainty in Solar Photovoltaic Generation at Multiple Timescales", NREL, TP-5500-5874, 2013.

87 Wind Energy The Facts, "Understanding Variable Output Characteristics of Wind Power", [online], wind-energy-the-facts.org, accessed 2020.

88 Jerald A. Caton, *An Introduction to Thermodynamic Cycle Simulations for Internal Combustion Engines*, Wiley, 2015.

89 M. E. Brokowski, "Design of an Otto Cycle", [online], qrg.northwestern.edu, 2002, accessed 2020.

90 David Chandler, "Explained: The Carnot Limit", [online], news.mit.edu, May 19, 2010, accessed 2020.
91 John Zactruba, "The Efficiency of Power Plants of Different Types", [online], brighthubengineering.com, 2010, accessed 2020.
92 Institut de Radioprotection et de Surete Nucleaire - IRSN, "Review of Generation IV Nuclear Energy Systems", 2015.
93 Steve Fetter, "How Long will the world's uranium supplies last?", Scientific American, January 26, 2009.
94 Global Energy Assessment, "Towards a Sustainable Future", Cambridge University Press, 2012.
95 Terra Power, "Resources", [online], terrapower.com, accessed 2020.
96 Richard A. Muller, *Physics for Future Presidents*, W.W. Norton, New York, 2009.
97 William Shockley and Hans J. Queisser, "Detailed Balance Limit of Efficiency of p-n Junction Solar Cells", Journal of Applied Physics, 32, 1961, 510.
98 David MacKay, *Sustainable Energy without the Hot Air*, UIT Cambridge, 2009.
99 US Department of Energy, "Wind Vision. A New Era for Wind Power in the United States", 2018.
100 IPCC, "Climate Change 2014: Impacts, Adaptation, and Vulnerability. Part A: Global and Sectoral Aspects", Contribution of Working Group II to the Fifth Assessment Report of the IPCC, Cambridge University Press, 2014, 670.
101 Charles A.S. Hall et al., "EROI of different fuels and the implications for society", Energy Policy, 64, 2014, 141-152.
102 Erik Mielke et al., "Water Consumption of Energy Resource Extraction, Processing and Conversion", Harvard Belfer Center, 2010.
103 Yi Jin et al., "Water use of electricity technologies: A global meta-analysis", Renewable and Sustainable Energy Reviews, 115, 2019, 109391.
104 International Renewable Energy Agency, "Costs", [online], irena.org, accessed 2020.
105 US Energy Information Administration, "Addendum: Updated Capital Cost and Performance Characteristic Estimates for Utility Scale Electricity Generating Plants in the Electricity Market Module of the National Energy Modeling System", Feb 5 2020, excel link tables.
106 Digvijay Bhusan, "The World's Biggest Under-Construction Power Plants by Capacity", April 3, 2019, [online], accessed 2020.
107 World Nuclear Association, "Economics of Nuclear Power", [online], world-nuclear.org, March 2020, accessed 2020.
108 Vincent Giles et al., "Global Nuclear Value in Life Extension", Credit Suisse, 21 November 2018.
109 PV Tech, "News", [online], pv-tech.org, accessed 2020.
110 Wind Power Monthly, "Tender Watch", [online], windpowermonthly.com, accessed 2020.
111 Renewables Now, "Latest Stories - Tenders", [online], renewablesnow.com, accessed 2020.
112 Florian Egli et al., "A dynamic analysis of financing conditions for renewable energy technologies", Nature Energy, 3, 2018, 1084-1092.
113 Bjarne Steffen, "Financing and Innovation in Energy Transitions", Energy Politics Group ETH Zurich, Nov 2019.
114 Business Wire, "kWh Analytics Releases 2020 Solar Generation Index and Issues First START Comps Reports to Address Biases in Solar Production Estimates", [online], businesswire.com, october 5 2020, accessed 2020.
115 Chris Goodall, *The Switch*, Profile Books, 2016.
116 Ryan Davis et al., "Process Design and Economics for the Production of Algal Biomass", NREL technical paper, TP-5100-64772, 2016.

Chapter 8: Energy Storage and Distribution – Building a Safe, Reliable, and Cost Competitive Network

1 Andrew J. Pimm et al., "The potential for peak shaving on low voltage distribution networks using electricity storage", Journal of Energy Storage, 16, 2018, 231-242.
2 Mark Z. Jacobson, *100% Clean, Renewable Energy and Storage for Everything*, 2020.
3 Alexander Zerrahn et al., "On the economics of electrical storage for variable renewable energy sources", European Economic Review, 108, 2018, 259-279.
4 US Energy Information Administration, "Monthly Energy Review", [online], eia.gov, accessed 2020.
5 World Energy Council, "Energy Storage Monitor", worldenergy.org, 2019.
6 ARUP, "Five Minute Guide to Electricity Storage", [online], arup.com, accessed 2020.
7 Deloitte, "Energy Storage: Tracking the technologies that will transform the power sector", 2015.
8 Cadex Electronics Inc, "Battery University", [online], batteryuniversity.com, accessed 2020.
9 IRENA, "Electricity storage and renewables: Costs and markets to 2030", 2017.
10 Gunnar Benjaminsson et al., "Power-to-Gas - A Technical Review", Swedish Gas Technology Centre, 2013.
11 Karlsruhe Institute of Technology, "Power-to-gas facility with high efficiency", [online], phy.org, 2018, accessed 2020.

12 Martin Lambert, "Power to Gas Linking Electricity and Gas in a Decarbonising World", The Oxford Institute for Energy Studies, 2018.
13 Spirit Energy, "Battery Storage Knowledge Bank: Understanding Batteries", [online], spiritenergy.co.uk, accessed 2020.
14 Andrew Blakers et al., "Global pumped hydro atlas", [online], re100.eng.anu.edu.au, accessed 2020.
15 O. Schmidt et al., "The future cost of electrical energy storage based on experience rates", Nature Energy, 2, 2017, 17110.
16 Neoen, 'Document De Base', 2018, p65.
17 James Thornhill, "Two Years On, Musk's big battery bet is paying off in Australia", Bloomberg, Feb 2020, [online], accessed 2020.
18 RWE Power, "ADELE - Adiabatic compressed - Air Energy Storage for Electricity Supply", [online], 2010, accessed 2020.
19 Jidai Wang et al., "Overview of Compressed Air Energy Storage and Technology Development", Energies, 10(7), 2017, 991.
20 I Arsie et al., "Optimal Management of a Wind/CAES Power Plant by Means of Neural Network Wind Speed Forecast", 2007.
21 Bushveld Minerals, "Energy Storage and Vanadium Redox Flow Batteries 101", [online], bushveld minerals.com, Nov 2018, accessed 2020.
22 John Fitzgerald Weaver, "World's largest battery: 200MW/800MWh vanadium flow battery-site work ongoing", [online], electrek.com, Dec 2017, accessed 2020.
23 USGS, "National Minerals Information Centre, Vanadium Statistics and Information", [online], usgs.gov, accessed 2020.
24 Energy Vault, [online], energyvault.com, accessed 2020.
25 Marko Aunedi et al., "The drive towards a low-carbon grid", [online], eonenergy.com, 2021, accessed 2021.
26 IRENA, "Thermal Energy Storage: Technology Brief", January 2013.
27 Ioan Sarbu and Calin Sebarchievici, "A Comprehensive Review of Thermal Energy Storage", Sustainability, 10(1), 2018, 191.
28 Lavinia Gabriela, "Seasonal Sensible Thermal Energy Storage Solutions", Leonardo Electronic Journal of Practices and Technologies, 19, 2011, 49-68.
29 Julian D. Hunt et al., "Global resource potential of seasonal pumped hydropower storage for energy and water storage", Nature Communications, 11(947), 2020.
30 HyWeb, "The Hydrogen and Fuel Cell Information System", [online], h2data.de, accessed 2020.
31 Brian David James et al., "Hydrogen Production Pathways Cost Analysis", DOE and OSTI.gov, 2016.
32 O. Schmidt,"Future cost and performance of water electrolysis: An expert elicitation study", International Journal of Hydrogen Energy, 42(52), 2017, 30470-30492.
33 DOE, "Technical Targets for Hydrogen Production from Electrolysis", [online], energy.gov, Hydrogen and Fuel Cell Technologies Office, accessed 2020.
34 Cision, "Nel ASA: Recieves purchase order from Nikola", Jun 3, 2020, [online], news.cision.com, accessed June 2020.
35 Mohamed El-Shimy, "Overview of Power-to-Hydrogen-to-Power (P2H2P) Systems Based on Variable Renewable Sources", The 5th International Conference on Electrical, Electronics, and Information Engineering, 2017.
36 Battelle Institute for DOE, "Manufacturing Cost Analysis of 100 and 250 kW Fuel Cell Systems for Primary Power and Combined Heat and Power Applications", Jan 2016.
37 The Milne Museum of Electricity, "Electricity in Brighton", [online], milnemuseum.org, accessed 2020.
38 Alfonzo Gomez-Rejon, "The Current Wars", Film, 2019.
39 Jeremy Rifkin, The Green New Deal, Macmillan, 2019.
40 Gagnon P, et al., "Rooftop solar photovoltaic technical potential in the United States: a detailed assessment", *National Renewable Energy Laboratory*, Jan 2016.
41 Katalin Bódis et al., "A high-resolution geospatial assessment of the rooftop solar photovoltaic potential in the European Union", Renewable and Sustainable Energy Reviews, 114, 2019, 109309.
42 Electric Power Research Institute, "Estimating the Costs and Benefits of the Smart Grid", Technical Report, March 2011.
43 CAISO, "Preliminary Root Cause Analysis, Mid August 2020 Heat Storm", October 6 2020.
44 Scott K. Johnson, "Report details causes of recent California rolling blackouts", [online], arstechnica.com, October 10 2020, accessed 2020.
45 Micah S. Ziegler and Jessika E. Trancik et al., "Storage Requirements and Costs of Shaping Renewable Energy Toward Grid Decarbonization", Joule, 3(9), 2019, 2134-2153.
46 IEA, "World Final Consumption 2018", [online], iea.org, accessed 2020.

Chapter 9: Re-engineering the Economy – Transport in a Net-Zero Future

1 EIA, "World Energy Outlook 2016, Transportation Sector Energy Consumption", Chapter 8, 2016.
2 BP, Statistical Review of World Energy, 2019, 68th edition.

3 IEA, "World Final Consumption 2018", [online], iea.org, accessed 2020.
4 David MacKay, *Sustainable Energy without the Hot Air*, UIT Cambridge, 2009.
5 ARUP, "Five Minute Guide to Electricity Storage", [online], arup.com, accessed 2020.
6 Cadex Electronics Inc, "Battery University", [online], batteryuniversity.com, accessed 2020.
7 DOE, "The History of the Electric Car", [online], energy.gov, accessed 2020.
8 Electric Vehicle Database, "Tesla Model 3 Standard Range", [online], ev-database.uk, accessed 2020.
9 Auke Hoekestra, "The Underestimated Potential of Battery Electric Vehicles to Reduce Emissions", 3(6), 2019, 1412-1414.
10 Continental, "Worldwide Emission Standards and Related Regulations", May 2019.
11 European Court of Auditors, "The EU's response to the 'dieselgate' scandal", Briefing Paper, February 2019.
12 P Jakubicek, "Semi Truck Size and Weight Laws in the United States and Canada", [online], bigtruckguide.com, Jan 2016, accessed 2020.
13 US Department of Transportation, "Freight Facts and Figures", 2013.
14 US DOE, "US Drive, Fuel Cell Technical Team Roadmap", Nov 2017.
15 US DOE, "Transportation Energy Futures Series, Commercial Trucks, Aviation, Marine Modes, Railroads, Pipeline, Off-Road Equipment", 2013.
16 Dan Ye et al., "Jet propulsion by microwave air plasma in the atmosphere", AIP Advances, 10, 2020.
17 Eviation, "All Electric-All ready, Meet Alice The World's first all-electric commuter aircraft", [online], eviation.co, accessed 2020.
18 Reuters, "easyJet partner starts developing engine for electric plane", [online], reuters.com, accessed 2020.
19 Ben Sampson, "Wright Electric awarded government grant for electric propulsion system", October, 2020.
20 Faig Abbasov, "Roadmap to decarbonising European Shipping", Transport and Environment, 2018.
21 Energy Transitions Commission, "Reaching Net-zero carbon emissions from harder-to-abate sectors: mission possible", 2018.

Chapter 10: Re-engineering the Economy – Industry in a Net-Zero Future

1 EIA, "World Energy Outlook 2016, Transportation Sector Energy Consumption", Chapter 8, 2016.
2 BP, Statistical Review of World Energy, 2019, 68th edition.
3 Pallav Purohit and Lena Hoglund-Isaksson, "Global emissions of fluorinated greenhouse gases 2005-2050 with abatement potentials and costs", Atmospheric Chemistry and Physics 17(4), 2017, 2795-2816.
4 Paul Griffin et al., "Industrial energy use and carbon emissions reduction in the chemicals sector: A UK perpective", Applied Energy, 277, 2018, 587-602.
5 Carbon Trust, "International Carbon Flows", [online], carbontrust.com, 2011, accessed 2020.
6 Thomas O. Wiedmann et al., "The material footprint of nations", PNAS, 112(20), 2015, 6271-6276.
7 Gregory P. Thiel, "To decarbonize industry, we must decarbonize heat", Joule, 2021.
8 Michael Ashby, "Materials Data Book", Cambridge University Engineering Department, 2003.
9 Tamaryn Brown et al., "Briefing Paper No 7, Reducing CO_2 emissions from heavy industry: a review of technologies and considerations for policy makers", Grantham Institute for Climate Change, Feb 2012.
10 World Steel Association, "World Steel in Figures, 2017", 2017.
11 Economics243, "BOF and EAF Steels: What are the differences?" [online], econ243.academic.wlu.edu, 2016, accessed 2020.
12 Industrial Efficiency Technology Database, "Iron and steel", [online], iipinetwork.org, accessed 2020.
13 S Dietz et al., "Carbon Performance Assessment of Steel Makers: Note on Methodology", Transition Pathway Initiative, 31 July 2017.
14 ArcelorMittal, "By-products, scrap and the circular economy", [online], corporate.arcelormittal.com, accessed 2020.
15 Plastics Europe, "World Plastics Production 1950-2015", [online], committee.iso.org, accessed 2020.
16 Borealis, "Petrochemical companies form Cracker of the Future Consortium and sign R&D agreement", [online], borealisgroup.com, Aug 2019, accessed 2020.
17 Statistica, "Cement: Global Production in 1990, 2000 and 2010, with forecasts for 2020 and 2030", [online], statistica.com, 2013, accessed 2020.
18 Christina Galitsky et al., "Energy Efficiency Improvement and Cost Saving Opportunities for Cement Making", Energy Analysis Department, Berkeley National Laboratory, 2008.
19 Industrial Efficiency Technology Database, "Cement", [online], iipinetwork.org, accessed 2020.
20 Karin Jager, "The rise of wooden skyscrapers", [online], dw.com, 2018, accessed 2020.
21 Fran Yanez, The 20 key technologies of Industry 4.0 and smart factories, 2017.
22 Michael Ashby, "Materials Data Book", Cambridge University Engineering Department, 2003.

23 Industrial Efficiency Technology Database, "Pulp and Paper", [online], iipinetwork.org, accessed 2020.

24 Ali Hasanbeigi and Lynn Price, "A Review of Energy Use and Energy Efficiency Technologies for the Textile Industry", Lawrence Berkeley National Laboratory, 2012.

25 Energy Transitions Commission, "Mission Possible, Reaching Net-zero carbon emissions from harder-to-abate sectors", November 2018.

26 Peter H. Pfromm, "Towards sustainable agriculture: Fossil-free ammonia", Journal of Renewable and Sustainable Energy, 9(034702), 2017.

Chapter 11: Re-engineering the Economy – Amenities in a Net-Zero Future

1 EIA, "World Energy Outlook 2016, Transportation Sector Energy Consumption", Chapter 8, 2016.

2 BP, Statistical Review of World Energy, 2019, 68th edition.

3 David MacKay, *Sustainable Energy without the Hot Air*, UIT Cambridge, 2009.

4 Bridget Mallon, "How big is the average house size around the world?", [online], elledecor.com, Aug 26 2015, accessed 2020.

5 Martin Armstrong, "Air Conditioning Biggest Factor in Growing Electricity Demand", [online], statistica.com, June 2018, accessed 2020.

6 IEA, "The Critical Role of Buildings, Perspectives for the Clean Energy Transition", April 2019.

7 IEA, "Tracking Buildings Report, Heating", 2020.

8 IEA, "Renewables 2018, Analysis and Forecasts to 2023", Oct 2018.

9 IEA, "Tracking Buildings 2020, Heat Pumps", [online], iea.org, June 2020, accessed 2020.

10 European Heat Pump Association, "Heat Pumps - Integrating Technologies to Decarbonise Heating and Cooling", ehpa white paper, November 2018.

11 Oliver Ruhnau et al., "Time series of heat demand and heat pump efficiency for energy system modeling", Nature Scientific Data, 6, 2019, 189.

12 IPCC, "Climate Change 2013: The Physical Science Basis", Contribution of Working Group I to the Fifth Assessment Report of the IPCC, Cambridge University Press, 2013, 690.

13 energy.gov, "Lumens and the Lighting Facts Label", [online], energy.gov, accessed 2020.

14 P. Pattison et al, "Solid State Lighting R&D Plan", DOE, 2016.

15 IEA, "Energy Access, Achieving modern energy for all by 2030 is possible", [online], iea.org, 2019.

16 United Nations, "Sustainable Development Goals, 6, Clean Water and Sanitation for all", [online], un.org, accessed 2020.

17 IPCC, "Climate Change 2014: Impacts, Adaptation, and Vulnerability. Part A: Global and Sectoral Aspects", Contribution of Working Group II to the Fifth Assessment Report of the IPCC, Cambridge University Press, 2014, Chapter 9, 692.

Chapter 12: Re-engineering the Economy – Agriculture in a Net-Zero Future

1 P. Smith et al., "Agriculture, Forestry and Other Land Use (AFOLU)", Contribution of Working Group III to the Fifth Assessment Report of the Intergovernmental Panel on Climate ChangeIPCC AR 5, Chapter 11, Cambridge University Press, 2014.

2 EIA, "World Energy Outlook 2016, Transportation Sector Energy Consumption", Chapter 8, 2016.

3 BP, Statistical Review of World Energy, 2019, 68th edition.

4 Food and Agriculture Organization of the United Nations, "Land Use", [online], fao.org/faostat, accessed 2020.

5 Food and Agriculture Organization of the United States, "FAOSTAT, Data", [online], fao.org, accessed 2020.

6 Mike Berners-Lee et al., "Current global food production is sufficient to meet human nutritional needs in 2050 provided there is radical societal adaptation", Elem Sci Anth, 6(1), 2018, 52.

7 Hope Jahren, *The Story of More*, Fleet, 2020.

8 Renewable Fuels Association, "2019 Ethanol Industry Outlook", Feb 2019.

9 University of Michigan, "Biofuels Factsheet", Center for Sustainable Systems, 2018, accessed 2020.

10 Mike Berners-Lee, *There is no Planet B*, Cambridge University Press, 2019

11 World Health Organisation, "Obesity and Overweight", March 2020, [online], who.int, accessed 2020.

12 World Hunger, 2018 World Hunger and Poverty Facts and Statistics, [online], accessed 2020.

13 Beyond Meat, "Products", [online], beyondmeat.com, accessed 2020.

14 Carmen Reinicke, "Beyond Meat extends its post-IPO surge", [online], markets.businessinsider.com, July 2019, accessed 2020.

15 Impossible, "Products", [online], impossiblefoods.com, accessed 2020.

16 Crunchbase, "Memphis Meats", [online], crunchbase.com, accessed 2020.
17 Footprint Coalition, "Aquaculture", [online], footprintcoalition.com, accessed 2021.
18 IPCC, "Climate Change 2014: Mitigation of Climate Change", Contribution of Working Group III to the Fifth Assessment Report of the IPCC, Cambridge University Press, Chapter 11, 833.
19 Lauren C. Ponisio et al., "Diversification Practices reduce organic to conventional yield gap", Proceedings of the Royal Society B, 2015.
20 OECD Data, "Meat Consumption", [online], data.oecd.org, accesssed 2020.

Chapter 13: The Net-Zero Economy – Social and Economic Benefits

1 IEA, 'Share of emissions reductions in 2050 by maturity category and scenario", IEA.org, accessed 2021.
2 IPCC, "Climate Change 2014: Mitigation of Climate Change", Contribution of Working Group III to the Fifth Assessment Report of the IPCC, Cambridge University Press, 62.
3 The Wall Street Journal, "Barrel Breakdown, Cost of producing a barrel of oil and gas", [online], graphics.wsj.com, April 2016, accessed 2020.
4 World Bank, "Oil Rents (% of GDP)", [online], data.worldbank.org, accessed 2020.
5 Max Wei et al., "Putting Renewables and energy efficiency to work", Energy Policy, 38(2), 2010, 919-931.
6 IPCC, "Climate Change 2014: Mitigation of Climate Change", Contribution of Working Group III to the Fifth Assessment Report of the IPCC, Cambridge University Press, 63.
7 Herman Bril, George Kell and Andreas Rasche, *Sustainable Investing: A Path to a New Horizon*, Routledge Publishing, 2021.
8 The Pacific Institute, "Water Data", [online], worldwater.org, 2014, accessed 2020.
9 USGS, "Phosphate Rock Statistics and Information", usgs.gov, [online], 2020, accessed 2020.
10 USGS, "Potash Rock Statistics and Information", usgs.gov, [online], 2020, accessed 2020.
11 The Open University, "Organisations, environmental management and innovation", [online], open.edu, accessed 2020.
12 David Anthoff and Richard Tol, "Schelling's Conjecture on Climate and Development: A Test", Economic and Social Research Institute, Papers WP390, 2011.
13 National Institute for Health and Care Excellence, "How NICE measures value for money in relation to public health interventions", September 2013.
14 Jurgen John, "Differential discounting in the economic evaluation of healthcare programs", Cost Effective Resource Allocation, 17(29), 2019.
15 Elizabeth A. Stanton, "Negishi Welfare Weights: The Mathematics of Global Inequality", Stockholm Environment Institute, Working Paper, WP-US-0902.
16 Voices for a Malaria Free Future, "Malaria", [online], Malariafreefuture.org, accessed 2020.
17 World Health Organisation, "Malaria control: the power of integrated action", [online], who.int, accessed 2020.
18 Elliot Marseille et al., "Thresholds for the cost-effectiveness of interventions: alternative approaches", Policy and Practice, 93(2), 2015, 118-124.
19 World Health Organisation, "Global costs and benefits of drinking-water supply and sanitation interventions to reach the MDG target and universal coverage", WHO, 2012.
20 Swinburn, Boyd A et al., "The Global Syndemic of Obesity, Undernutrition, and Climate Change: The Lancet Commission report", The Lancet, 393(10173), 2019, 791 - 846.
21 Guy Hutton and Mili Varughese, "The costs of meeting the sustainable development goal targets on drinking water, sanitation, and hygiene", World Bank, Jan 2016.
22 Center for the Evaluation of Value and Risk in Health, CEA Registry, [online]
23 Jack Houston, "Why organ transplants are so expensive in the US", [online], businessinsider.com, Sept 12 2019, accessed 2020.
24 Michael Fehlings et al., "Is surgery for cervical spondylotic myelopathy cost-effective? A cost-utility analysis based on data from the AOSpine North America prospective CSM study", Journal of Neurosurgery Spine, 2012, 89-93.
25 Ian Tab, "Quality Adjusted Life Years", [online], slideserve.com, accessed 2020.
26 Elliot Marseille, Bruce Larson, Dhruv S Kazi, James G Kahn and Sydney Rosen, World Health Organisation, "Thresholds for the cost-effectiveness of interventions: alternative approaches", Volume 93, Number 2, February 2015, 118-124.
27 Amory Lovins, *Soft Energy Paths: towards a Durable Peace*, Pelican, 1977.
28 Global Energy Assesment Writing Team, "Global Energy Assessment, Towards a Sustainable Future", Cambridge University Press, 2012.

29 Johan Rockstrom and Jeffrey D. Sachs et al., "Sustainable Development and Planetary Boundaries", Sustainable Devlopment Solutions Network, May 2013.

30 Jason Channell et al., "Energy Darwinism 2, Why a Low Carbon Future Doesn't Have to Cost the Earth", Citi Global Perspectives and Solutions, August 2015.

31 Christopher T.M. Clack et al., "Future cost-competitive electricity systems and their impact on US CO_2 emissions", Nature Climate Change, 6, 2016, 526–531.

32 REN21, "Renewables Global Futures Report", 2017.

33 Jesse D. Jenkins et al., "Getting to Zero Carbon Emissions in the Electric Power Sector", Joule, 2(12), 2018, 2498-2510.

34 Jesse Jenkins et al., "Net-Zero America: Potential Pathways, Infrastructure, and Impacts, interim report, Princeton University", Princeton, December 15, 2020.

35 Sven Teske, "Achieving the Paris Climate Agreement Goals", Springer International Publishing, 2019.

36 IPCC Global Warming of 1.5 C special report, Oct 2018, IPCC, 2019: IPCC Special Report on the Ocean and Cryosphere in a Changing Climate, 2018.

37 IEA, "World Energy Model, Sustainable Development Scenario", [online], iea.org, 2019.

38 Mark Z. Jacobson et al., "Impacts of Green New Deal Energy Plans on Grid Stability, Costs, Jobs, Health, and Climate in 143 Countries", One Earth, 1(4), 2019, 449-463.

39 Johannes Emmerling et al., "The WITCH 2016 Model - Documentation and Implementation of the Shared Socioeconomic Pathways", Working Paper 42.2016, June 17, 2016.

40 BloombergNEF, "The New Energy Outlook", [online], bnef.com, accessed 2021.

41 Carbon Brief, "How integrated assessment models are used to study climate change", [online], carbonbrief.org, 2 October 2018, accessed 2020.

42 Krey Volker et al., "Looking under the hood: A comparison of techno-economic assumptions across national and global integrated assessment models", Energy, 172, 2019, 1254-1267.

43 O. Y. Edelenbosch, "Decomposing passenger transport futures: Comparing results of global integrated assessment models", Transportation Research Part D: Transport and Environment, 55, 2017, 281-293.

44 Michael Grubb and Claudia Wieners, 'On the Need for Dynamic Realism in DICE and other Equilibrium Models of Global Climate Mitigation', Working paper 112, Jan 25, 2020.

45 Frank Ackerman and Elizabeth A. Stanton, "Climate Economics: The State of the Art", Stockholm Environment Institute-US Center, 2011.

46 Robert S. Pindyck, "The Use and Misuse of Models for Climate Policy", Review of Environmental Economics and Policy, 11(1), 2017, 100–114.

47 Jeffrey D. Sachs, *The Ages of Globalization*, Columbia University Press, 2020, p13.

Chapter 14: A World Trying to Transition – Progress and Hurdles to Action

1 Spencer Weart, *The Discovery of Global Warming*, Harvard University Press, 2008.

2 Thomas R. Anderson et al., "CO_2, the greenhouse effect and global warming: from the pioneering work of Arrhenius and Callendar to today's Earth System Models", Endevour, 40(3), 2016, 178-187.

3 Svante Arrhenius, "On the Influence of Carbonic Acid in the Air upon the Temperature of the Ground", Philosophical Magazine and Journal of Science, 5(41), 1896, 237-276.

4 G. S. Callendar, "The Artificial Production of Carbon Dioxide and its Influence on Temperature", Quarterly Journal of the Royal Meteorological Society, 64(275), 1938.

5 American Chemical Society, "The Keeling Curve: Carbon Dioxide Measurements at Mauna Loa", [online], acs.org, 2015, accessed 2020.

6 J. S. Sawyer, "Man-made Carbon Dioxide and the "Greenhouse" Effect", Nature, 239, 1972, 23–26.

7 CIA document, "A Study of Climatological Research as it Pertains to Intelligence Problems", August 1974.

8 Wallace S. Broecker, "Climate Change: Are we on the brink of a pronounced Global Warming", Science, 189(4201),1975, 460-463.

9 Pawan Kumar Bhartia and Richard D. McPeters, "The discovery of the Antarctic Ozone Hole", Comptes Rendus Geoscience, 350(7), 2018, 335-340.

10 World Health Organisation, "Health Effects of the Chernobyl accident and special health care programmes", 2006.

11 Philip Shabecoff, "Global Warming Has Begun, Expert Tells Senate", New York Times, June 24, 1988.

12 United Nations Environment Programme, "Climate Change Information Sheet 17", UNFCCC, [online], accessed 2020.

13 IPCC, "Climate Change: The IPCC Scientific Assessment", Cambridge University Press, 1990.

14 United Nations, "What is the United Nations Framework Convention on Climate Change?", [online], unfccc.int, accessed 2020.

15 IPCC, "Climate Change 1995: The Scientific of Climate Change. Contribution of Working Group I to the Second Assessment Report of the Intergovernmental Panel on Climate Change", Cambridge University Press, 1995.

16 IPCC, "Climate Change 2001: The Scientific Basis. Contribution of Working Group I to the Third Assessment Report of the Intergovernmental Panel on Climate Change", Cambridge University Press, 2001.

17 IPCC, "Climate Change 2007: Synthesis Report. Contribution of Working Groups I, II and III to the Fourth Assessment Report of the Intergovernmental Panel on Climate Change", Switzerland, 2007.

18 United Nations Foundation, "Copenhagen Climate Change Conference - December 2009", [online], unfccc.int, accessed 2020.

19 Department of Energy and Climate, "Government Response to the House of Commons Science and Technology Committee 8th Report of Session 2009 to 10", Crown Copyright, 2010.

20 United Nations Foundation, "COP21", [online], un.org, accessed 2020.

21 NDC Partnership, "Pocket Guide to NDCs", [online], ndcpartnership.org, accessed 2020.

22 UN Environment Programme, "Emissions Gap Report 2020", Dec 2020.

23 BP, Statistical Review of World Energy, 2019, 68th edition.

24 IEA, "World Final Consumption (2018)", [online], iea.org, accessed 2020.

25 BSRIA, "World Heat Pump Market Study 2019", 2019.

26 Kingsmill Bond and Carlo. M. Funk, "2020 Vision: The Coming Era of Peak Fossil Fuels", Carbon Tracker Initiative, 2020.

27 Jack Weatherford, *The History of Money*, Penguin Random House, 1998.

28 Umbra Capital, "History of Merchant Banking", [onnline], umbracapital.com, 23rd July 2019, accessed 2020.

29 Tim Gould and Zakia Adam, "Low fuel prices provide a historic opportunity to phase out fossil fuel consumption subsidies", IEA, [online], iea.org, 2 June 2020, accessed 2021.

30 Oil Change International, "Fossil Fuel Subsidies Overview", [online], priceofoil.org, accessed 2020.

31 IEA, "World Energy Investment 2019", May 2019.

32 Mariana Maqzqzucato and Gregor Semieniuk, "Financing renewable energy: Who is financing what and why it matters", Technological Forecasting and Social Change, 127, 2018, 8-22.

33 OECD, *Green Investment Banks*, OECD Publishing, 2016.

34 David G. Rose, "The Raising Project Finance Handbook", 2020.

35 Rob Macquarie et al., "Updated View on the Global Landscape of Climate Finance 2019", Dec 2020, Climate Policy Initiative.

36 Blackrock, "Sustainability: The tectonic shift transforming investing", Blackrock Investment Institute, February 2020.

37 Bloomberg NEF, "Sustainable Debt Breaks Annual Record Despite Covid-19 Challenges", [online], about.bnef.com, Jan 2021, accessed 2021.

38 Credit Suisse, "The Global Wealth Report 2020", Research Institute, 2020.

39 Allianz, "Allianz Global Wealth Report 2020", Sept 2020.

40 IMF, "A Global Picture of Public Wealth", [online], blogs.imf.org, June 2019, accessed 2020.

41 Thomas Piketty, "Capital in the 21st Century", Harvard University Press, 2014.

42 Boston Consulting Group, "Global Asset Management 2019, Will These '20s Roar?", 2019.

43 OECD, "Pensions Markets in Focus 2019", OECD publishing, 2019.

44 Elke Weber, "Perception and expectation of climate change: Precondition for economic and technological adaptation", Psychological Perspectives to Environmental and Ethical Issues in Management, Jossey-Bass, 314 – 341.

45 William Forster Lloyd, "Two Lectures on the Checks to Population", Oxford University, 1833.

46 Garrett Hardin, "The Tragedy of the Commons", Science, 162 (3859), 1968, 1243-1248.

47 G. Janssens-Maenhout etal., "Fossil CO_2 and GHG emissions of all world countries", Publications Office of the European Union, Luxembourg, 2017.

48 Roberto Roson and Martina Sartori, "Estimation of Climate Change Damage Functions for 140 Regions in the GTAP 9 Data Base", Journal of Global Economic Analysis, 1(2), 78-115.

49 Jenny Chase, "Solar Power Finance Without the Jargon", World Scientific, 2019, p38.

50 Lave, Lester B., An Empirical Description of the Prisoner's Dilemma Game. Santa Monica, CA: RAND Corporation, 1960.

51 Nicholas Stern, "The Economics of Climate Change", The Stern Review, Cambridge, 2007, Chapter 17.

52 Amitav Ghosh, *The Great Derangement*, The University of Chicago Press, 2016, p115.

53 Dominic Roser and Christian Seidel, *Climate Justice*, Routledge, 2017.

54 CDIAC, "Fossil-Fuel CO_2 Emissions by Region", [online], cdiac.ess, accessed 2020.

55 George Monbiot, *HEAT*, Penguin Books, 2006, p25.

56 Naomi Oreskes and Erik M. Conway, *Merchants of Doubt*, Bloomsbury, 2010.

57 Riley E Dunlap, Peter J Jacques, 'Climate Change Denial Books and Conservative Think Tanks: Exploring the Connection', Am Behav Sci., June, 2013.
58 Timo Busch, Lena Judick, 'Climate Change – that is not real! A comparative analysis of climate-sceptic think tanks in the USA and Germany', Climatic Change, 164 (18), 2021.
59 Suzanne Goldenberg and Helen Bengtsson, "Biggest US coal company funded dozens of groups questioning climate change", [online], theguardian.com, June 13, 2016, accessed 2020.
60 Gretchen Goldman, 'Peabody Energy Discloses Extensive Payments to Climate Denial Groups', Union of Concerned Scientists, June 30, 2016.
61 Peabody Energy Corporation, et al., Certificate of Service Chapter 11, Case No. 16-42529, 2016.
62 ExxonMobil, 'Worldwide Giving Report', 2006.
63 ExxonMobil, 'Dimensions', 1998.
64 Exxon Mobil, "Citizenship Report 2007", page 39.
65 Reuters, "Exxon again cuts funds for climate change skeptics", [online], reuters.com, May 23 2008, accessed 2020.
66 Cato Institute, '25 years at the CATO Institute', 2001.
67 Cato Institute, 'CATO Institute 2019 Annual Report', 2019.
68 Suzanne Goldenberg, 'Big donors ditch rightwing Heartland Institute over Unabomber billboard', The Guardian, 9th May 2012.
69 Mike Berners-Lee and Duncan Clark, *The Burning Question*, Profile Books Ltd, 2013.
70 Craig Idso, Robert M. Carter, S. Fred Singer, 'Why Scientists Disagree About Global Warming', 2015, online, heartland.org.
71 Leah Cardamore Stokes, *Short Circuiting Policy*, Oxford University Press, 2020.
72 Michael Liebreich, "How to Save the Planet: Be Nice, Retaliatory, Forgiving and Clear", Liebreich Associates, Sept 2012.
73 Robert Axelrod, *The Evolution of Cooperation*, Basic Books, 1984, p124.
74 Michael E. Mann, *The New Climate War*, PublicAffairs, 2021.
75 Peabody Energy Corporation, Form 8-k, March 16, 2017
76 Scott Waldman, 'Cato closes its climate shop; Pat Michaels is out', May 29, 2019.
77 Michael E. Mann, The New Climate War, Scribe, 2021.
78 CDIAC, "Fossil-Fuel CO_2 Emissions by Region", [online], cdiac.ess, accessed 2020.

Chapter 15: Plotting a Good Course – The Politics of Change

1 World Meteorological Organization, "Statement on the State of the Global Climate in 2019", WMO, 2020.
2 World Economic Forum, "Climate Change is a planetary emergency - how can we avert disaster?", [online], weforum.org, Jan 2020, accessed 2020.
3 Climate Emergency Declaration, [online], climateemergencydeclaration.org, 7 Feb 2021, accessed 2021.
4 Extinction rebellion contributors, *This is Not a Drill*, Pengiun Randon House, 2019.
5 BBC News, "Our Planet Matters: What's the BBC plan all about?", [online], bbc.co.uk, Jan 2020, acessed 2020.
6 Ceres, "Climate Action 100+", [online], ceres.org, accessed 2021.
7 350.org, "Stop Fossil Fuels. Build 100% Renewables", [online], 350.org, accessed 2021.
8 The Institutional Investors Group on Climate Change, "Our Mission", [online], accessed 2021.
9 Principles for Responsible Investment, [online], unpri.org, accessed 2021.
10 United Nations, 'Net-Zero Emissions Must be Met by 2050', un.org, 12 Nov 2020, accessed 2021.
11 UNFCCC, 'Climate Ambition Alliance: Net Zero 2050, climateaction.unfccc.int, accessed April 2021
12 Yuval Noah Harari, *Sapiens: A Brief History of Humankind*, Harper, 2014.
13 The Editors of Encyclopaedia Britannica, "Citizenship", [online], britannica.com, accessed 2020.
14 Hal Harvey, *Designing Climate Solutions*, Island Press, 2018.
15 Climate Policy Info Hub, "Non-Market Based Climate Policy Instruments", [online], climatepolicyinfohub.eu, accessed 2021.
16 Climate Policy Info Hub, "Market Based Climate Policy Instruments", [online], climatepolicyinfohub.eu, accessed 2021.
17 Christina Nunez, "Ozone depletion, explained", [online], nationalgeographic.com, April 2019, accessed 2021.
18 Leah Stokes, *Short Circuiting Policy*, Oxford University Press, 2020.
19 Corrie E. Clark, "Renewable Energy R&D Funding History: A Comparison with Funding for Nuclear Energy, Fossil Energy, Energy Efficiency, and Electric systems R&D", Congressional Research Service, June 2018.
20 Jenny Chase, *Solar Power Finance Without the Jargon*, World Scientific, 2019.
21 Carbon Disclosure Project, "Home", [online], cdp.net, accessed 2021.

22 Jinqiang Chen, "The Challenges and Promises of Greening China's Economy", Harvard Kennedy School Belfer Center, January 2017.

23 John Fialka, "Why China is Dominating the Solar Industry", [online], scientificamerican.com, Dec 2016, accessed 2020.

24 Northern Ireland Audit Office, "Report by the Comptroller and Auditor General for Northern Ireland – CAG-04051", April 2017.

25 Rory Carrol, "Cash-for-Ash fiasco: Northern Ireland's Enron on Craggy Island", The Guardian, 28 Sept 2018.

26 Martin McGuinness MLA, 'Letter from the Executive Office', 9 Jan, 2017.

27 Peter L Singer, "Federally Supported Innovations: 22 Examples of Major Technology Advances That Stem From Federal Research Support", The Information Technology and Innovation Foundation, Feb 2014.

28 Mariana Mazzucato, *The Value of Everything*, Penguin, 2018, p194.

29 Bea Perks, "Goodenough rules", Royal Society of Chemistry, [online], chemistryworld.com, Dec 2014.

30 Dana Hull, "The man behind Solyndra's rise and fall: Chris Gronet", [online], mercurynews.com, Nov 2011, accessed 2020.

31 Report to Congressional Committees, "Further Actions Are Needed to Improve DOE's Ability to Evaluate and Implement the Loan Guarantee Program", United States Government Accountability Office, July 2010.

32 Robert P. Murphy, "Lessons From Solyndra", [online] econlib.org, Feb 2012, accessed 2020.

33 Tesla, "Tesla repays department energy loan nine years early", [online], May 2013.

34 Richard Schmalensee and Robert N. Stavins, "Lessons Learned from Three Decades of Experience with Cap and Trade", Review of Environmental Economics and Policy, 11(1), 2017, 59–79.

35 ember-climate.org, "Carbon Price Viewer", [online], ember-climate.org, accessed 2020.

36 European Environment Agency, "Trends and projections in Europe reports", 2019.

37 IPCC, "Climate Change 2014: Mitigation of Climate Change", Contribution of Working Group III to the Fifth Assessment Report of the IPCC, Cambridge University Press, 450.

38 IPCC Contributors, "Annex III: Technology-specific cost and performance parameters. In: Climate Change 2014: Mitigation of Climate Change. Contribution of Working Group III to the Fifth Assessment Report of the Intergovernmental Panel on Climate Change", Cambridge University Press, 2014.

39 IPCC, "Climate Change 2014: Mitigation of Climate Change. Contribution of Working Group III to the Fifth Assessment Report of the IPCC", Cambridge University Press, 466.

40 OECD, "Taxing Energy Use 2019: Using Taxes for Climate Action", OECD Publishing, 2019.

41 IEA, "Energy subsidies, Tracking the impact of fossil-fuel subsidies", [online], iea.org, accessed 2020.

42 Government Offices of Sweden, "Sweden's carbon tax", 11 Jan 2021, [online], accessed 2021.

43 Danny Cullenward and David G. Victor, *Making Climate Policy Work*, Polity, 2021, p175.

44 Climate Change Committee, "Reducing UK Emissions: Progress Report to Parliament", CCC, 2020, 82.

45 Office for National Statistics, "Environmental Accounts", [online], ons.gov.uk, 2019.

46 Katy Hargreaves et al., "The Distribution of Household CO_2 Emissions in Great Britain", Joseph Rowntree Foundation, March 2013.

47 Office for National Statistics, "Personal and household finances", [online], ons.gov.uk, accessed 2020.

48 William D Nordhaus, *The Climate Casino*, Yale University Press, 2015.

49 Peter Drucker, *The Practice of Management*, Butter-Worth Heinemann, 1955.

50 Malcolm S. Salter, "Rehabilitating Corporate Purpose", Harvard Business School NOM Unit Working Paper No. 19-104, April 2019.

51 Milton Friedman, *Capitalism and Freedom*, The University of Chicago Press, 1962.

52 George Serafeim et al., "Corporate Sustainability: First Evidence on Materiality", The Accounting Review, 91(6), 1697-1724.

53 Gordon L. Clark et al., "From the Stockholder to the Stakeholder: How Sustainability Can Drive Financial Outperformance", SSRN, March 2015.

54 Herman Bril, George Kell and Andreas Rasche, *Sustainable Investing: A Path to a New Horizon*, Routledge Publishing, 2021, 90.

55 Arabesque, "Home", [online], arabesque.com, accessed 2020.

56 Michael Aklin and Johannes Arpelainen, *Renewables: The Politics of a Global Energy Transition*, The MIT Press, 2018.

57 Noam Chomsky and Robert Pollin, *Climate Crisis and the Global Green New Deal*, Verso, 2020.

58 Ray Dalio, *Principles for Navigating Big Debt Crises*, Bridgewater Associates, 2018.

59 Trading Economics, "United States Initial Jobless Claims", [online], tradingeconomics.com, accessed 2020.

60 Investing.com, "Crude Oil WTI Futures", [online], uk.investing.com, accessed 2020.

61 Carbon Brief, "Tracking how the worlds green recovery plans aim to cut emissions", [online], carbonbrief.org, June 2020, accessed 2021.

62 Kate Larsen et al., "It's Not Easy Being Green: Stimulus Spending in the World's Major Economies", [online], September 2020, accessed 2021.

63 Rhodium Group, "Green Stimulus and Recovery Tracker", [online], rhg.com, accessed 2021.

64 Dr Fatih Birol, "What the 2008 financial crisis can teach us about designing stimulus packages today", [online], iea.org, 19 April 2020, accessed 2020.

65 Credit Suisse, "Global Wealth Report", Research Institute, 2019.

Chapter 16: Betting the Planet – The Future of Humanity

1 Herman Daly Chapter, *Scarcity and Growth Reconsidered*, Johns Hopkins, 1979, p 79.

2 Steven Hawking, "Life in the Universe", Public Lecture Transcript, [online], web.archive.org, 12 July 2000, accessed 2021.

3 Ramez Naam, *The Infinite Resource*, University Press of New England, 2013.

4 Michelle Kunimoto and Jaymie M. Matthews, "Searching the Entirety of Kepler Data. II. Occurrence Rate Estimates for FGK Stars", The Astronomical Journal, 159 (6), 2020, 248.

5 Milan M. Cirkovic, *The Great Silence*, Oxford University Press, 2018.

6 Gino Segre and Bettina Hoerlin, *The Pope of Physics*, Picador, 2016.

7 Robin Hanson, "The Great Filter - Are we Almost Past it?", [online], web.archive.org, Sept 15, 1996, accessed 2021.

Afterword: Hedgehogs and Foxes – Understanding the Climate Divide

1 Isaiah Berlin, *The Hedghog and the Fox: An Essay on Tolstoy's View of History*, Weidenfeld & Nicolson, 1953.

2 Philip E Tetlock and Dan Gardner, *Superforcasting: The Art and Science of Prediction*, Crown Publishers, 2015.

3 David Epstein, Range: *How Generalists Triumph in a Specialized World*, Pan Books, 2020.

4 Friedrich Hayek, *The Road to Serfdom*, George Routledge & Sons, 1944, p57.

Printed in Great Britain
by Amazon